PROBLEMS IN ORGANIC CHEMISTRY

PROBLEMS IN ORGANIC CHEMISTRY

By
Dr. Shardendu Kislaya

DISCOVERY PUBLISHING HOUSE PVT. LTD.
NEW DELHI-110 002

Published by:
Tilak Wasan
DISCOVERY PUBLISHING HOUSE PVT. LTD.
4383/4B, Ansari Road, Darya Ganj
New Delhi-110 002 (India)
Phone : +91-11-23279245, 23253475, 43596065
E-mail : discoverybooksindia@gmail.com
discoverypublishinghouse@gmail.com
web : www.discoverypublishinggroup.com

Reprinted: **2019**

First Published: **2011**

ISBN: 978-81-8356-814-2

Problems in Organic Chemistry

Printed at:
Infinity Imaging Systems
Delhi

Preface

The book "Problems in Organic Chemistry" has been written to meet the requirements of graduate students of all Indian Universities. The matter of this book has been detail theory and sufficient solved examples. We have been selected sufficient problems from various Universities examination papers.

One of salient course syllabi features of the present edition of the book is the inclusion of a large number of typical worked out problems which will elucidate various abstract principles and theories discussed in the text. It covers the complete syllabus of physics prescribed by Technical Universities.

We have tried to our best to keep the book free from the misprint. The author shall be grateful to the readers who point out errors and omissions which inspite of all care might have been there.

We shall indeed be very thankful to our colleague for their recommendations this book of for their students. The present book will be warmly received by the students and teachers.

—Author

Contents

1

Basic Problems of the Organic Chemistry

INTRODUCTION

Another set of reactions that has received considerable attention is that in which optically active coordination compounds, especially tris (chelate) compounds, racemise. The racemisation of these compounds is often rapid compared to that of the carbon compounds. However, there are also coordination compounds which racemise very slowly. Generally, complexes which are labile racemise rapidly and are difficult or impossible to resolve, and conversely, non-labile complexes racemise slowly and are usually resolvable. However, the solution of certain labile complexes such as

$Al(C_2O_4)_3^{2-}$, $Ga(C_2O_4)_3^{2-}$, $Fe(C_2O_4)_3^{3-}$, $Cd(en)_3^{2+}$

and $Zn(en)_3^{2+}$ has been reported.

The rates of racemisation of octahedral complexes vary from extremely rapid to extremely slow. In many cases, the rates are such that they can be followed by conventional techniques.

MECHANISM

There are two mechanism by which optically active octahedral complexes may racemise. These are called the intermolecular and intramolecular processes. In some cases, racemisation occurs by one mechanisms in preference to other while occasionally racemisation occurs by both mechanisms simultaneously.

We shall now discuss the two mechanisms separately.

1. Intermolecular Mechanism

According to this mechanism, racemisation involves a ligand interchange process as represented by the equilibria:

$$D-M(AA)_3 \underset{+AA}{\overset{-AA}{\rightleftarrows}} M(AA)_2 \underset{-AA}{\overset{+AA}{\rightleftarrows}} L-M(AA)_3$$

In the above equilibria, the solvent is not included but in water solution the intermediate would surely be $M(AA)_2\ (H_2O)_2$.

The intermolecular mechanism of racemisation has been established in several cases. For example, the rates of water exchange of cis-$Co(en)_2(H_2O)_2{}^{3+}$ and of cis-$Co(en)_2\ NH_3H_2O^{3+}$ are approximately sixty five and forty times larger than are their rates of racemisation. This reveals that most of the water interchange process occurs with retention. However, such an exchange takes places with rearrangement.

The rate of loss of optical activity of methanol solution of (+)-$Co(en)_2Cl_2{}^+$ has been found to be rate of radiochlorine exchange of one chloro group. The simplest mechanisms consistent with this observation is that the complex dissociates to a symmetrical five-coordinated intermediate; thus each dissociation is leading to a loss of optical activity.

Thomas (1929) suggested that the racemisation of $M(C_2O_4)^{3-}$ takes place by an intermolecular process involving a planar intermediate $M(C_2O_4)^{3-}$; the fact that such an intermediate is symmetrical would then account for the loss in the optical activity.

Kinetic studies on the racemisation of (+)-cis-$Co(en)_2Cl_2{}^+$ reveals that the loss of optical rotation does not result from the observation intermediate but rather is due to trans isomerisation. This inference is drawn from the observation that the acid hydrolysis of cis-$Co(en)_2Cl_2{}^+$ gives a mixture of cis-and trans-$Co(en)_2\ (H_2O)\ Cl_2{}^+$. On this basis, the probable mechanism of racemisation of (+)-cis-$Co(en)_2Cl_2$ is as follows:

$$(+)\text{-cis-}Co(en)_2Cl_2{}^+ \xrightarrow{k} (+)\text{-cis-}Co(en)_2\ (H_2O)Cl^+ \xrightarrow{k'} \text{trans-}Co(en)_2\ (H_2O)Cl_2{}^+$$

2. Intramolecular Mechanism

For the first time, it was Werner who suggested that complex ions may racemise by an intermolecular mechanism. Later on, it was shown that this type of mechanism has been found in the systems trioxalatocohalt (III) and-chromium (III) ions, tris (acetyl acetonato) cobalt

(III) and-chromium (III), ethylene diamine tetracctato cohaltitate (III) and ion and also in part for tris (1, 10-phenanthroline) an tris(2,2′-bipyridinc) iron (II) ions.

Various workers studied the oxalato complexes and most of them postulated intramolecula mechanism for their racemisation. However, Johnson et. al. (1933) suggested that these compound racemise by an intermolecular mechanism and they cited the following evidence against the intramolecula mechanism.

(i) There could not he any dissociation of $Co(C_2O_4)_3^{2-}$ or $Cr(C_2O_4)_3^{2-}$ in aqueous solution.

(ii) These complex ions undergo racenusation even in the solid state and

(iii) The rate of racemisation is not decreased by the presence of oxalate ions.

The above evidence can not be accepted as proof in favour of intermolecular mechanism. However this controversy was settled on the rate of oxalate ion exchange in the systems $Co(C_2O_4)_3^{2-} * - C_2^*O_4^{2-}$ and $Cr(C_2O_4)_3^{-} - *C_2O_4^{3-}$. As the rate of oxalate ion exchange is much slower than the rate of racemisation, it is indicative of the fact that racemisation of these ions takes place by an intramolecular mechanism

Intramolecular mechanism has been observed in $Fe(phen)_3{}^{2+}$ and $Fe(bipy)_3{}^{2+}$ ions. The rates of racemisation of these ions have been found to he greater than the rates of dissociation. From this it could be concluded that racemisation must, at least in part, involve an intramolecular process. However, the kinetic studies on the racemisation and dissociation of the Fe (II) complexes suggest a dual mechanism.

It is interesting to note that very similar Fe (II) and Ni (II) complexes should racemise by different mechanisms. The reason is that although Fe (II) and Ni (II) complexes have the same charge and size but differ in stability and bond type, *i.e.*, the Fe (II) complexes are diamagnetic low-spin dx system while Ni (II) complexes are paramagnetic high-spin d^* systems.

The acceptable explanation is that the intramolecular racemisation of the Fe (II) complex may result from a process of expansion which allows loss of activity. Thus, there may occur an increase in the inter-atomic distances between donor atom and metal due to the excitation of the low-spin to a high-spin state which then might well rearrange before returning to the original state. On the basis of this interpretation, the racemisation may he represented by the equilibria:

$$Fe(AA)_3{}^{2+} \underset{k_2}{\overset{k_1}{\rightleftarrows}} Fe...(AA)_3{}^{2+} \underset{k_4}{\overset{k_3}{\rightleftarrows}} Fe...(AA)_2 + AA$$

From the above equilibria, the following points are to he noted:

(i) If $k_2 > k_3$, the racenusation will take place by an intra-molecular process.

(ii) If $k_2 < k_3$, the racemisation will take place by an inter-molecular process.

(iii) If $k_2 = k_3$, both mar contribute to the observed rate of isomerisation.

The intramolecular mechanism is believed to proceed through a symmetrical trigonal-prism inter mediate due to distortion of all the ligands by a twisting motion.

HOW RACEMISATION TAKES PLACE?

The exact nature of the isomerisation is not known. There several ways in which such a process can happen. However, some of which can he eliminated by appropriate experiments.

For example, one possible way would be complete dissociation of one chelate ring with formation of a square planar complex of a trans diaquo complex Fig. 1.

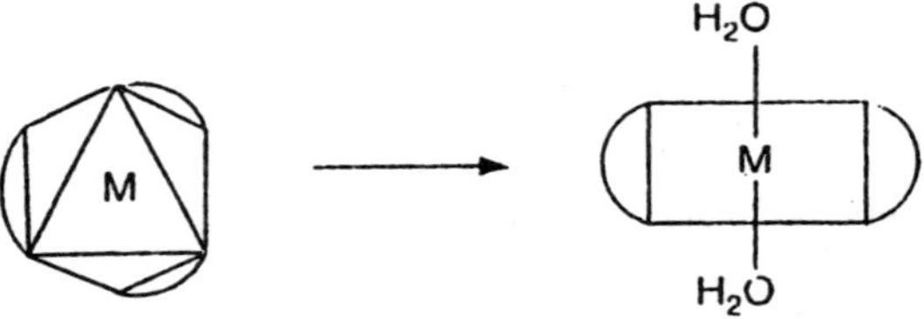

Fig. 1

The asymmetry has been lost and if the chelate ring is reformed, it will have a 50-50 chance of forming either the Δ or ^ isomer. But the rate determining step in the racemisation (k_r) must be equal to the rate of dissociation (k_d).

For example, tris(phenathraline) nickel (II) racemises at the same rate ($K_r = 1.5 \times 10^{-4}$ sec^{-1}) as it dissociates ($k_d = 6.7 \times 10^{-4}$ sec^{-1}). If racemisation occurs faster than dissociation [*e.g.*, tris (phenanthroline) iron (II), $k_r = 6.7 \times 10^{-4}$; $k_d = 0.70 \times 10^{-4}$ sec^{-4}], this mechanism can be eliminated.

The more reasonable way is that only on end of the chelate ring will be detached, thereby forming a 5-coordinate complex. This complex might undergo a Berry pseudorotation with scrambling of the positions. Reforming of the chelate ring would then form a racemic mixture of Δ or isomers. In order to account for racemisation, twist mechanisms have been proposed. These do not require bond rupture. The earliest mechanism of that sort had been proposed by Ray and Dutt and is known as the rhombic twist [Fig. 2(a)]. Later Bailar postulated a trigonal twist mechanism [Fig. 2(b)].

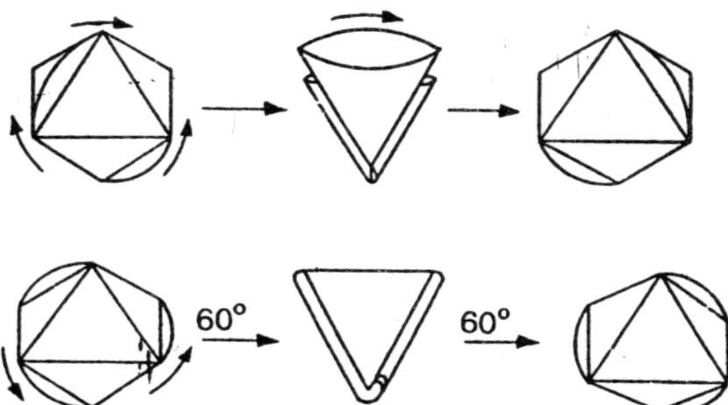

Fig. 2(a) and (b) : (a) Rhombic or Ray-Duff twist. (b) Trigonal or Bailar twist.

Due to experimental difficulties, there is no firm experimental evidence in favour or against any of the twist mechanisms. It appears that rigid chelates and those with small bite angles favour a trigonal twist. Rigidity in the chelate ring inhibits one-ended dissociation. Small bite angles are known to stabilise trigonal prismatic coordination and might be expected to reduce the energy barrier to the twist.

ELECTRON-TRANSFER REACTIONS OR REDOX REACTIONS

These are the reaction in which the oxidation states of some atoms may change by the transfer of an electron from one atom to the others. These reactions can be devided into two main types:

(a) Outer sphere reactions

(b) Inner sphere reactions or ligand bridged activated complex reactions.

We shall discuss these one by one.

(a) Outer Sphere Reactions

These are the reactions in which the coordination sphere of ions undergoing redox reaction is not altered. The simplest examples of oxidation-reduction reactions are those between species which differ only in charge, such as

$$[Fe(CN)_6]^{4+} \text{ and } [Fe(CN)_6]^{3-} \text{ or } [IrCl_6]^{3-} \text{ and } [IrCl_6]^{2-}.$$

Such reactions have been investigated by tracer methods using flow methods and rapid quenching if necessary, for very fast reactions, the broadening of e.s.r. or n.m.r. spectra can sometimes be utilised. In one novel method, rate can be determined by the more recent technique of dynamic n.m.r., for example using ^{17}O n.m.r. of the following reaction.

$$[MnO_4]^- + [MnO4_4]^{2-} \rightarrow [MnO_4]^{2-} + [MnO_4]^-$$

In another novel method, the rate of loss of activity on mixing solutions of a D-complex of on oxidation states and the L-complex of another oxidation state gives the rate of electron transfer by the following reaction :

$$D\text{-}[Os(dipy)_3]^{2+} + L\text{-}[Os(dipy)_3]^{3+} \rightleftarrows L\text{-}[Os(dipy)_3]^{2+} + D\text{-}[Os(dipy)_3]^{3+}$$

(b) Inner Sphere Reactions

These are the reactions in which electron transfer takes place through a bridging group common to the coordination shells of both metal ions. An example of inner sphere reactions is that between $[Co(NH_3)_5Cl]^{2+}$ and $[Cr(H_2O)_6]^{2+}$ in acidic solution, first investigated by Taube. The net reaction has been found to be,

$$[Co(NH_3)_5Cl]^{2+} + [Cr(H_2O)_6]^{2+} + 5H_2O^+ \rightarrow$$
$$[Co(H_2O)_6]^{2+} + [Cr(H_2O)_5Cl]^{2+} + 5NH_4^+$$

Similar reactions occur when the chloride in the above reaction is replaced by other halide ions, sulphate, phosphate, acetate, oxalate, succinate and maleate. Inner sphere reactions have been studied by many standard physical and chemical methods.

Mechanism of Outer-Sphere Reaction of Outer-Sphere Mechanism

The mechanism of outer sphere reactions involves-direct electron transfer from the outside of a ligand in one coordination sphere over the outside of a second sphere through an outer-sphere activated complex. In the electron transfer, the ligands are simply acting as electron-conduction media.

The above mechanism has been confirmed by the recent work on electron spin resonance of paramagnetic complexes. This work further reveals that unpaired electrons spend part of their time on ligands.

In principle, it is possible to recognise an outer sphere reaction on the basis of following criteria:

(i) The rate low should he of first in both the reactants.

(ii) The activated complex will contain the intact coordination shells of both metal ions.

(iii) The coordination shell of either metal ion should be sufficiently inert to substitution so that the rate of electron transfer is faster than the rate of substitution.

(iv) Electron transfer is expected to be fast when no change in the molecular dimensions takes place.

(v) The rate constant depends upon the cation present in solution; ion-pair formation decreases the activation energy by reducing the electrostatic repulsion energy.

(vi) Reactions involving large-size difference proceed slowly.

The exchange of $[Fe(CN)_6]^{3-}$ and $[Fe(CN)_6]^{4-}$ denotes a typical example of a process which takes place by direct electron transfer through an outer-sphere activated complex. The rate of exchange, studied by isotopic labelling, is very rapid ($k \sim 10^3 M^{-1}s^{-1}$ at 4°C). Ferrocyanide (a low spin d^6 system) and ferricyanide (a low spin d^7 system) are both inert to substitution.

When Franck-condon principle is applied to this outer-sphere reaction, it has some interesting repercussions. This principle governs the probability of transition between different electronic states of a molecule. This principle states that an electronic transition occurs so rapidly that during the transition the nuclei in the molecule remain static.

In order to understand the effect of Franck-Condon principle, the transfer of an electron from $[Fe(CN)_6]^{4-}$ to $[Fe(CN)_6]^{3-}$ is considered. The rate of this redox reaction has been studied by labelling either of the complexes with a radioactive isotope of Fe or with ^{14}C.

$$\underset{\substack{\text{Fe–C bonds}\\ \text{log er}}}{[\overset{*}{Fe}(CN)_6]^{4-}} + \underset{\substack{\text{Fe–C bonds}\\ \text{shorter}}}{[Fe(CN)_6]^{3-}} \rightarrow \underset{\substack{\text{Fe–C bonds}\\ \text{log er}}}{[\overset{*}{Fe}(CN)_6]^{3-}} + \underset{\substack{\text{Fe–C bonds}\\ \text{shorter}}}{[Fe(CN)_6]^{4-}}$$

The above reaction has been found to be fast and its second-order rate constant is $\sim 10_5$ at 25°C. Further, no heat change takes place because the same products are obtained after electron transfer. The $[Fe(CN)_6]^{4-}$ is a low spin d^6 $(t_{2g}^6\ e_g^0)$ system and the $[Fe(CN)_6]^{3-}$ is high spin d^5 $(t_{2g}^5\ e_g^0)$ system. As both ions are inert, the loss or exchange of CN^- of any substitution reaction is very fast. No possibility exists for electron transfer through a bridged activated complex because the formation of the activated complex amounts to a substitution reaction.

During this reaction, there is no change in the configuration of the atoms, Fe, C or N. Consequently the bond length of Fe-C bond in the newly formed $[Fe(CN)_6]^{3-}$ ion will be longer than the equilibrium value and the length of Fe-C bond in the product $[Fe(CN)_6]^{4-}$ ion will be shorter than the equilibrium value. Thus if an election is to he transferred between the anions in their ground state equilibrium configuration by the Franck-Condor, principle there would be the expansion of the product $[Fe(CN)_6]^{3-}$ by increasing the length of Fe-C bond compression of $[Fe(CN)_6]^{4-}$ by shortening Fe-C bonds. For this to be so, the energy has to be added to the system *i.e.*, the products would be of higher energy than the reactants.

However, this contradicts the fact that no heat change takes place in electron transfer reaction. It means that an electron transfer reaction is only possible when the vibrations within two anions have made them of identical geometries and electronic configurations.

Under such a situation, the products and reactants in the electron-transfer process would be equivalent and then no energy change takes place due to electron transfer. But both the anions are similar.

Therefore, the addition of a relatively small amount of energy, *i.e.*, activation energy, would make the anions alike and consequently electron transfer reaction occurs rapidly.

If both the reactants are inert as in the above example, a close approach of the metal atoms is not possible and therefore the electron transfer must take place by a tunneling or outer-sphere mechanism. In this mechanism, each complex is retaining its full coordination shell in the activated complex so that no ligand is common to each central metal atom and an electron is supposed to pass through both the coordination shells. This mechanism is expected to be correct if both the reactants are exchanging their ligands more slowly.

For an isotopic change, the equilibrium constant is unity and ΔG° is nearly zero. Therefore, activation energy is required to overcome the electrostatic repulsion between ions of like charge, to distort the coordination of both species and to modify the solvent structure around both species. It is interesting to note that the outer-sphere electron-transfer reactions are more rapid for complexes having ligands like CN^- and o-phe-nanthroline than for corresponding complexes having ligands H_2O or NH_3.

The electron-transfer between $[Co(NH_3)_6]^{3+}$ and $[Co(NH_3)_6]^{2+}$ is slow.

$$\underset{\text{High-spin}}{[\overset{*}{C}o(NH_3)_6]^{2+}} + \underset{\text{Low-spin}}{[Co(NH_3)_6]^{3+}} \rightarrow \underset{\text{Co-N=2.11-0.35=1.76Å}}{[\overset{*}{C}o(NH_3)_6]^{3+}} +$$

$$\text{Co} - \text{N} = 2.11\text{Å} \quad \text{Co} - \text{N} = 1.96\text{Å} \quad Co^{3+} \rightarrow d_6(t_{2g}^6\ e_g^1)$$

$$\underset{\text{Co-N=1.96+0.35=2.31Å}}{\left[Co(NH_3)_6\right]^{2+}}$$

$$Co^{2+} \rightarrow d^7\,(t_{2g}^5\ e_g^2)\ Co^{3+} \rightarrow d^6(t_{2g}^6\ e_g^0)\ Co^{2+} \rightarrow d^7(t_{2g}^6\ e_g^1)$$

Both the complexes are octahedral and do not differ greatly in size. Therefore, the electron transfer in these complexes should be very fast. Actually this reaction is slow. This has been explained as follows.

The two complexes have been different electronic configurations and have also different Co-N bond lengths. For example, $[Co(NH_3)_6]^{2+}$ is a high-spin d^7 $(t_{2g}^5\ e_g^2)$ system in which Co-N bond length is 2.11Å $[Co(NH_3)_6]^{3+}$ is a low spin d^6 $(t_{2g}^6\ e_g^0)$ system in which Co-N bond length is 1.96A. When electron transfer occurs, the configurations become t_{2g}^6, e_g^1 and $t_{2g}^6\ e_g^1$ respectively. Both of these are not in ground-slate configurations but both are electronically excited. This excess of energy will he lost either by radiation or in the form of thermal energy. Thus this exchange reaction has higher activation energy. Therefore, it is expected that the rate of electron transfer reaction between $[Co(NH_3)_6]^{2+}$ and $[Co(NH_3)_6]^{3+}$ is much slower than that between $[Fe(CN)_6]^{3-}$ and $[Fe(CN)_6]^{4-}$.

Alkali metal ions have been found to catalyse the $[Fe(CN)_6]^{4-}$ – $[Fe(CN)_6]^{3-}$ exchange reaction. This effect has been found to be greater for cesium and smallest for lithium. These results are indicative of the fact that a partly desolated cation is accelerating exchange repulsion by helping to overcome electrostatic repulsion by forming a transition state like,

$$[Fe(CN)_6]^{4-} \ldots\ldots M^+ \ldots\ldots [Fe(CN)_6]^{3-}$$

The alkali metal ion has also been found to catalyse the MnO_4^{2-} to MnO_4^- exchange reaction. The order of effectiveness is same as for the $[Fe(CN)_6]^{4-}$ - $[Fe(CN)_6]^{3-}$ reaction.

Outer-sphere reactions involving complexes of different metals have been found to he faster than outer-sphere exchange reactions between different oxidation states of the same element. An example is

$$[Os(dipyr)_3]^{2+} + [Mo(CN)_8]^{3-} \rightarrow [Os(dipy)_3]^{3+} + [Mo(CN)_8]^{4-}$$

The decrease in energy for such reactions will appear as the free energy of the reactions when excited states of products are converted into ,round states. This means that for such reaction the structure of the transaction state is more like that of the reactants; hence the activation energy is lowered and the rate is increased.

A number of electron transfer reactions are known in which the electron transfer occurs by both inner and outer-sphere mechanisms. The examples of such reactions are between $[Co(NH_3)_4X]^{2+}$ and $[Co(CN)_5]^{3-}$ where X = F^-, CN^-, NO_3, NO, etc.

2. Inner Sphere Mechanism

In inner-sphere reactions, substitution of the coordination shell of one of the metal ions occurs prior to electron transfer. In such reactions, an intimate contact between oxidant and reductant is required. This requirement is fulfilled if a bridged activated complex is formed due to the attachment between the oxidant and reductant. In this activated complex, there must he one ligand which would he common to the coordinate sphere of both the reacting complexes and this ligand forms a bridge between them.

The formation of activated complex is followed by dissociation and electron transfer. In simple words, the inner-sphere mechanism proceeds through the formation of a bridged intermediate followed by dissociation and the electron transfer.

Taube *et al.* (1959) studied the oxidation of aqueous Cr(II), $[Cr(H_2O)_6]^{2+}$ by pentammine cobaltic(III) chloride $[Co(NH_3)_5Cl]^{2+}$ in acidic medium. This reaction may he represented as follows :

$$\underset{\text{Oxidnat}}{[Co(NH_3)_5Cl]^{2+}} + \underset{\text{Reductant}}{[Cr(H_2O)_6]^{2+}} + 5H_3O^+ \rightarrow \underset{\text{Reduced product}}{[Co(H_2O)_6]^{2+}} + \underset{\text{Oxidised product}}{[Cr(H_2O)_5Cl]^{2+}} + 5NH_4^+$$

$$\underset{\text{Law-spin and inert}}{Co^{3+} \rightarrow 3d^6} \quad \underset{\text{High-spin and labile}}{Cr^{2+} \rightarrow 3d^6} \qquad \underset{\text{High-spin and labile}}{Co^{2+} \rightarrow 3d^7} \quad \underset{\text{Low-spin and inert}}{Cr^{3+} \rightarrow 3d^8}$$

The above reaction can he explained by the mechanism given below:

(i) $[\overset{III}{Co}(NH_3)_5Cl]^{2+} + [\overset{II}{Cr}(H_2O)_6]^{2+} \rightarrow [(NH_3)_5\overset{III}{Co}-Cl-\overset{II}{Cr}(H_2O)_6]^{4+} + H_2O$

(ii) $[(NH_3)_5\overset{III}{Co}-Cl-\overset{II}{Cr}(H_2O_6)]^{4+} \underset{\text{trasfer}}{\overset{\text{Electron}}{\rightarrow}} [(NH_3)_5\overset{II}{Co}]^{2+} + [\overset{II}{Cr}(H_2O)_5Cl]^{2+}$

(iii) $[(NH_3)_5\overset{III}{Co}]^{2+} + H_2O \rightarrow [(NH_3)_5\overset{II}{Co}(H_2O)]^{2+}$

(iv) $[(NH_3)_5\overset{III}{Co}(H_3O)]^{2+} + 5H_2O^+ \rightarrow [Co(H_2O)_6]^{2+} + 5NH_4^+$

From the above mechanism it is obvious that the above reaction proceeds through the following steps:

(i) The $[Cr(H_2O)_6]^{2+}$ loses a molecule of water to form an activated bridged intermediate with $[Co(NH_3)_5Cl]^{2+}$. This activated bridged intermediate contains Co^{3+} and Cr^{2+} ions link through Cl^- ion which acts as a bridge between the two coordinate spheres. The bridging Cl^- ion is provided to the activated intermediate by the inert reactant, $[CO(NH_3)_5Cl]^{2+}$.

This has been proved by the addition of radioactive Cl^- to the above reaction when no labelled Cl atoms are formed in the chlorochromium(III) complex; this experiment also indicates that no ionisation of complexes takes place.

(ii) In the activated intermediate, there occurs the transfer of an electron front Cr^{2+} to Co^{3+} through the bridging Cl^- ion to convert Cr^{2+} to Cr^{3+} and Co^{3+} to Co^{2+} . Then, the intermediate dissociates to give a 6-co-ordinated Cr(III) and a 5-co-ordinated Co(II) complex.

(iii) The 5-co-ordinated Co(II) complex then picks up the additional eater molecule from the medium to develop into a 6- coordinated Co(II) complex.

(iv) The 6-coordinated Co(II) complex formed in step (iii) is unstable. Therefore it undorgoes complete aquation give to hydrated Co(II) ion, thereby losing five ammonia molecules in the form of NH_4^+ ions.

The reaction studied has been found to be of first order with respect to the oxidant and reductant *i.e.*,

$$v = k\ [\text{Oxidant}]\ [\text{Reductant}]$$

Similar mechanism has been suggested when the chloride in the above example is replaced by other halide ions, sulphate, phosphate, acetate, succinate, oxalate, etc.

$$[\overset{III}{Co}(NH_3)_5X]^{2+} + [\overset{II}{Cr}(H_2O)_6]^{2+} + 5H_3O^+ \rightarrow [\overset{II}{Co}(H_2O)_6]^{2+} + [\overset{III}{Cr}(H_2O)_5X]^{2+} + 5NH_4^+$$

Here X = F^-, Cl^-, Br^-, I^-, SO_4^{2-}, NCS^-, PO_4^{3-}, $P_2O_7^{4-}$, CH_3COO^-, oxalate, succinate, etc.

Among halide ions, the effectiveness for bridges purposes, as measured by relative reaction rates, is $F^- < Cl^- < Br^- < I^-$. This order is in accordance with the expected order of ability to transmit an electron and undergo covalent bond-cleavae.

Among organic ions, oxalate and maleate are considerably more effective than acetate and succinate. Some redox reactions are known in which multiple bridges are formed. For example, in the following reaction three oxygen atoms of EDTA serve as bridges.

$$[\overset{II}{Co}(EDTA)]^- + \overset{II}{C}r^{2+} \rightarrow \overset{II}{C}o^{2+} + [\overset{II}{Cr}(H_2O)(EDTA)]^-$$

In reactions having low bridging groups, there occurs transfer of two ligands but only one electron. Conversely, in various $\overset{II}{Pt} \mid \overset{IV}{Pt}$ exchanges, there occurs transfer of one ligand but two electrons.

Certain reactions are known in which electron transfer by an inner-sphere mechanism is not accompartied by ligand transfer. For example,

$$[\overset{II}{Co}(EDTA)]^- + [\overset{III}{Fe}(CN)_6]^{3-} \rightarrow [\overset{III}{Co}(EDTA)]^{-1} + [\overset{II}{Fe}(CN)_6]^{4-}$$

Certain reactions are known in which inner and outer-sphere mechanisms occur together. An example of these is the reaction between cyanide ions and $Co(NH_3)_5X$. In this two distinct patterns of behaviour were noted, *i.e.*,

(i) When X (the ligand present in the complex) is Cl^-, N_3^-, NCS^- or OH^-, the reaction conforms to

$$Co(NH_3)_5X + 5CN^- \rightarrow Co(CN)_5X + 5NH_3$$

and the rate law of this equation is

$$v = k_1 [Co(NH_3)_5X][Co(CN)_5^{3-}]$$

(ii) When X= PO_4^{3-}, CO_3^{2-}, SO_4^{2-}, NH_3, OAc^- or other carboxylates, a different stoichiometry rate law would be obtained.

$$Co(NH_3)_5X + 6CN^- \rightarrow [Co(CN)_5X]^{3-} + 5NH_3 + X$$

$$v = k'_0 [Co(NH_3)_5X][Co(CN)_5^{3-}](CN^-)$$

In case (i), the substitution takes place by an inner sphere electron transfer between $[Co(CN)_5]^{2-}$ and $\overset{III}{Co}(NH_3)_5X$ through the bridged intermediate.

$$[(CN)_5\overset{II}{Co}-X-\overset{III}{Co}(NH_3)_5]$$

In case (ii), $[Co(CN)_5]^{3-}$ in the presence of cyanide is in equilibrium with a low concentration of $[Co(CN)_6]^{4+}$ and that this reacts with $[Co(NH_3)_6]^{3+}$ by the outer sphere mechanism. As $[Co(CN)_5]^{3-}$ is rapidly formed from Co^{2+} (aq) or $[Co(NH_3)_6]^{2+}$ and cyanide, it may he seen that Co^{2+} (aq) will catalyse the reactions. However, the mechanism is as follows:

$$[\overset{(III)}{Co}(CN)_5]^{3-} \underset{k}{\overset{+Co^+}{\rightleftarrows}} [\overset{(II)}{Co}(CN)_6]^{2-}$$

$$[\overset{(II)}{Co}(CN)_5]^{3-} + \overset{(III)}{Co}(NH_3)X_5 \xrightarrow{k_1} [(CN)_5\underset{(III)}{\overset{(II)}{Co}}-X-\overset{(III)}{Co}(NH_3)_5] \rightarrow \overset{(III)}{Co}(CN)_5X$$

$$[\overset{(II)}{Co}(CN)_6]^{4-} + \underset{(II)}{Co}(NH_3)_5X_5 \xrightarrow{k'} [(CN)_5\overset{(III)}{Co}(CN)(X)\overset{(III)}{Co}(NH_3)_5] \rightarrow [\overset{(III)}{Co}(CN)_6]^{3-}$$

The observed rate constant for the outer sphere mechanism has been found to be $k'_0 = kK$. From the kinetic data, it is evident that the outer-sphere process is largely independent of the nature of the ligand X.

TYPES OF ELECTRON-TRANSFER REACTIONS

These are as follows:

(a) **One equivalent-One equivalent Reactions :** These are the electron reactions in which there occurs the transfer of one electron from one species to the other. These simple reactions serve as models for more complicated systems and their study has proved invaluable in developing an understanding of the electron transfer in solution.

Examples of these reactions are as follows:

(i) $Ce^{3+} + Co^{3+} \rightarrow Co^{2+} + Ce^{4+}$

(ii) $Fe^{2+} + Co^{3+} \rightarrow Fe^{3+} + Co^{2+}$

The kinetic data for the oxidation of cerium (III) by cobalt (III) were obtained by following the disappearance of Co(III) at its absorption maximum of 605mμ. At this wave length, there is no interference from Ce (IV) or Ce (III). This reaction has been found to be of second order.

The kinetic data for the oxidation of iron (II) by cobalt (III) in perchloric acid are obtained by making the concentrations of reactant equal and then the reactant solution was sampled and run into a quenching solution made up of ammoniacal 2, 2′-bipyridine. Under the conditions employed, the rate law.

$$\frac{-d[Fe(III)]}{dt} = \frac{d[Co(III)]}{d} t = k'[Co(III)][Fe(II)]$$

has the integrated from

$$\frac{I}{Fe(II)} = k't + \frac{i}{[Fe(II)]_0}$$

Where k′ is related to half-life of reaction by

$$k' = \frac{I}{[Fe(II)]_0 t\sqrt{2}}$$

(b) **Two-equivalent-One-equivalent Reactions :** Many examples of these reactions are known. The reactions of Fe^{2+} with two equivalent oxidants are complex. These reactions must involve unstable states of iron or the oxidant. For example, two successive one electron transfers are possible for the oxidation of Fe(II) by Tl (III) in aqueous perchloric acid solution in which both Fe (II) and Fe (III) are competing for the Tl(II) intermediate.

$$Tl(III) + Fe(II) \underset{k_2}{\overset{k_1}{\rightleftarrows}} Tl(II) + Fe(III)$$

$$Tl(II) + Fe(III) \overset{k_3}{\rightleftarrows} Tl(I) + Fe(III)$$

A mechanism similar to the Tl(III)-Fe(II) has been demonstrated for the oxidation of V (IV) by Tl(III).

$$\text{Tl(III)} + \text{V(IV)} \rightleftarrows \text{Tl(II)} + \text{V(V)}$$

$$\text{Tl(II)} + \text{V(IV)} \rightarrow \text{Tl(I)} + \text{V(V)}$$

The rate data for the oxidation of Tl (I) by Ce (IV) in 6.2M nitric acid at 54°C indicate the following mechanism.

$$2\text{Ce(IV)} \rightleftarrows [\text{Ce(IV)}]_2$$

$$\text{Ce(III)} + \text{Ce(IV)} \rightleftarrows [(\text{Ce(III)}\,\text{Ce(IV)}]$$

$$\text{Ce(IV)} + \text{OH}^- \rightleftarrows \text{Ce(III)} + \text{OH}$$

$$\text{Tl(I)} + \text{OH} \rightleftarrows \text{Tl(II)} + \text{OH}^-$$

$$\text{Tl(I)} + \text{Ce(IV)} \rightleftarrows \text{Tl(II)} + \text{Ce(III)}$$

$$\text{Tl(II)} + \text{Ce(IV)} \rightleftarrows \text{Tl(III)} + \text{Ce(III)}$$

(c) **Two-equivalent-Two-equivalent Reactions :** An interesting example of these reactions is

$$\text{U(IV)} + \text{Tl(III)} \rightarrow \text{U(VI)} + \text{Tl(I)}$$

The acceptable mechanism of this reaction is as follows:

$$\text{U(IV)} + \text{Tl(III)} \rightarrow \text{U(V)} + \text{Tl(II)}$$

$$\text{U(IV)} + \text{Tl(II)} \rightarrow \text{U(V)} + \text{Tl(I)}$$

$$\text{U(V)} + \text{Tl(III)} \rightarrow \text{U(VI)} + \text{Tl(II)}.$$

ISOMERISATION OF OCTAHEDRAL COMPLEXES

As early as 1889, Jorgensen for the first time recognised the isomerisation of geometrical isomers of complex compounds. Although these arrangements are not among square-planar complexes, they have often been recognised among octahedral complexes. For example, the cis salts of $[Co(en)_2NO_2Cl]$ Cl, $[Co(en)_2(NO_2)_2]$ NO_3, $K_3[Ir(C_2O_4)2Cl_2]$ and $K_3[Rh(C_2O_4)_2Cl_2]$ on prolonged boiling or evaporation to dryness yield the corresponding trans isomers.

Isomerisation has been studied in considerable detail in few systems only. We shall now attempt to study this one by one.

(a) Dichlorobis (ethylenediamine) Cobalt (III) Ion

The cis-trans isomerisation most extensively studied is that between the praseo, trans-$Co(en)_2Cl_2^+$ and violeo, cis-$Co(en)_2Cl_2^+$ ions.

When an aqueous solution of green trans-[Co(en)$_2$Cl$_2$]Cl is concentrated on a steam both, violet crystals of cis-[Co(en)$_2$Cl$_2$]Cl are obtained. The hydrochloric acid solution of the latter salt on evaporation again yields green crystals of trans-[Co(en)$_2$Cl$_2$]Cl.HCl.2H$_2$O.

When the above system was studied by using radiochloride ion; it was concluded that isomerisation may occur as a result of an intermolecular process and also there is no direct replacement of the coordinated chloride by chloride ion.

The particular isomer separating from solution is mainly determined by the relative solubilities of the isomeric salts. For example, the less soluble cis-[Co(en)$_2$Cl$_2$]Cl is obtained from aqueous solution whereas the still less soluble trans-[Co(en)$_2$Cl$_2$](H$_2$O)Cl is separated from hydrochloric acid solution.

Thus, the main role of hydrochloric acid is that of a precipitant and not that of opening of the chelate ring. This inference has been confirmed by using trans-[Co(en*)$_2$Cl$_2$]Cl. After dissolving this salt in cold water, it is again precipitated by the addition of HCl acid. Hydrogen chloride liberated at 110°C by this hydrochloride salt was not radioactive.

In order to explain the above observations, the following mechanism has been postulated:

(i) $\text{cis-Co(en)}_2\text{Cl}_2^+ + H_2O \rightleftarrows \text{cis-Co(en)}_2(H_2O)Cl_2$
$\uparrow\downarrow Cl^-$

(ii) $\text{trans-Co(en)}_2\text{Cl}_2^+ + H_2O \rightleftarrows \text{trans-Co(en)}_2(H_2O)Cl_2^+ + Cl^-$

In the above mechanism, step (i) is correct but (ii) is not because the equation of trans-Co(en)$_2$Cl$_2^+$ yields directly a mixture of 35% cis- and trans-Co(en)$_2$ H$_2$O Cl$_2^+$.

On the basis of dissociation mechanism, it is evident that the cis and trans isomers may have a common trigonal pyramidal intermediate which readily provides a path for isomerisation Fig. 3.

$$X\text{–}M(AA)_2\text{–}X \underset{+X}{\overset{-X}{\rightleftarrows}} X^{(4)}\text{–}M\ (2)(3) \rightleftarrows X\text{–}M\text{–}X$$

Fig. 3 : Isomerisation of cis-trans isomers of the type M(AA)$_2$X$_2$ by a dissociation mechanism.

The above mechanism is not limited to any one set of cis-trans isomers but to those in which X can be either chloride ion or water.

(b) Diaquo cis (ethylenediamine) Cobalt (III) Ion

When the absorption spectra of aqueous solutions of cis- and trans-$Co(en)_2\ (H_2O)_2^{3+}$, are recorded separately, their spectra change rapidly to the same spectrum, indicating that an equilibrium mixture of cis- and trans-isomers is obtained. Later on it was found that this change is much slower in acid solution than in water.

This means that the hydroxoaquo species, $Co(en)_2\ (H_2O)\ OH^{2+}$, undergoes isomerisation more readily than do the diaquo complexes. Further, ΔH^0 for isomerisation varies with temperature.

These observations are indicative of the fact that more than one mechanism are operating in the isomerisation of $Co(en)_2\ (OH)_2^+$. One of the paths is corresponding to Co-O bond cleavage while the other to the Co-N bond rupture.

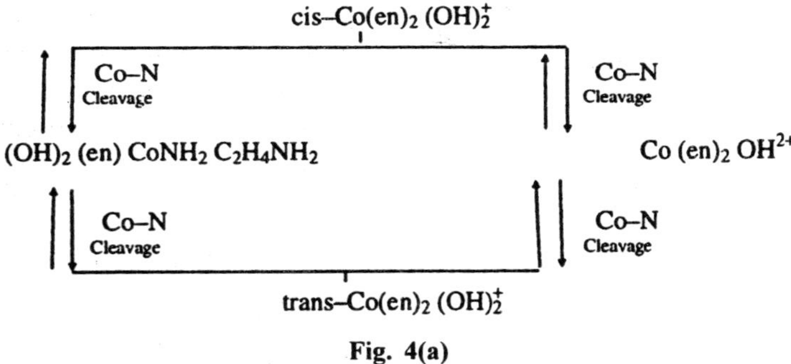

Fig. 4(a)

It is quite probable that the two processes should be able to compete.

(c) Aquoamminebis (ethylene) Diamine Cobalt (III) Ions

As the isomerisation of trans-$Co(en)_2\ NH_3OH^{2+}$ into the cis isomer takes place without oxygen exchange, this reveals that an intra-molecular mechanism must be involved.

This mechanism may occur by the opening and reclosing of one of the ethylene diamine chelate rings (Fig. 5).

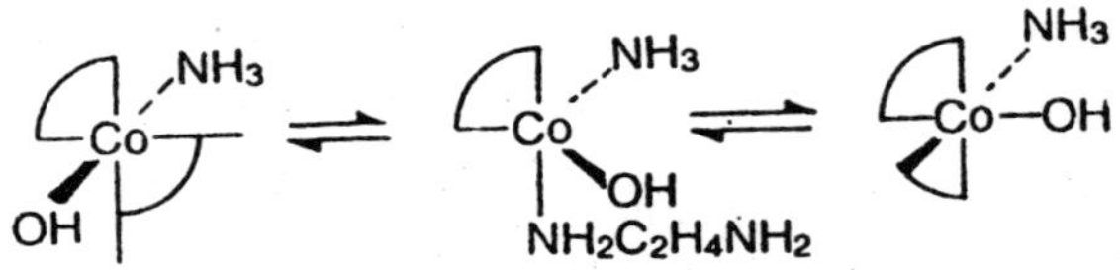

Fig. 5 : Isomerisation of cis- and trans- $Co(en)_2NH_3OH^{2+}$ by an intramolecular mechanism involving.

(d) 1.10-Bis(salicylidieneamino)-4,7 dithiodecane cobalt (III) cation

Dwyer and Lions prepared many coordination compounds using sexadentate ligands of the type:

$$\text{—CH} = \text{N}\,(\text{CH}_2)_x\ \text{S}\,(\text{CH}_2)_y\ \text{S}\,(\text{CH}_2)_z\ \text{N} = \text{CH—}$$
$$\text{—OH} \qquad\qquad\qquad\qquad\qquad\qquad \text{OH—}$$

Where x, y and z may be either two or three.

When x = y = z = 2, a green Co (III) complex is isolated whereas when x = y = z = 3, a brown product is obtained.

As the C=N-C grouping be linear, this is possible in the green compound having rigid five membered chelate ring (Fig. 6). However, the greater flexibility of the six membered chelate rings permits the existence of structure which is assigned to the brown form (Fig. 6).

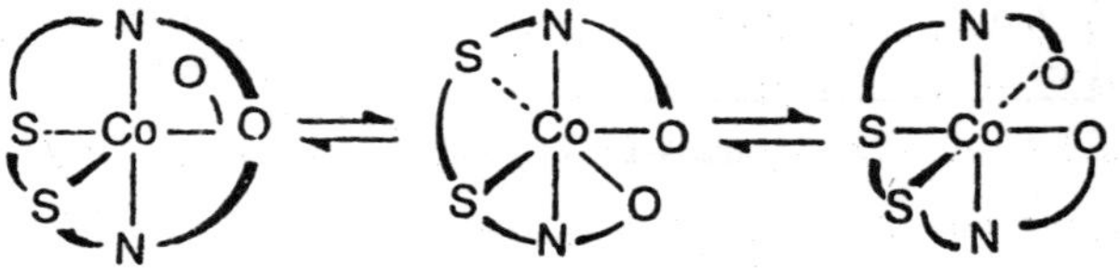

Fig. 6 : Intramolecular isomerisation of 1, 10-bis (salicyldeneamine)-6, 7-dithiodecae-cobalt (III) ion by a disociation machanism.

Excellent proof that the isomerisation in green and brown form takes place by an intramolecular process is offered by the observation that these optically active complexes isomerise readily without any loss of optical rotation.

KINETIC ASPECTS OF COMPLEXES

Liability, Inertness, Stability and Instability : Depending upon their reactivity, Taube classified the complexes into two types:

(A) Labile or Non-inert Complexes

These are the complexes which undergo ligand replacement reactions rapidly.

(B) Inert Complexes

These are the complexes which undergo ligand replacement reactions slowly. In order to make this distinction quantitatively, Taube suggested that complexes which undergo ligand replacement within one minute or so at 25°C and 0.1 M reactant concentration are arbitrarily termed labile or non-inert complexes, other less reactive complexes are referred to as inert.

The four-coordinated complexes of Cu^{2+} are labile while the six-coordinated complexes of Rh^{3+} are inert.

$$[Co(H_2O)_4]^{2+} + 4NH_3 \rightarrow [Cu(NH)_4]^{2+} + 4H_2O$$

$$[RhCl_4]^{3-} + C_2O_4^{2-} \rightarrow [Rh(C_2O_4)_3]^{3-} + 6Cl^-$$

It is interesting to note the inertness has no relation with stability which is determined thermodynamically. For example, although the complexes like $[Ni(CN)_4]^{2-}$, $[Mn(CN)_6]^{3-}$ and $[Cr(CN)_6]^{3-}$ are, having high stability constants yet the rate of exchange of CN^- by labelled $^{14}CN^-$ gives half time period as 30s, lh and 24 days, respectively. Hence, $[Ni(CN)_4]^{2-}$ is labile whereas $[Mn(CN)_6]^{3-}$ and $[Cr(CN)_6]^{3-}$ are inert.

Half life

$$[Ni(CN)_6]^{2-} + 4\ {}^{14}CN^- \rightleftarrows [Ni({}^{14}CN)_4]^{2-} + 4CN^- \quad \sim 30s \quad ...(1)$$

$$[Mn(CN)_6]^{3-} + 6\ {}^{14}CN^- \rightleftarrows [Mn({}^{14}CN)_6]^{3-} + 6CN^- \quad \sim Ih \quad ...(2)$$

$$[Cu(CN)_6]^{3-} + 6\ {}^{14}CN^- \rightleftarrows [Cr({}^{14}CN)_6]^{3-} + 6CN^- \quad \sim 24d \quad ...(3)$$

It is important to distinguish between instability and lability. From simple rate theory, the liability of complex is dependent upon the activation energy, *i.e.*, the difference in energy between the reactants and activated complex; a low activation energy result in a fast reaction.

Instability is generally decided by the difference between the free energies of the reactants and products (loosely by the beat of reaction).

The $[Ni(CN)_4]^{2-}$ complex ion is a good example of thermodynamically stable complex which is kinetically labile. The classic example of the kinetically inert complex which is thermodynamically unstable, is the

$[Co(NH_3)_6]^{3+}$ in acid solution. One might expect it to decompose as follows:

$$[Co(NH_3)_6]^{3+} + 6H_3O^+ \rightarrow [Co(H_2O)_6]^{3+} + 6NH_4^+ \quad ...(4)$$

The tremendous thermodynamic driving force of six basic ammonia molecules attaching to six protons results in an equilibrium constant for the above reaction (4) of 10^{25}. There is another way for expressing the difference between stable and inert complexes. For instance, stable complexes have large positive free energies of reaction. ΔG while inert complexes merely have large positive free energies of activation ΔG^*. According to Taube, the substitution reactions of the inert complexes might be studied by the classical methods of changes in absorbance, pH, gas evolution, etc., and followed directly while the substitution reactions of the labile complexes might be studied by modern techniques like flow systems, electronic recorders, and relaxation effects.

Kinetic and Reaction Rates of Octahedral Substitution

Using MOT treatment for the kinetics of exchange of water from the aqua cation, Gray and Langford (1968) classified the metal ions into the following four categories.

1. Class I.

This includes the alkali metals and larger alkaline earth metals. The complexes of these metal ions are bound by essentially purely electrostatic forces. The Z^2/r ratio for these ions ranges upto about $10C^2m^{-1}$. The exchange of water for the complexes is extremely fast. In general, even the fastest kinetic techniques fail to follow the reactions. First order rate constants have been found to be of the order of 10^8 sec^{-1}.

2. Class II.

This includes the dipositive transition metals, Mg^{2+} and tripositive ions. The bonding in the complexes of there ions is somewhat stronger than that of complexes in class I, but LFSE'S are somewhat smaller. The Z^2/r for these ions ranges from about to 10 to $30C^2m^{-1}$. The exchange of water has been found to be fast. First order rate constants for exchange of water have been found to range from 10^5 to 10^8 sec^{-1}.

These reactions have been studied by the fastest techniques in which the equilibrium is perturbed by a fast variation of a physical parameter like pressure (ultrasonic or "P-Jump" methods) or temperature (T-Jump" method) and the response of the system has been used to estimate rate of reaction.

3. Class III.

This includes the tripositive transition metal ions (stabilised to some extent by LFSE), and two very small ions Be^{2+} and La^{3+}. The Z^2/r ratios for these ions have been found to be greater than about 30 C^2m^{-1}. When compared to classes I and II, the exchange of water has been found to be relatively slow although the first order rate constants have been found to range from 1 to $10^4 sec^{-1}$. It is possible to follow the reaction by traditional kinetic technique but by augmenting with flow techniques.

4. Class IV.

This includes metal ions Cr^{3+}, Co^{3+}, Pt^{2+}. The sizes of the ions of this class are comparable to class III ions. Also these ions exhibit considerable LFSE : Cr^{3+}, (d^3), Co^{2+} (low-spin d^6), Pt^{2+} (low-spin d^8). The complexes of these metals ions are only inert complexes. The exchange of water has been found to be slow. The first order rate constants may range from 10^{-1} to $10^{-9} sec^{-1}$.

The above classification of metal ions is summarised in Table 1.

Table 1 : Classification of the metal ions as labile and innert

Class	*Rate of Exchange of water*	*First order rate constant (s^{-1})*	*Metal ions*
I	Extremely fast	$>10^8$	Alkali and alkaline earth metals
II	Fast	$10^5 - 10^8$	Dipositive transition metals, Mg^{2+} and tripositive lanthanides
III	Slow	$1 - 10^4$	Tripositive transition metal ions (stabilised by LFSE to some extent only) and two very small ions, Be^{2+} and Al^{3+}.
IV	Extremely slow	$10^{-9} - 10^{-1}$	Ions with high CFSE Cr^{3+} (d^3), Co^{3+} (low-spin d^6), Pt^{2+} (low-spin d^8).

Ligand Field Effects and Reactions Rates

Ligand filed theory has been applied with moderate success to the problem of reactivity in octahedral complexes by Basolo and Pearson (1967). These workers carried out calculations which were based upon the concept of ligand-field stabilisation energy-(LFSE). From these calculations, these workers found a pattern of kinetic behaviour in terms of the electronic configuration of the central metal ion. For example,

(i) If the LFSE of the activated complex is greater than that of the coordinate compound, then on reaction via. the transition state, there will be a gain of LFSE which will result in a low activation energy and a rapid reaction.

(ii) If the activated complex is less LFSE stabilised than the original substrates, there will be an increase in activation energy and a rate of reaction.

Table 2 : Changes in LFSE (unit dq) upon changing A 6-coordinate complex to a 5-coordinate (square pyramidal) of a 7-coordinate (Pentagonal Bypyramidal) species

Systems	*High* CN = 5	*Spin* CN = 7	*Low* CN = 5	*Spin* CN = 7
d^0	0	0	0	0
d^1	+0.57	+1.28	+0.57	+1.28
d^2	+1.14	+2.56	+1.14	+2.56
d^3	–2.00	–4.26	+1.14	+2.56
d^4	+3.14	–1.07	–1.43	–2.98
d^5	0	0	–0.86	–1.70
d^6	+0.57	+1.28	–4.00	–8.52
d^7	+1.14	+2.56	+1.14	–5.34
d^8	–2.00	–4.26	–2.00	–4.26
d^9	–3.14	–1.07	+3.14	–1.07
d^{10}	0	0	0	0

By taking the five-coordinated square pyramid and seven-co-ordinated pentagonal bipyramid structures as approximations to the transition states, adopted by an octahedral complex reacting via a dissociative and associative mechanism respectively, the results collected in Table 2 are obtained.

From Table (2), it may be seen that systems which suffer a loss in LFSE, on either mechanism, are the either d^3, d^8 and low spin (strong field) d^4, d^5 and d^6. Such configurations give rise to inertness whereas other configurations give rise to liability. This line of demarcation has been verified experimentally.

For tripositive metal ions, the order to liability follows from the magnitude of LFSE losses and the liability is expected to increase as follows,

$$Co(III) < Cr(III) < Mn(III) < Fe(III) < Ti(III) < Ga(III) < Sc(III)$$

The comparison of dipositive with tripositive species is difficult because the rate is also affected by the charge on the central metal ion which strengthens the metal-ligand bonds. For instance, in the non-transition metal series the liability decreases in the order shown with SF_6 being exceptionally inert.

$$[AlF_6]^{3-} > [SiF_6]^{2-} > [PF_6]^{-} > SF_6,$$

Due to greater liability of dipositive metal, octahedral Ni(II) is considerably more labile than corresponding tripositive ions in line with the exchange rate shown in Eqs. (1) to (3). In case of dipositive metal ions, it has been found that other ions like Mn(II), Fe(II), Co(II) and Cu(II) that possess less LFSE are even more labile.

The differences between Pt(II) and Ni(II) complexes can be attributed to the (i) higher LFSE for Pt(II); (ii) availability of antibonding d2 orbital for accepting an electron pair of the fifth ligand by the associated mechanism.

Mechanisms of Nucleophilic (ligand) Substitution Reactions (SN Reactions) in Octahedral Complexes

The study of substitution mechanism has occupied many works for a long time and continues to be an active area of investigation. The generalised substitution reaction may be written as follows.

$$MX_nY + Z \rightarrow MX_nZ + Y$$

where M is the metal ion, Y is the leaving group, Z is the entering ligand and X_n includes all the other ligands which are not undergoing substitution. The value of n is 5 for an octahedral complex.

For racemisation reaction, Z is the solvent molecule which may of may not be present in the final product.

For acid hydrolysis reaction, Z is water or proton. For base hydrolysis reaction, Z is OH^-.

For anation reaction, Z is anion. These reactions are known as anion substitution reactions. These reactions proceed in two steps involving water molecules.

For exchange reactions, Z and Y are same; Z is usually a labelled species for exchange reactions.

Two different mechanisms have been proposed for substitution reactions in octahedral complexes:

1. Dissociative S_{N^1} Mechanism

S_{N^1} indicates substitution (S) nucleophilic (N) unimolecular of first order reaction (1). The reaction is nucleophilic as the ligand is seeking a positive centre, *i.e.,* the metal ion.

Dissociative S_{N^1} mechanism is supposed to involve the following two step:

(i) In the first step, the octahedral complex MX_5Y is dissociated to Y and MX_5. The latter is a five-coordinated intermediate activated complex which has either square pyramidal or trigonal pyramidal shape. This activated complex, MX_5 is an electron-deficient intermediate.

$$\underset{(CN=6)}{MX_5Y} \xrightarrow{\text{Slow}(-Y)} \underset{(CN=6)}{MX_5 + Y}$$

The above dissociative step is a metal ligand bond breaking step and the reaction occurring in this step is unimolecular because this step involves only one reactant species, MX_5Y. Further this step is slow and therefore it is the rate determining step.

(ii) In the second step, the nucleophilic reagent Z attacks the short-lived pentacoordinated intermediate activated complex MX_5 of very limited stability to give the complex, MX_5Z.

$$\underset{(CN=5)}{MX_5} \xrightarrow{+Z} \underset{(CN=6)}{MX_5Z}$$

This step is fast and bimolecular. But the activation energy for the first step is high while the activation energy for the second step is low. Therefore, the rate of overall reaction will not depend upon the concentration of Z but on that of MX_5Y. Hence, the rate of reaction is of first order with respect to MX_5Y and of zero order with respect to Z. Thus,

$$\text{Rate of reaction} = k_1[MX_5Y]$$

The formation of MX_5Z can be shown by combining the steps (i) and (ii).

$$\underset{(CN=6)}{MX_5Y} \xrightarrow[\substack{\text{Slow;}\\ \text{unimolecular}\\ \text{and rate}\\ \text{determining step}}]{-Y} \underset{\substack{\text{5-Coordinate}\\ \text{intermediate}\\ \text{complex}\\ (CN=5)}}{MX_5} \xrightarrow[\text{Fast}]{-Z} \underset{(CN=6)}{MX_5Z}$$

The formation of MX_5Z form MX_5Y may be shown diagrammatically as in Fig. 7.

For the S_{N^1} mechanism, the following is expected

(i) As there occurs the dissociation of the ligand completely from the octahedral complex, it is expected that the trans effect of the ligands would not be operative. The evidence for the presence of trans effect in octahedral complex is not abundant.

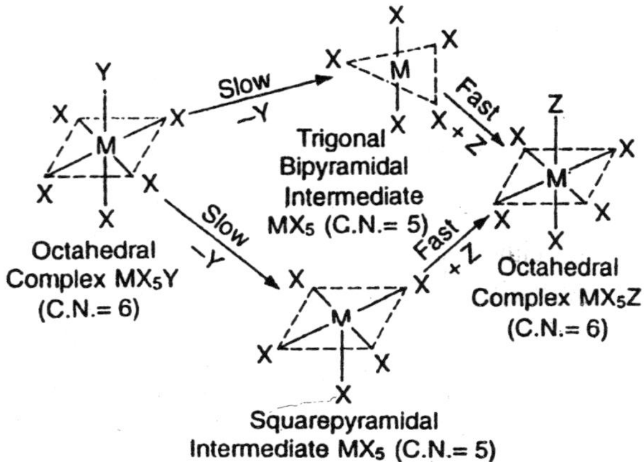

Fig. 7 : The dissociation S_{N^1} mechanism.

However, there is definite evidence for a trans influence of the ground state of cobalt (III) complexes of the type $[Co(NH_3)_5X]^{2+}$.

Significant length is found in the Co–N bond trans to the nitrosyl group and shortening trans to a chloride ion. In metal carbonyls, trans effect is operative to the extent that substitution takes place readily as long as carbonyl groups are trans to each other.

(ii) The rate of S_{N^1} substitution should be inversely proportional to the strength of Co–Y bond. Further, this rate should also depend upon charge, steric factors and chelation effects of the leaving group, Y.

(iii) The value of k_1 has been found to be independent of the nature of Z as well as its concentration except for the OH^- group for which the reaction is of second order.

(iv) If there occurs an increase in the electron density on metal atom by the electron donors in X_n, this should assist M–Y bond breaking. This may be seen from the k_1 values for the acid hydrolysis of $[Co(en)_2(Y–p)ClYJ$ (where Pk and k_1 are Y = H : 8.82, $1.1 \times 10^{-5} s^{-1}$: m-$CH_3$8.19, $1.3 \times 10^{-5} s^{-1}$:p- CH_3:7.92, $1.4 \times 10^{-6} s^{-1}$: p – OCH_3,7.53, $1.5 \times 10^{-5} s^{-1}$).

(v) Cis effect. If ligands are having another pair of electrons like CNS^- or OH^- this increases the rate of hydrolysis of complex about ten fold when present cis to Y as compared to the rate if they are present trans to Y.

This rate increase has been attributed to the stabilisation of the square pyramidal complex by the electron pair donated by OH^- or CNS^- along the cis position through p-d-pi bonding. The product is 100 per cent cis isomer. However, the ligands, which are either not having an extra pair of electrons (NH_3) or are themselves π acceptors (NO_2^-, CO, NO, etc..) do no show the is effect (Fig. 8).

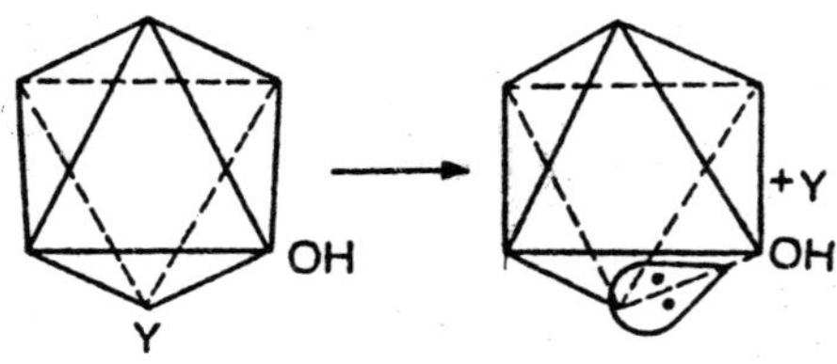

Fig. 8 : The cis effect of the hydroxyl group in stabilising the square pyramidal intermediate.

2. Associative or Displacement, S_{N^2} Mechanism

In this mechanism, SN2 indicates substitution (S) nucleophilic (N) bimolecular or second order (2) reaction. This mechanism also involves the following two steps:

(i) In the first step, the incoming nucleophile Z attaches MX_5Y to form a seven coordinated unstable transition state intermediate which is probably pentagonal bipyramidal in shape. Eventually, it is a metal-ligand bond making step.

$$\underset{(CN=6)}{MX_5Y} \xrightarrow{+Y} \underset{\substack{\text{Unstable seven coordinated} \\ (CN=7)}}{MX_5YZ}$$

As this step involves the association of Z with MX_5Y to form MX_5YZ, hence the name association mechanism is given. Further this step is slow. it is rate determining. Also, this step is bimolecular because this step involves two reactions viz, MX_5Y and Z. Therefore, the rate is of first order with respect to the complex MX_5Y and also of first order with respect to the entering ligand, *i.e.*,

Rate of reaction = $k_2\,[MX_5Y][Z]$

(ii) In the second step, Z is added to MX_5Y at the same time or shortly thereafter and Y leaves rapidly to give MX_2Z. This is a fast step and is not a rate determining step.

$$\underset{(CN=7)}{MX_5YZ} \xrightarrow{-Y} \underset{(CN=6)}{MX_5Z}$$

Both the steps (i) and (ii) may he represented as follows:

$$\underset{(CN=6)}{MX_5X} \xrightarrow[\substack{\text{Bimolecular and rate} \\ \text{determining step}}]{\text{slow}(+Z)} \underset{\substack{\text{Seven coordinated} \\ \text{unstable intermediate} \\ (CN=5)}}{MX_5YZ} \xrightarrow{\text{Fast}(-Y)} \underset{(CN=6)}{MX_5Z}$$

Both the steps (i) and (ii) are shown diagrammatically in Fig. 9.

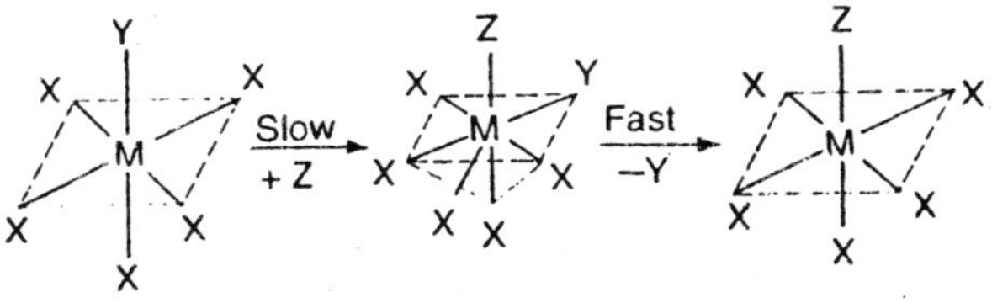

Slow +Z → Fast −Y →

Octahedral Compex MX_5Y (C.N.= 6) | Seven Coordinated Unstable Intermediate MX_5YZ (C.N.= 7) | Octahedral Compex MX_5Z (C.N.= 6)

Fig. 9 : S_{N^2} mechanism for the substitution reaction.

Comparison between S_{N^1} and S_{N^2} Mechanisms

The various differences between these mechanisms are as follows:

(i) In the S_{N^1} mechanism, the rate determining step is metal-ligand bond breaking step because the coordination number of 6 of the complex MX_5Y is reduced to 5 which is the coordination number of the intermediate, MX_5.

In the S_{N^2} mechanism, the determining step is metal-ligand forming step because the coordination number ion 6 of the complex MX_5Y is increase to 7 which is the coordinate number of intermediate MX_5YZ.

(ii) The rate of S_{N^1} mechanism has been found to be of first order with respect to MX_5Y, *i.e.*, it means that the rate determining in S_{N^1} mechanism is unimolecular.

The rate determining step for S_{N^2} mechanism has been found to be bimolecular, *i.e.*, it means that its rate of reaction is of second order, of first order with respect to MX_5Y and also of first order with respect to Z. Hence,

$$\text{Rate of } S_{N^1} = k_1[MX_5Y]$$

$$\text{and Rate of } S_{N^2} \text{ mechanism} = k_2[MX_5Y][Z]$$

Anation Reactions

The replacement of water from an aqua complex by an anionic group is called anation reaction. This reaction is reverse of an acid hydrolysis reaction, *i.e.*, aquation. An example of anation reactions is

$$[CoX_5(H_2O)]^{3+} + Z^- \rightarrow [CoX_5Z]^{2+} + H_2O$$

Mechanism

Although some anation reactions are of second order yet this does not indicate an associative mechanism. However, the following dissociative mechanism is more likely:

(i) $$[CoX_5(H_2O)]^{3+} \underset{k_2}{\overset{k_1}{\rightleftarrows}} [CoX_5]^{3+} + H_2O$$

(ii) $$[CoX_5]^{3+} + Z^- \rightarrow [CoX_5Z]^{2+}$$

Rate of formation of $[CoX_5]^{2+}$ is given as:

$$\text{Rate} = \frac{d[CoX_5Z]^{2\cdot}}{dt} = K_3\ [CoX_5]^{3+}\ [Z^-] \qquad ...(1)$$

On applying steady statement treatment to steps (i) and (ii), we get

$$[CoX_5]^{3\cdot} = \frac{k_1[CoX_5(H_2O)]^{3}}{k_2 + k_3[Z^-]} \qquad ...(2)$$

On substitution equation (2) in (1), we get

$$\text{Rate} = \frac{d[CoX_5Z]^{2\cdot}}{dt} = \frac{k_1k_3[CoX_5(H_2O)]^{3\cdot}[Z^-]}{k_2 + k_3[Z^-]} \qquad ...(3)$$

If $[Z^-]$ is low and $k_2 >> k_3\ [Z^-]$, equation (3) modifies to

$$\text{Rate} = \frac{k_1k_3}{k_2}[CoX_5(H_2O)]^{3}\ [Z^-] \qquad ...(4)$$

$$\text{Rate} = \frac{k_1k_3}{k_2 + k_3}[CoX_5(H_2O)]^{3+}[Z^-]$$

If $[Z^-]$ is high and $k_2 >> k_3\ [Z^-]$, equation (3) modifies to

$$\text{Rate} = k_1\ [CoX_5(H_2O)]^{3+} \qquad ...(5)$$

From equations (4) and (5), it follows that the rate of anation reaction, depending upon the conditions applying may or not be dependent upon the concentration of the entering substituent. In principle, there occurs a gradual change from second to first-order kinetics as the concentration of substituent increases. Once the anation reaction becomes of first order, its rate should be independent of the nature of the incoming group and its observed rate constant should be equal to that for the isotopic exchange between water and aquo complex.

When attempts were made to detect the above mentioned effects in reactions of cobalt (III) amines with azide, thiocyanate and shiphate, these proved futile because in these reactions there also occurs outer-sphere association between the oppositely charged reactions at the high anionic concentrations so as to achieve the anion- independent limit. Under these conditions, there occurs competition between an ion path of second order and a reaction via a dissociative process. Thus,

(i) $$[CoX_5(H_2O)]^{3+} + Z^- \underset{k_2}{\overset{k_1}{\rightleftarrows}} [CoX_5(H_2O)]^{3+}\ Z^-$$

(ii) $[CoX_3(H_2O)]^{3+} Z^- \xrightarrow{k_3} [CoX_5Z]^{2+} + H_2O$

When steady-state approximation is applied to steps (i) and (ii), the rate of formation of $[CoX_5Z]^{2+}$ is given by

$$\frac{d}{dt}[CoX_5Z]^{2+} = \frac{k_1k_3}{k_2+k_3}\left[CoX_5(H_2O)\right]^{3+}\left[Z^-\right]$$

In order to avoid the complication of ion-pairing. Haim and Wilmarth used a cationic complex Co(III) in place of its anionic complex. With this complex, these workers were successful in demonstrating a dissociative mechanism in the reaction of aquopenta cyano cobaltate (III), $[Co(CN)_5(H_2O)]^{2-}$ with N_3^-, and SCN^- ions. Their kinetic results were interpreted by the following scheme:

$$[Co(CN)_5(H_2O)]^{2-} \xrightarrow{k_1} [Co(CN)_5]^{2-} + H_2O$$

$$[Co(CN)_5]^{2-} + Z^- \xrightarrow{k_2} [Co(CN)_5Z]^{3-}$$

When the concentration of Z^- become very high, the reaction becomes of second order in $[Co(CN)_5(H_2O)]^{2-}$, *i.e.*,

$$\frac{d\left[Co(CN)_5Z\right]^{-3}}{dt} = k_{obs}\,[Co(CN)_5]^{2-} + H_2O$$

where $$k_{obs} = \frac{k_1k_3\left[Z^-\right]}{k_2+k_3\left[Z^-\right]}$$

The values of k_2/k_3 reveal that azide is a better nucleophile to $[Co(CN)_s]^{2-}$ than thiocyanate. When similar type of work is extended to other ligands on the $[Co(CN)_5]^{2-}$, the following order of reactivity has been reported.

$$OH^- > N_3^- > SCN^- > I^- > BR^- > S_2O_3^{2-} > CNO^- > H_2O$$

It is interesting to note that a cyanide group, *i.e.*, CN^-, favours an associative mechanism S_{N^2} in organic systems while it favours a dissociative mechanism in inorganic systems.

The difference in the apparent stability of $[Co(CN)_5]^{2-}$ and $[Co(NH_3)_5]^{3-}$ lies in the special nature of the cyano group. This may he seen as follows:

(i) Due to strong electron donation from the CN^- group to the metal in the a bond, there occurs an accumulation of negative charge on the metal atom which is only partly removed by back π-bonding. This weakens the Co–H_2O bond and promotes an S_{N^i} mechanism.

(ii) In the high energy five coordinated species, p-bonding becomes more important.

(iii) Also an increase in bond angle would be expected to stabilise the five-coordinated intermediate $[Co(CN)_5]^{2-}$.

Substitution Reactions Without Breaking Metal-ligand Bond

The substitution reactions discussed above involve the cleavage of metal-ligand bond. However, there are substitution reactions in which ligand exchange takes place without cleavage of metal-ligand bond, *i.e.*, these substitution reactions occur with preservation of metal-ligand bond. An interesting example of such reactions is found in preparative coordination chemistry which is the rapid conversion of carbonate ammine cobalt (III) complexes like $[CO(NH_3)5CO_3]^+$ into the corresponding aquo complexes by the addition of excess acid.

$$[(NH_3)_5Co{-}O{-}CO_2]^+ + 2H^+ \rightarrow [(NH_3)_5Co{-}OH_2]^{3+} + CO_2$$

When the above reaction is performed in the presence of ^{18}O-labelled water, it is found that it is the O–O bond, rather than the Co–O bond which breaks, as both products do not show any uptake of ^{18}O. This result has been further confirmed from the observation that reactions involves fission of Co–O bonds are invariably slow.

$$\underset{\text{(Caronatocomplex)}}{\left[(NH_3)_5Co-O-CO_2\right]^+} + 2H_3{}^{18}O^+ \rightarrow \underset{\text{(Aquocomplex)}}{\left[(NH_3)_5Co-O-OH_2\right]^{3+}} + CO_2 + 2H_2{}^{18}O^+$$

Mechanism

Kinetic study reveals that the reacting species is the bicarbonate complex, $[Co(NH_3)5CO_3H]^{2+}$ which is believed to be formed by the attack of proton on the oxygen atom bonded to Co.

$$\left[(NH_3)_5Co-O-C\begin{matrix} \diagup O \\ \diagdown O \end{matrix}\right] + H^+ \overset{\text{fast}}{\rightleftarrows} \left[(NH_3)_5Co-O-C\begin{matrix} \diagup O \\ \diagdown OH \end{matrix}\right]^{2+}$$

$$\uparrow\downarrow \text{ Fast}$$

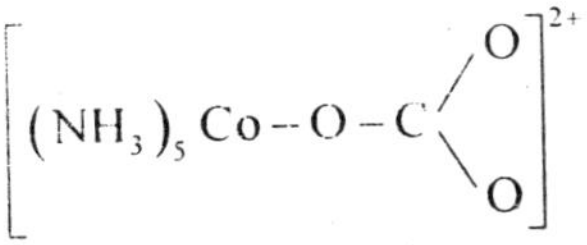

$$\left[(NH_3)_5\,Co-O-C\begin{matrix}\diagup O\\ \diagdown O\end{matrix}\right]^{2+}$$

Decarboxylation of the above formed bicarbonate complex then occurs as the rate-determining step.

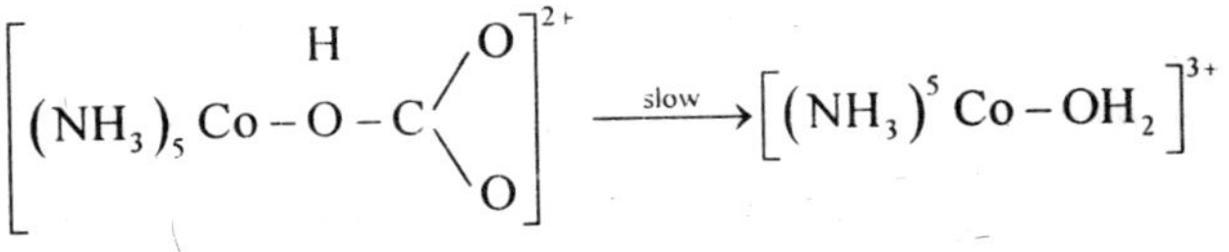

$$\left[(NH_3)_5\,Co-\overset{H}{O}-C\begin{matrix}\diagup O\\ \diagdown O\end{matrix}\right]^{2+} \xrightarrow{\text{slow}} \left[(NH_3)^5\,Co-OH_2\right]^{3+}$$

followed by

$$[(NH_3)_5\ Co–OH]^{2+} \rightleftarrows [NH_3)_5\ Co–OH_2]^{3+}$$

Thus, this is a decarboxylation reaction rather than an acid hydrolysis reaction.

Another reaction which does not involve the cleavage of metal-ligand bond is the formation of nitro complex $[Co(NH_3)_5\ NO_2)^{2+}$ from its corresponding chloro complex $[Co(NH_3)_5Cl]^{2+}$ by reaction with NO_2^-.

$$[Co(NH_3)_5\ Cl]^{2+} + NO_2^- \rightleftarrows [Co(NH_3)_5NO_2]^{2+} + Cl^-$$

This reaction has been studied in considerable detail. This has been confirmed that the formation of nitro complex occurs indirectly through the following sequence:

Chloro complex → aquo complex → nitrito complex → nitro complex.

By tracer experiments, it has been established that the Co–O bond of the aquo complex does not undergo cleavage in the formation of the nitrito form because the following reaction takes place with complete retention of ^{18}O.

$$\underset{\text{Aquo complex}}{\left[(NH_3)_5\,Co-{}^{18}OH_2\right]^{3+}} + NO_2^- \xrightarrow{H_2O} \left[(NH_3)_5\,Co-{}^{18}ONO\right]^{2+} + H_2O$$

The formation of nitro complex from chloro complex involves following mechanistic scheme:

(i) $[(NH_3)_5\ Co–Cl]^{2+} + H_2O \rightarrow [(NH_3)_5\ Co–OH_2]^{3+} + Cl^-$

(ii) $[(NH_3)_5\ Co–OH_2]_3 \rightleftarrows [(NH_3)_5\ Co–OH]^{2+} + H^+$

(iii) $2HNO_2 \rightleftarrows N_2O_3 + H_2O$

(iv) $[(NH_3)_5 Co-OH]^{2+} + N_2O_3 \rightleftarrows (NH_3)_5Co-O...H^+$
$|$
$O-N...NO_2^-$

$\uparrow\downarrow$

$[(NH_3)_5 Co-ONO]^{2+} + HNO_2$

(v) $[(NH_3)_5 Co-ONO]^{2+} \rightleftarrows [(NH_3)_5 Co-NO_3]^{2+}$

In the above mechanism, step (iv) is the rate determining step.

Hydrolysis Reactions

There are certain substitution reactions in which a ligand is replaced by a water molecule or by OH^- group. Such reactions are known as hydrolysis reactions.

$$[MX_5Y]^{n+1} + H_2O \rightarrow [MX_5H_2O]^{n+1} + Y^-$$

$$[MX_5Y]^{n+1} + OH^- \rightarrow [MX_5OH]^{n} + Y^-$$

Hydrolysis reactions are of two types:

(a) Acid Hydrolysis or Aquation Reactions

These are the reactions in which an aquo complex is formed as a result of the replacement of a ligand by H_2O molecules. These reactions occur in acid or neutral solutions. Examples of acid hydrolysis reactions are as follows:

$$[Co(NH_3)_5Cl]^{2+} + H_2O \rightarrow [Co(NH_3)_5H_2O]^{3+} + Cl$$

$$[Co(en)_2ACl]^{+} + H_2O \rightarrow [Co(en)_2A(H_2O)]^{2+} + Cl^-$$

(Where A = OH^-, Cl^-, NCS^-, NO_2^-)

(b) Base Hydrolysis Reactions

These are the reactions in which hydroxo complex is formed by the replacement of ligand by OH^- group. These reactions occur in basic solutions.

An example of base hydrolysis reactions is as follows:

$$[Co(NH)_2Cl]^{2+} + HO \rightarrow [Co(NH_3)_5OH]^{2+} + Cl^-$$

Acid-Hydrolysis Reactions of Cobalt Complexes

These reactions may be studied under two headings:

1. One type includes hydrolysis reactions which take place under acid conditions but are not dependent upon the concentration of acid provided pH of an acid solution is less than 4. These reactions are studied under simple acid hydrolysis reactions.

2. Another type includes hydrolysis reactions whose rate of hydrolysis is acid catalysed. These reactions are studied under acid catalysed aquation reactions.

1. Simple acid Hydrolysis Reactions

These reactions show no acid dependence. However, the simple acid hydrolysis reaction of type.

$$[MX_5Y]^{n+} + H_2O \rightleftarrows [MX_5(H_2O)]^{(n-1)-} + Y^-$$

follows the first order kinetics. The observed rate of this hydrolysis reaction is Rate = $K[MX_5Y]^{n-}$

This form of rate law is compatible with either mechanism (a) or (b).

(a) $$[MX_5Y]^{n-} \underset{fast}{\overset{k_a,\ slow}{\rightleftarrows}} [MX_5]^{(n+n)+} + Y$$

$$[MX_5]^{(n+1)+} + H_2O \xrightarrow{fast} [MX_5(H_2O)]^{(n-1)-}$$

$$Rate = k_a\ [MX_5Y]^{n-}$$

(b) $$[MX_5Y]^{n+} + H_2O \overset{k_b\ slow}{\rightleftarrows} [MX_cY.H_2O]^{n+}$$

$$[MX_5Y.H_2O]^{n+} \xrightarrow{fast} [MX_5.H_2O]^{(n+1)+} + Y^-$$

$$Rate = k_b\ [MX_5Y]^{n+}\ [H_2O]$$

The rate laws given above do not indicate whether these reactions proceed by dissociation or association mechanism. However, a study as to how the following factors affect the rate constant of them reactions can give us an information about the nature of the mechanism by which these reactions proceed.

The greater bulk of work performed has been concerned with acid hydrolysis of Co (III) complexes because it is known that most substitution processes of cobalt (III) in aqueous solution proceed indirectly via reaction with the solvent.

(A) Steric Factors

These factors may decide about the type of mechanism. For instance, the crowding of a reaction, centre might favour a dissociation process by effectively alleviating some degree of steric but reduce the chance of association by discouraging attack of an approaching group.

In order to study the steric factors, the acid hydrolysis of Co(III) complexes of the type trans $[Co(AA)_2\ Cl2]^+$ (where AA = a bidentate derived from ethylene diamine) is investigated and data art recorded in Table 3.

Table 3 : Rate Constants for the Acid Hydrolysis of trans-$[Co(AA)_2Cl_2]^+$ at pH and 25°C $[Co(AA)_2\ Cl_2]^+ \rightarrow [Co(AA)_2(H_2O)Cl]^{2+} + Cl^-$

AA	*10^3k (m^{-1})*
$NH_2 - CH_2 - CH_2 - NH_2$(en)	1.9
$NH_2 - CH_2 - CH(CH_3) - NH_2$(pn)	3.7
dl $- NH_2 - CH(CH_3) - CH(CH_3) - NH_2$(dl − bn)	8.8
meso $- NH_2 - CH(CH_3) - CH(CH_3) - NH_2$(m- bn)	250
$NH_2 - C(CH_3)_2 - C(CH_3)_2 - NH_2$(tetramine)	very fast
$NH_2 - CH_2 - CH_2 - CH_2 - NH_2$(tn)	600

From the above Table it can be seen that there exists a correlation between rate and increased C-methyl substitution on ethylene diamine.

When the size of the chelate ring is increased from five to six by triethylene diamine in place of diethylene diamine, the rate increase drastically with the increase in size of chelate ring. This rate increase favours a dissociative mechanism in which the increasing steric hindrance is conductive to the formation of an intermediate of reduced coordination number.

From the kinetic results of hydrolysis of $[Co(en)_2LCl]^+$, some interesting facts about steric factors are revealed (Table 4). For instance, for the ligands OH^-, Cl^- and NCS^-, the cis-isomers react more rapidly than trans. Further cis isomers but not trans isomers react with retention of configuration. These results have been explained in terms of π-bonding theory.

Table 4 : Rate Constants and Isomeric Products for the Acid Hydrolysis of Chlorobis-(Ethylene-diamine) (Ligand) Cobalt (III) Complexes at 25°C

Ligand L	*cis isomer Rate Constant cis isomer $k_c(sec^{-1})$*	*% cis isomer in the product*	*trans isomer Rate Constant trans isomer $k_t(sec^{-1})$*	*% trans isomer in the product*	$\frac{k_c}{k_t}$
NCS	1.1×10^{-5}	100	5.0×10^{-8}	30–50	220
OH	1.3×10^{-2}	100	1.4×10^{-2}	25	9.2
Cl	2.4×10^{-4}	100	3.2×10^{-5}	65	7.5
N_3	2.5×10^{-4}	100	2.4×10^{-4}	80	1.05
NH_3	5.0×10^{-7}	84	4.0×10^{-7}	17	1.25
NO_2	1.1×10^{-4}	100	1.0×10^{-2}	100	0.11

According to Pearson and Basolso, those ligands that exhibit a strong cis effect are those that are having unshared pairs of electrons (in addition to the pair used in the sigma dative bond). The unshared pair of electrons is available to be donated to the metal in a p-d π-bond (Fig. 10).

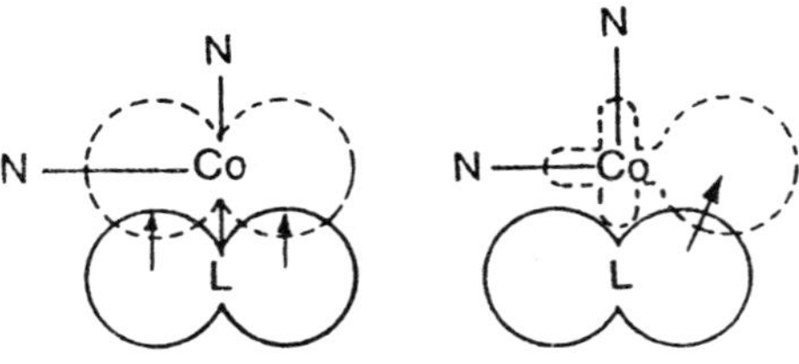

Fig. 10 : (a) and (b) overlapping of filled p-orbital of L with vacent (a) p-orbital or (b) d^2sp^3 hybrid orbital of cobalt in a 5-coordinated square pyramidal activated complex resulting from the dissociation of Cl^- from cis $[Co(en)_2LCl]^+$. The two N ligands not shown are above and below the plane of paper.

This pi bonding by a cis substituent can stabilise a square pyramidal activated complex by decreasing the positive charge on the metal. This allows the reaction to proceed without extensive rearrangement and the product obtained is 100% cis isomer.

If the same ligands are present in the trans position, no orbital is available for overlap unless the complex undergoes rearrangement to yield a trigonal bipyramidal structure (Fig. 11). This increases the energy of activation and makes substitution reactions more difficult and results in mixtures of isomers in the product. These ligands which fail to exhibit a positive cis effects are of two types:

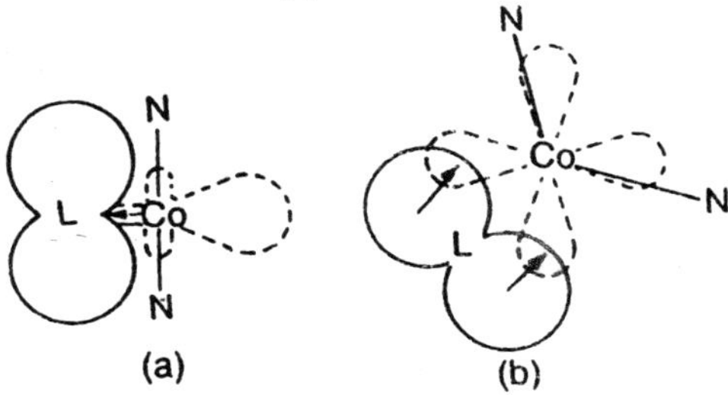

Fig. 11 : (a) and (b) no overlapping of filled p-orbital of L with vacent (a) p-orbital or (b) d^2sp^3 hybrid orbital of cobalt in a square pyramidal activated complex resulting from the dissociation of Cl^- from trans-$[Co(en)_2LCl]^+$. (b) Efficient overlapping with vacent $dx^2- y^2$ orbital if there is rearrangement of a trigonal bibyramidal. The two N ligands not shown are above and below the plane of paper.

(i) Those ligands which are lacking another pair to donate to the metal. An example is ammonia.

(ii) Those ligands which are pi acceptors. An example is the nitro group. These ligands have tendency to withdraw electrons via it-bonding and hence can not operate through mechanism described above.

(b) Inductive Effects

Rate increase with the increased C-methyl substitution of the acid hydrolysis of Co(III) complexes of the trans-$[Co(AA)_2Cl_2]^+$ may he due to inductive effects, by increasing alkyl substitution, the electron density is distorted towards the cobalt atom, thereby assisting the dissociation of the chloride ion.

The influence of inductive effects has been understood from studying the rate constants for the acid hydrolysis of $[Co(en)_2(X - py)Cl]^{2+}$ at 0°C. (Table 5).

From the Table 5 it follows that an increasing inductive effect which is operating in this is evidenced by the increasing basic strength of the pyridine molecule. From the rate and base dissociation constant, it is seem that a 20-fold increase in basic strength will result in a 40 percent increase in rate. But the basic strengths of the substituted ethylene diamines vary by a factor of about 1.5. Therefore it may be concluded that the marked rate increase is due to other influences than inductive effects.

Table 5 : Rate Constants for Acid Hydrolysis of $[Co(en)_2(X-py)Cl]^{2+}$ at 50°C

X-Py	pK_b	$10 \times ks^{-1}$
pyridine	8.82	1.1
3-methyl pyridine	8.19	1.3
4-methyl pyridine	7.92	1.4
4-methoxy pyridine	7.53	1.5

(c) Solvent Effect

This gives the most acceptable explanation.

When a dissociation reaction proceeds, it occurs via a transition state by separating charges. The greater the solvation of the activated complex, the greater will be the seperation of charges. Therefore, the factors responsible for poor solvation of the transition state result in lower rates of reactions.

2. Acid-Catalysed Aquation Reactions

In some systems the rate of hydrolysis has been found to be acid-catalysed and a protonated species is formed as a reactive intermediate. These systems fall into two categories:

(a) Complexes having ligands derived from weak acids, *e.g.*, $[M(NH_3)_5Y]^{n+}$ where

$Y = F^-$, N_3^-, $RCOO^-$, CO_3^{2-} or NO_2^-

$[Co(en)_2X_2]^+$ where $X = NO_2^-$ or N_3^-

(b) Complexes having polydentate ligands, *e.g.*,

$[M(EDTA)]^{n-}$ where M = Ni(II) or Fe(III)

$[M(C_2O_4)_3]^{3-}$ where M = Co(II), Cr(III) or Rh(III)

$[M(bipy)_3]^{2+}$ where M = Fe(II) or Ni(II) and

$[Ni(en)_3]^{2+}$

Mechanism

In complexes of category (a) acid catalysis is attributed to the tendency of strongly basic ligand to attach a proton. For instance,

$$[M(NH_3)_5(RCOO)]^{2+} + H^+ \overset{k}{\rightleftarrows} [M(NH_3)_5RCOOH]^{3+} + \text{(fast)}$$

The above reaction may be followed by aquation which takes place through the alternative paths:

$$[M(NH_3)_5(RCOO)]^{2+} + H_2O \overset{k_1}{\longrightarrow} [M(NH_3)_5(H_2O)^{3+} + RCOO$$

$$[M(NH_3)_5(RCOOH)]^{3+} + H_2O \overset{k_2}{\longrightarrow} [M(NH_3)_5(H_2O)]^{3+} + RCOOH$$

On applying the steady state treatment of the above mechanism by keeping the idea in mind that the concentration of protonated species is small even under strongly acid conditions, the following rate law containing only two terms is obtained.

$$\text{Rate} = k_1\,[M(NH_3)_5(RCOO)^{2+}] + k_{H^+}\,[M(NH_3)_5(ROO)^{2+}]\,[H^+]$$

where $k_{H^+} = k_2k$,

The observed rate constant is then given by

$$k_{obs} = k_1 + k_2k[H^+]$$

If k_{obs} is plotted against $[H^+]$, a linear plot is obtained. This has an intercept equal to k_1 and a gradient of k_2k.

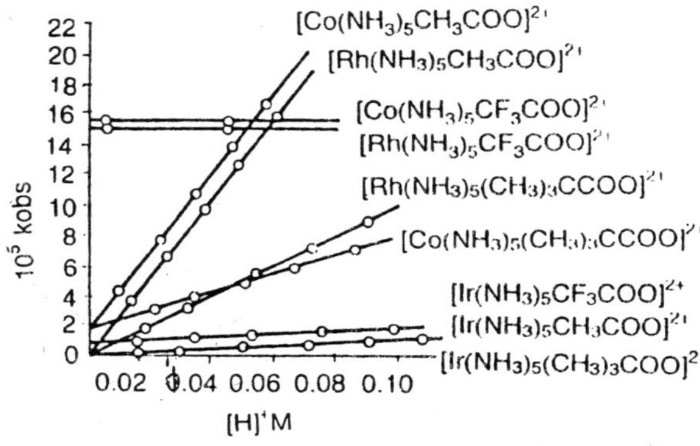

Fig. 12 : Rates of hydrolysis of $[M(NH_3)_5RCOO]^{2+}$ versus H^+ at 80° and π = 0.10 M.

This type of curve is obtained in many acid catalysed equation of series of pentammine complexes of Co(III). Rh(III) and Ir(III) (Fig. 12), thereby supporting the mechanism given above.

The mechanism given above undergoes variation in the acid-catalysed aquation of azidopenta cyano cobaltate (III). If SCN^- is absent, the reaction mechanism of hydrolysis of this complex may he as follows:

$$[Co(CN)_5N_3]^{3+} + H^+ \underset{slow}{\overset{k}{\rightleftarrows}} [Co(CN)_5N_3H]^{2-} \text{ (Fast)}$$

$$[Co(CN)_5N_3H]^{2-} + H_2O \rightarrow [Co(CN)_5OH_2]^{2-} + HN_3$$

However, if SCN^- is present, the reaction mechanism of hydrolysis is assumed to occur through pentacyano cobaltate (III) intermediate.

$$[Co(CN)_5N_3H]^{2-} \overset{slow}{\rightarrow} [Co(CN)_5]^{2-} + HN_3$$

$$[Co(CN)_5]^{2-} + H_2O \rightarrow [Co(CN)_5OH_2]^{2-}$$

in competition with

$$[Co(CN)_5]^{2-} + SCN^- \rightarrow [Co(CN)_5NCS_2]^{3-}$$

For complexes of category (b), the mechanism similar to given above could operate. For instance, the acid dependence in the dissociation of $[Fe(bipyr)_3]^{2+}$ could he attributed to the formation of a protonated intermediate.

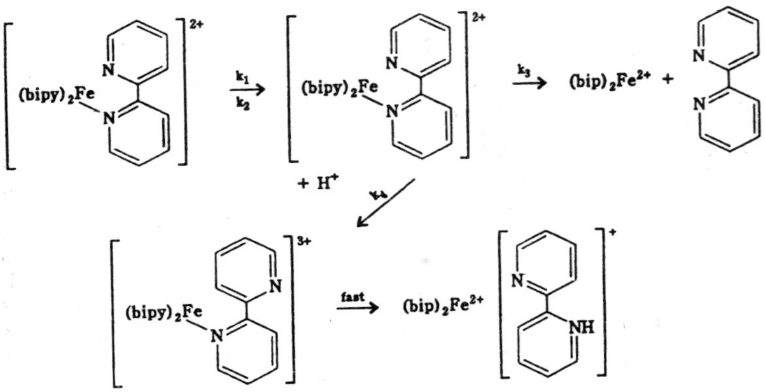

Fig. 13

However, the weak basicity of the bipyridyl group does not favour appreciable concentrations of such entities and this also does not explain the limiting rate obtained at high acid concentration which means the completion of protonation. An acceptable mechanism, in which there occurs an opening up of the chelate ring which is followed by protonation of free end and then loss of the protonated ligand, may be given as follows:

When the steady state approximation is applied to the concentration of the intermediate, $[Fe(bipy)_2H]^{3+}$, the observed rate constant obtained is as follows:

$$k_{obs} = \frac{k_1\left(k_3 + k_4\left[H^+\right]\right)}{k_2 + k_3 + k_4\left[H^+\right]}$$

From this expression, the following two limiting cases arise:

(i) If hydrogen ion concentrations are very low, then

$$k_{obs} = \frac{k_1 k_3}{k_2 + k_3}$$

(ii) If hydrogen ion concentrations are very high *i.e.*, $K_4[H^+]$ $(k_2 + k_3)$, then

$$k_{obs} = k_1$$

From the ratio of these limiting rates [(i) and (ii)], an approximate value of 0.2 for $k_4 - (k_2 + k_3)$ was obtained. This result indicates that, on the average, 80 percent of the occasions on which the first the metal-nitrogen bond breaks, result in the reformation of the bond, and 20 percent result in the breaking of bond, causing dissociation. A similar type of mechanism has been postulated for the dissociation of $[Ni(en)_3]^{2+}$ and $[Ni(bipy)_3]^{2+}$.

For the acid catalysed aquation of trioxalato chromium (III),

$$[Cr(C_2O_4)_3]^{3-} + 2H_3O^+ \rightarrow [Cr(H_2O)_2(C_2O_4)_2]^- + H_2C_2O_4$$

the rate law takes the following form:

$$-\frac{d\left[Cr(C_2O_4)_3\right]^{3-}}{dt} = k'[H_3O^+][Cr(C_2O_4)_3]^{3-} + k''[H_3O^+]^2[Cr(C_2O_4)_3]^{3-}$$

The mixed order dependence on the hydrogen ion concentration has been explained on the basis of a rapid pre-equilibrium protonation of the complex.

$$[Cr(C_2O_4)_3]^{3-} \overset{k}{\rightleftarrows} [Cr(H_2O)(C_2O_4)_2(NC_2O_4)]^{2}$$

followed by the parallel rate-determining steps;

$$[Cr(H_2O)(C_2O_4)_2(HC_2O_4)]^{2-} + H_2O \overset{k_1}{\rightarrow} [Cr(H_2O)_2(C_2O_4)_2]^{-} + HC_2O_4^{-}$$

and

$$[Cr(H_2O)(C_2O_4)_2(HC_2O_4)]^{2-} + H_3O^{+} \rightarrow [Cr(H_2O)_2(C_2O_4)_2]^{-} + H_2C_2O_4$$

where $k' = k_1k$ and $k'' = k_2k$

In some reactions like the aquation of trans-dichlorotetra aquo-chromium (III), an inverse acid-dependence has been noted:

$$[CrCl_2(H_2O)_4]^{+} + H_2O \rightarrow [CrCl(H_2O)_5]^{2+} + Cl$$

The mechanism of this aquation is as follows:

$$[CrCl_2(H_2O)_4]^{+} \overset{fast}{\rightleftarrows} [CrCl_2(H_2O)_3OH] + H^{+}$$

$$[CrCl_2(H_2O)_3OH] + H_2O \overset{slow}{\rightarrow} [CrCl(H2O)4OH]^{+} + Cl$$

$$H^{+} + [CrCl(H_2O)_4OH]^{+} \overset{fast}{\rightleftarrows} [CrCl(H_2O)_5]^{2+}$$

The above mechanism has been found to be in agreement with the following empirical rate law:

$$k_{obs} = k' + k'' [H^{+}]$$

In the above expression, the first term denotes the direct rate of aquation of trans-$[CrCl_2(H_2O)_4]^{+}$.

Base Hydrolysis Reactions

The greater bulk of work performed has been concerned with the base hydrolysis of cobalt (III) ammine complexes. For ammine complexes of Co(III) containing N-H bonds, the rate of base hydrolysis has been found to be 10^6 times faster the corresponding rate of hydrolysis.

The base hydrolysis reactions have been discussed by using $[Co(NH_3)_5Cl]^{2+}$ as an example

$$[Co(NH_3)_5Cl]^{2+} + OH^{-} \rightarrow [Co(NH_3)_5(OH)]^{2+} + Cl^{-}$$

The above reaction may proceed by any of the following mechanisms:

(a) S_{N^2} Associate Mechanism

The probable mechanism according to this may be as follows:

$$[Co(NH_3)_5Cl]^{2+} + OH^- \underset{}{\overset{slow}{\rightleftarrows}} [Co(NH_3)_5(OH)Cl]^+ \overset{fast}{\rightarrow} [Co(NH_3)_5(OH)]^{2+} + Cl^-$$

In the above mechanism, the slow step is the rate-determining step and is of second order, first order with respect to complex and first order with respect to base, *i.e.*,

$$\text{Rate of reaction} = k\,[\text{Complex}][\text{Base}]$$
$$= k\,[Co(NH_3)_5Cl^{2+}][OH^-]$$

The above mechanism fails to explain quite a few observations:

(i) The 7-co-ordinate complexes are not very stable.

(ii) Although, the value of k6 has been found to be nearly 10^4 times higher than k_a, why do the hydroxyl ions possess the exceptionally high nucleophilic activity as compared to the similar anions ?

(b) S_{N^1} CB Dissociation Mechanism

This mechanism has been given by Garrick (1937). This mechanism involves a rapidly established pre-equilibrium in which the hydroxyl ion abstracts a proton from one of the ammine nitrogen and forms an amino conjugate base. The rate-determining step is the dissociation of the conjugate base to yield a penta-coordinated intermediate which then adds solvent to form the hydroxo product.

(i) $[Co(NH_3)_5Cl]^{2+} + OH^- \overset{k}{\rightleftarrows} [Co(NH_3)_4NH_2Cl]^+ + H_2O$ (fast)

(ii) $[Co(NH_3)_4NH_2Cl]^+ \underset{k}{\rightarrow} [Co(NH_3)_4NH_2]^{2+} + Cl^-$

(Rate-determining)

(iii) $[Co(NH_3)_4NH_2]^{2+} + H_2O \rightarrow [Co(NH_3)_5OH]^{2+}$ (fast)

This mechanism has become known by the symbol S_{N^1} CB (where CB = conjugate base). When the steady-state approximation is applied to above mechanism, the rate follows as:

$$-\frac{d\left[Co(NH_3)_5Cl\right]^{2+}}{dt} = k[Co(NH_3)_4NH_2Cl]^+$$

$$= kK[Co(NH_3)_5Cl^{2+}][OH^-]$$

$$= \frac{kK_A}{k_w}[Co(NH_3)_5Cl^{2+}][OH^-]$$

In the above rate expression, K_A is the dissociation constant of the complex and kW is the ionic product of water.

If the rate expression (I) is true, the following conditions are required to be fulfilled:

(i) The establishment of the pre-equilibrium should be rapid comparison with the rate of the overall process.

(ii) The concentration of the amido conjugate-base should be low.

It is interesting to note that both condition have been found to be fulfilled. These have been verified experimentally. For instance,

(i) From the exchange studies of $[Co(NH_3)_5Cl]^{2+}$ with D_2O in basic solution, it is been proved that the exchange rate is faster than the release of the chloride ion.

(ii) The cobalt, (III) ammine complexes are such weak acids that the concentration of base form is very low even in strongly alkaline solution.

In addition to the direct evidences (i) and (ii), there are indirect evidences which are as follows

(i) When chelation is increased, there should occur an increase in the base hydrolysis. This has been found to be so in number of cobalt(III) ammine complexes and also in the case of the analogous platinum (IV) complexes. Further, complexes like traps-$[Co(py)_4Cl_2]^+$ and $[Co(CN)_5X]^{3-}$, which are not having acidic protons, undergo slow base hydrolysis at rates which are independent of hydroxide ion concentration.

(ii) Taube and Green (1963) proved that the final hydroxo product, $[Co(NH_3)_5OH]^{2+}$ must drive from the reaction of intermediate, $[Co(NH_3)_4NH_2]^{2+}$ with water and not with the hydroxide ion because the $^{18}O - ^{16}O$ ratio in the product has been found to be identical with the ratio in water.

(iii) If S_{N^1} CB mechanism above is applied to the base hydrolysis of the complexes of the type $[Co(en)_2AX]^{n+}$, the concentration of the isomeric product should be independent of the nature of A provided X is constant-because reaction, occurs via a common intermediate This has been found to be true in the number of complexes in which there occurs agreement between the proportion of cis-product from different parents Table 6.

Table 6 : Isomeric Products of Base Hydrolysis of cis- and trans$[Co(en)_2AX]^{n+}$ (A = non replaced ligand, X = replaced ligand).

Reactant	*Product*	*% cis-isomer in product*	
		For cis-reactant	For trans-reactant
$[Co(en)_2ClBr]^+$	$[Co(en)_2Cl(OH)]^+$	30	5
$[Co(en)_2ClCl]^+$	$[Co(en)_2Cl(OH)]^+$	37	5
$[Co(en)_2(NCS)N_3]^+$	$[Co(en)_2NCS(OH)]^+$	70	70
$[Co(en)_2(NCS)Cl]^+$	$[Co(en)_2NCS(OH)]^+$	80	76

Limitations : If it becomes possible to detect the conjugate-base intermediate, it would he regarded as the direct proof of the mechanism. Although there is some evidence for this five-co-ordinated species in dimethylsulphoxide solvent yet the attempts to detect a similar intermediate in aqueous solution by spectrophotometry are unsuccessful due to the formation of ion pairs.

(c) **Ion-pair mechanism :** This mechanisms avoids the postulation of altered coordination number. However, this postulates the formation of ion-pair. This mechanism has been applied to the base hydrolysis of Co(III)-ethylenediamine complexes of the type $Co(en)_2L$, where L is hydroxylamine.

$$[Co(en)_2LCl]^{2+} + OH^- \overset{k}{\rightleftarrows} [Co(en)_2LCl]^{2+}.OH^- \text{ (ion pair)}$$

$$[Co(en)_2LCl]^{2+} + H^- \overset{k}{\rightarrow} [Co(en)_2L(OH)]^{2+} + Cl^-$$

In order to he compatible with the observed second order kinetics it is necessary that the equilibrium concentration of the ion- pair is low. Then, the rate law becomes as:

$$\text{Rate} = k[\text{ion-pair}] = kK[\text{complex}][OH^-]$$

In order to he more precise, ion-pairing of this type may be considered in terms of solvation changes of the complex ion in which the hydroxide ion is replacing a solvent molecule in the inner solvation shell;

$$\{[Co(en)_2LCl](H_2O)_n\} + OH^- \overset{k}{\rightleftarrows} \{[Co(en)_2LCl]OH(H_2O)_{n-1}\}^+ + H_2O$$

$$\{[Co(en)_2LCl]OH(H_2O)_{n-1}\}^+ \overset{k}{\rightarrow} \{[Co(en)_2LOH]Cl(H_2O)_{n-1}\}^+$$

Square brackets are indicating the species enclosed in the coordination shell while { } are indicating the species enclosed in the inner solvation shell. Similarly, the aquation may be regarded as a rearrangement between the solvation and co-ordination shells.

$$\{[Co(en)_2LCl](H_2O)_n\}^{2+} + H_2O \overset{k}{\rightleftarrows} \{[Co(en)_2LCl]H_2O(H_2O)_{n-1}\}^{2+} + H_2O$$

$$\{[Co(en)_2LCl](H_2O)_{n-1}\}^{2+} \overset{k}{\rightleftarrows} \{[Co(en)_2L\,H_2O]Cl(H_2O)_{n-1}\}^{2+}$$

The attack of the hydroxide ion on the inner co-ordination shell may well proceed by a Grotthus chain-transfer. The solvent shell which consists of tightly-bound and well-ordered solvent molecules, is acting as a physical barrier against attack by an incoming group. However, the hydroxide ion can generate a new hydroxide ion by removing a proton from a solvent molecule and in this way the process is repeated. Eventually, attack on the metal is taking place without penetration of the barrier.

COMPLEMENTARY REDOX REACTIONS

These are the redox reactions in which the oxidant and reductant are changing their oxidation states by the same number. A consequence of such reactions is that only one molecule of each reactant is required to react. Examples of complementary reactions are:

$$Sn(II) + Tl(III) \rightarrow Sn(IV) + Tl(I)$$

$$\underset{\text{Reductant}}{Sn(II)} + \underset{\text{Oxidant}}{Hg(II)} \rightarrow \underset{\text{Oxidant}}{Sn(IV)} + \underset{\text{Reductant}}{Hg(O)}$$

NON-COMPLEMENTARY REDOX REACTIONS

These are the radox reactions in which the oxidant and reductant change their oxidation states by different numbers. The result is that the different number of molecules of oxidant and reductant is that the different number of molecules of oxidant and reductant in the stoichiometric equation are to be written, *e.g.*,

$$2Fe(III) + Sn(II) \rightarrow 2Fe(II) + Sn(IV)$$

Non-complementary reactions are often slow compared to complementary reactions.

MECHANISM OF SUBSTITUTION REACTIONS IN SQUARE-PLANAR COMPLEXES

Square-planar complexes are formed by the d^8 systems of Pt (II), Pd (II), Ni (II), Au (III), Rh (I) and Ir (I). Out of these, the kinetics of ligand of the inert and stable complexes of platinum (II) have been studied in greater detail.

The kinetis of the reactions such as

$$PtA_2LX + Y \rightarrow PtA_2LY + X$$

are described by a two term rate law which has the form

$$\text{Rate} = K_S \text{ (complex)} + K_Y \text{ (complex) (Y)}.$$

The composite nature of the rate law implies that such reactions may proceed by bimolecular associative (S_{N^2}) mechanisms involving either the solvent or the entering ligand as the nucleophilic reagent. These two mechanisms are as follows:

(i) Solvent as a Nucleophilic Agent

$$PtA_2LX + H_2O \overset{\text{Slow, } K_s}{\rightleftarrows} PtA_2L(H_2O) + X$$

$$PtA_2L(H_2O) + Y \overset{\text{Fast}}{\rightleftarrows} PtA_2LY + H_2O$$

(ii) Entering Ligand as a Nucleophilic Agent

$$PtA_2LX + Y \overset{\text{Slow, } K_Y}{\rightleftarrows} \left[A_2Pt \begin{matrix} X \\ Y \end{matrix}\right] \rightarrow PtA_2LY + X$$

Sufficient evidence is available which shows that both mechanisms involve bimolecular (S_{N^2}) mechanism. It will be suffice at this stage to cite two sorts of evidence;

(i) Firstly, variation of the charge of the complex has only a slight effect on the rate. Insensitivity of the rate to the charge on the complex is a characteristic of associative mechanism in which bond making and bond breaking play roles of compatable importance.

This has been best illustrated by studying the kinetics of the four complexes. viz. $[Pt(Cl_4)]^{2-}$, $[Pt(NH_3)Cl_3]^-$, $[Pt(NH_3)_2Cl_2]^0$ and $[Pt(NH_3)_3Cl]^+$. The rates of hydrolysis reaction of these complexes vary by a factor of two (quite a number) although the charge on the substance is varying from –2 to +1.

(ii) Secondly, an increase in steric hindrance in the complex is accompanied by a decrease in reactivity. This is expected if there occurs an increase in co-ordination number as in bimolecular associative mechanism.

This has been verified experimentally by studying the rates of reactions of a series of cis- and trans-$[Pt(PEt_3)_2RCl]$ complexes with pyridine in ethanol solution. In these reactions, there occurs a marked decrease in rate as the group R is varied from phenyl to o-tolyl to mesityl; relative rates for the cis and trans isomers are respectively, 100,000 : 200 : 1 and 30 : 6 : 1.

We shall now discuss the two mechanisms of substitution reactions separately.

1. Solvent as a Nucleophilic Reagent

Three mechanisms involving solvent as nucleophilic reagent have been proposed, hereby leading to an intermediate aquo complex which is assumed to have a tetragonal structure:

(a) An association type mechanism in which solvent participates and a trigonal bipyramidal transition state.

(b) A purely dissociative type mechanism which does not involve solvent in the transition state.

(c) A dissociative process in which solvent participates and a square pyramidal transition state is formed.

The above three possible mechanism are outlined as follows (Fig. 14).

The evidence against scheme (b) may be summarised as follows:

(i) The rate constant k_s is not affected by changes in the net charge of the complex.

(ii) The values of k_s have been found to decrease with increase in steric hindrance.

(ii) The values of k_s in different solvents do not bear any relationship to the dielectric constant of the medium but exhibit a correlation with the coordinating power of the solvent.

Similarly, there is no direct or indirect proof for the scheme (c). Thus, a trigonal bipyramid is only the reasonable transition state for the solvent path as given in scheme (a).

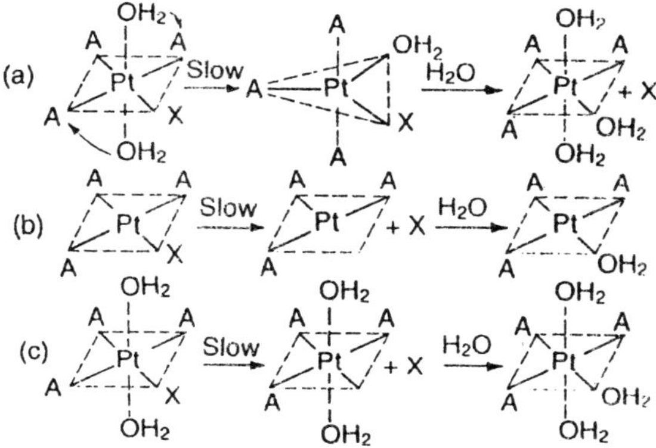

Fig. 14 : Three possible mechanisms involving solvent as a nucleophilic reagent.

2. Entering Ligand As A Nucleophilic Agent

The mechanism of this type of substitution also involves trigonal bipyramidal structure as the transition state as shown below (Fig. 15).

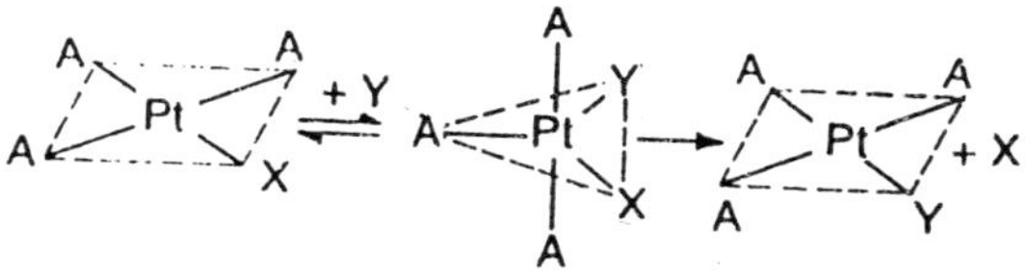

Fig. 15

Similarly, substitution reactions of cis- and traps-$PtAl_2LX$ with Y to yield $PtAl_2LY$ are also explained on the basis of nucleophilic attack of Y through trigonal bipyramidal structure. This process has been found to be entirely stereoscopic, *i.e.*, cis trans product (Fig. 16(a) and (b))..

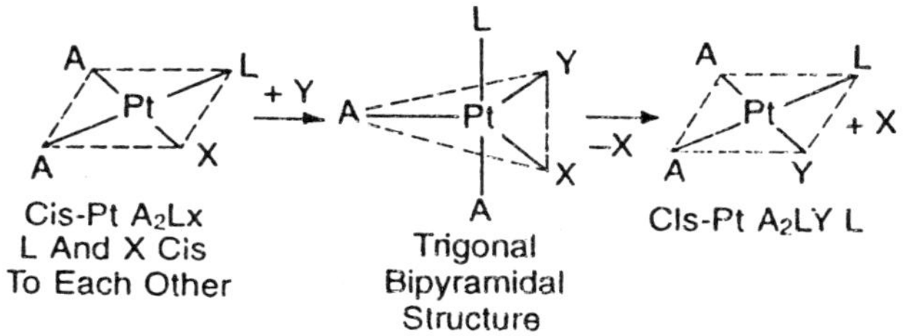

Cis-Pt A_2Lx
L And X Cis
To Each Other

Trigonal
Bipyramidal
Structure

CIs-Pt A_2LY L

(a)

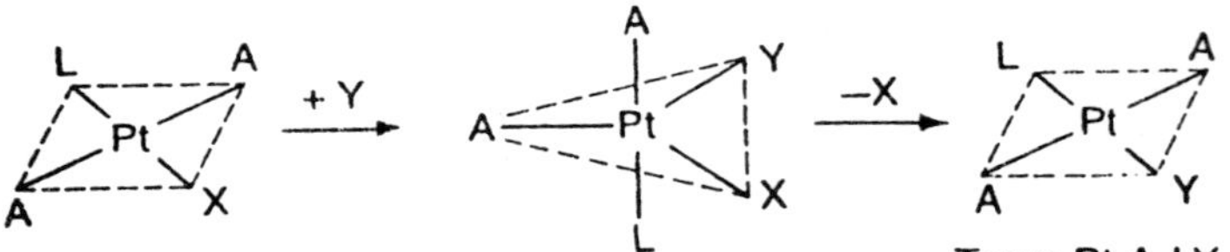

Trans-Pt A_2LX (L And X Trans To Each Other) $\xrightarrow{+Y}$ Trigonal Bipyramidal Structure $\xrightarrow{-X}$ Trans-Pt A_2LY

(b)

Fig 16

Factors Affecting the Rates of Substitution Reactions in Square Planar Complexes

Sufficient kinetic data are available which deal with the effect of various factors on the rates of substitution reactions in platinum (II) complexes. Some of these factors are:

(a) **Trans effect.** In order to observe the effect of ligand L trans to the leaving group Cl^- in some Pt (II) square-planar complexes, the relative rates of the following reaction have been studied:

$$\text{trans}-\left[\begin{matrix} H_3N & & Cl \\ & Pt & \\ L & & Cl \end{matrix}\right]^0 + Pt \rightarrow \text{trans}-\left[\begin{matrix} H_3N & & py \\ & Pt & \\ L & & Cl \end{matrix}\right] + Cl^-$$

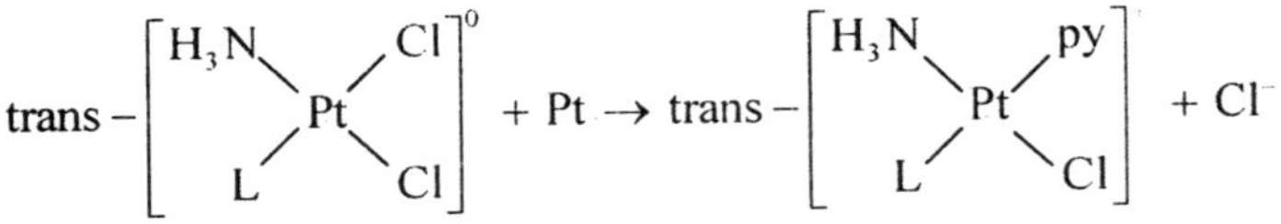

(where L = C_2H_4, NO_2^-, Br^-, Cl^-)

The relative rates of above reaction have been found to decrease with the decrease of relative trans-effect of these groups whereas the activation energy E_a have been found to increase in this order.

Relative				
Trans-effect	:	$C_2H_4 > > NO_2$	$> Br^-$	$> Cl^-$
Relative rates	:	$> 100 > 90$	> 3	> 1
Ea (Kcal/more)	:	$- < 11$	< 17	< 19

In order to observe the trans-effect of some usual ligands, the relative rates of the following reaction have been studied in ethanol at 25°C. In this reaction the ligand L trans, to the leaving Cl^- group is replaced

$$\text{trans}-\left[\begin{array}{ccc} Et_3P & & Cl \\ & Pt & \\ L & & PEt_3 \end{array}\right]^0 + Pt \rightarrow \text{trans}-\left[\begin{array}{ccc} Et_3P & & Cl \\ & Pt & \\ L & & PEt_3 \end{array}\right]^+ + Cl^-$$

replaced by py to yield $[Pt(PEt)_2pyL]^+$. In this reaction, L is changed by groups like H^-, CH_3^-, $C_6H_5^-$ etc. The observed rate constant for this reaction is defined by the following equation:

$$k_{obs} = k_1 + k_2\,[Y]$$

In the above expression, k_1 and k_2 are rate constants for forward and backward reactions. Both these are contributing to over all observed rates. The values of k_1 and k_2 are determined by trans-directing ability of group L trans to Cl which is leaving group.

Trans-directing ability of group L trans to Cl which is leaving group	H^- > methyl > phenyl - p-chlorophenyl - p - methoxy phenyl > biphenyl > Cl^-
k_1 (min^{-1})	$1.1 > 1 \times 10^{-2}$ $> 2 \times 10^{-3}$ -1.7×10^{-3} $> 1 \times 10^{-3}$ $> 6 \times 10^{-5}$
k_2(min^{1})	$2.5 \times 10^{2} > 5$ $> 9.5 \times 10^{-1}$ $- 9 \times 10^{-1}$ -7.9×10^{-1} $> 5.8 \times 10^{-1}$ $> 2.4 \times 10^{-2}$

From the above results it follows that the effect on the rates of reaction of different L groups trans to Cl^- which is being replaced by py has been found to decrease in the following order.

H^- > methyl > phenyl - p-chlorophenyl

The above order is also the order of trans-effect of these groups.

(b) **Effect of labile groups :** A number of investigations have been carried out to study the effect of replaced ligand X on the rates of some nucleophile Y in $[Pt(dieu)X]^+$ complexes. Recently, it has been

$$[Pt(dien)X]^+ + Y \rightarrow [Pt(dien)Y]^+ + X$$

shown that the ease of replacement of X follows the order: Cl^- ~, Br^- ~, $I^- \geq N_3 > SCN^- > NO_2^- > CN^-$. From this order, it follows that ligands

Cl^-, Br^- and I^- are replaced with comparable ease while, N_3^-, SCN^-, NO_2^-, CN^- are more difficult to replace and also the rate of replacement has been found to be dependent on the nature of the ligand. It is interesting to note that the behaviour of the halide groups is remarkable because the strength of the metal-halide bond increases in the following order:

$$Cl^- < Br^- < I$$

(c) **Effect of solvent :** Considerable evidence has been cited to prove that in the solvent paths the solvent replaces X^- directly. Therefore, it is expected that with the increase in the coordination ability of the solvent, the contribution made by this path to the overall rate of reaction should also increase. This has been demonstrated experimentally by studying the effect of solvent on the chloride exchange of trans $[Pt(py)_2Cl_2]$.

$$\text{trans-}[Pt(py)_2Cl_2] + Cl^- \rightarrow \text{trans-}[Pt(py)_2Cl_2] + Cl^-$$

When moderately low concentration of Cl^- is used, the solvents has been divided into two categories:

(i) One category includes those solvents whose rate of exchange does not depend, upon the concentration of Cl^-. These solvents provide almost entirely a solvent path for exchange. The solvents included in this group are good coordinating solvents. It has been found experimentally that the solvent effect on the value of k_s is in the following order:

$$DMs > H_2O, MeNO_2 > ROH$$

(ii) Another category includes those solvents whose rate of exchange depends on the concentration of Cl^-. These solvents possess poor coordinating power towards Pt(II) and contribute little to k_s. In these solvents, exchange proceeds through a reagent-dependent path (k_s). In this category of solvents, the order reactivity is as follows:

Carbon tetrachloride > benzene > m-cresol > t-butanol > ethyl acetate > acetone > dimethyl formamide (DMF).

The nature of the solvent (acetone, DMF, and methanol) does not affect the nucleophilicity order for trans-$[Pt(PEt_3)_2Cl_2]$.

(d) **Effect of charge on the complex :** For this see under "Mechanism of substitution reactions".

THE CAUSE OF OPTICAL ACTIVITY

Two important points that arise from the property of optical activity are: What types of structure give rise to optical activity, and why? Fresnel (1822) suggested the following explanation for optical activity in crystalline substances such as quartz, basing it on the principle that any simple harmonic motion along a straight line may be considered as the resultant of two opposite circular motions. Fresnel assumed that plane-polarised light, on entering a substance in a direction parallel to its optic axis, is resolved into two beams of circularly polarised light, one right-handed (dextro-) and the other left-handed (laevo-) and both having the same frequency.

If these two component beams travel through the medium with the same velocity, then the issuing resultant beam suffers no rotation of its plane of polarisation (Fig. 17a). If the velocity of the left-circularly polarised component is, for some reason, retarded, then the resultant beam is rotated through some angle to the right (in the direction of the faster circular component; Fig. 17b). Similarly, the resultant beam is rotated to the left if the right-circularly polarised component is retarded (Fig. 17c).

Fresnel tested this theory by passing a beam of plane-polarised light through a series of prisms composed alternately of dextroand laevorotatory quartz (Fig. 18). Two separate beams emerged, each circularly polarised in opposite senses; this is an agreement with Fresnel's explanation. Fresnel suggested that when planepolarised light passed through an optically active crystalline substance, the plane of polarisation was rotated because of the retardation of one of the circular components. Stated in another way, Fresnel's theory requires that the refractive indices for right- and left-circularly polarised light should be different for optically active substances.

It has been shown mathematically that only a very small difference between these refractive indices gives rise to fairly large rotations, and that if the refractive index for the left-circularly polarised light is greater than that for the right component the substance, will be dextrorotatory. The difficulty of Fresnel's theory is that it does not explain why the two circular components should travel with different velocities. It is interesting to note, however, that Fresnel (1824) suggested that the optical activity of quartz is due to the structure being built up in right- and left-handed spirals.

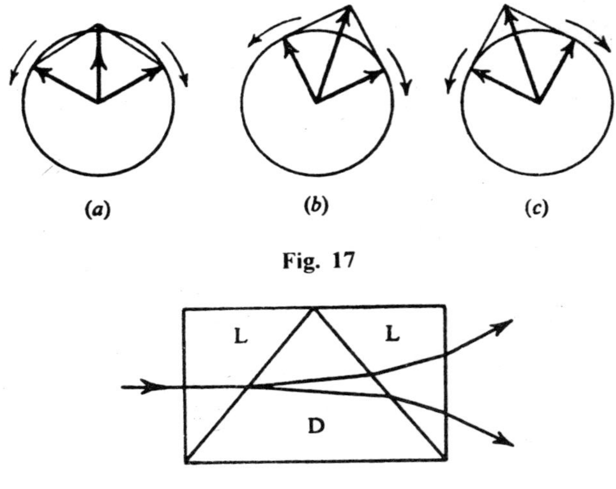

Fig. 17

Fig. 18

If n_L and n_R respectively represent the refractive indices for left- and right-circularly polarised light then, if these are different, the substance is said to exhibit circular birefringence. If $n_L > n_R$, the emergent linearly-polarised resultant is rotated to the right. Since refraction and absorption of light are interconnected, the implication is that if $n_L > n_R$ in the long wavelength region where the optically active substance is transparent, then $\varepsilon_L > \varepsilon_R$ (where ε is the molar absorptivity) for the shorter wavelength region where light is absorbed. This effect, *i.e.*, when ε_L and ε_R are unequal, is known as circular dichroism.

The combined phenomenon of circular dichroism and unequal velocity of travel of left- and right-circularly polarised light is known as the Cotton effect. Now, ORD and CD curves are studied in the region of maximum absorption of the optically active compound, i..e., in the region of an optically active chromophore. Such a chromophore is either inherently asymmetric, *e.g*, twisted biphenyls, or inherently symmetric, *e.g*, a carbonyl group.

In the latter, when this is in an asymmetric environment, it then behaves as an asymmetric chromophore, *i.e.*, optical activity is induced in the chromophore by the environment. Because of this, the carbonyl chromophore is referred to as an inherently symmetric, but asymmetrically perturbed, chromophore. The amplitudes of the ORD curves of compounds containing inherently asymmetric chromophores are usually much greater than those containing asymmetrically perturbed symmetric chromophores.

Drude (1900) showed that if a molecule possessed a structure such that when light is absorbed an electron is displaced along a right-handed helical path, then the result is a positive circular dichroism ($\varepsilon_L > \varepsilon_R$), and the molecule is dextrorotatory at longer wavelengths where $n_L > n_R$. In the enantiomer, the electron is displaced along a left-handed helical path when light is absorbed, the result being a negative circular dichroism ($\varepsilon_R > \varepsilon_L$) and an optical laevorotation ($n_R > n_L$) at longer wavelengths.

This theory of optical rotatory power has been modified by quantum mechanics treatment. A helical motion is the resultant of two components, a linear and a circular displacement. A linear and a circular charge-displacement produce an electric and magnetic dipole moment, respectively. If a transient electric dipole and a magnetic dipole are produced by absorption of light, the molecule is circularly dichroic. If the two moments are parallel, the circular dichroism is-positive; if the two moments are antiparallel, the circular dichroism is negative. If the two moments are mutually perpendicular, then no circular dichroism results.

Now let us consider the problem of optical activity of substances in solution. In this case the optical activity is due to the molecules themselves, and not to crystalline structure. Any crystal which has a plane of symmetry but not a centre of symmetry rotates the plane of polarisation, the rotation varying with the direction in which the light travels through the crystal. No rotation occurs if the direction of the light is perpendicular or parallel to the plane of symmetry. If we assume that molecules in a solution (or in a pure liquid) behave as individual crystals, then any molecule having a plane but not a centre of symmetry will also rotate the plane of polarisation, provided that the light travels through the molecule in any direction other than perpendicular (or parallel) to the plane of symmetry. Let us consider the molecule Ca_2bd (Fig. 19).

This has a plane of symmetry, and so molecule (I) and its mirror image (II) are superimposable. Now let us suppose that the direction of plane-polarised light passing through molecule (I) makes an angle $0°$ with the plane of symmetry, and that the resultant rotation is $+\alpha°$. Then if the direction of the light through molecule (II) also makes an angle

θ° with the plane of symmetry, the resultant rotation will be $-\alpha^\circ$. Thus the total rotation produced by molecules (I) and (II) is zero. In a solution of compound Ca_2bd there will be an infinite number of molecules in random orientation.

Statistically one can expect to find that whatever the angle θ is for molecule (I), there will always be molecule (II) also being traversed by light entering at angle θ. Thus, although each individual molecule rotates the plane of polarisation by an amount depending on the value of θ, the statistical sum of the contributions of the individual molecules will be zero.

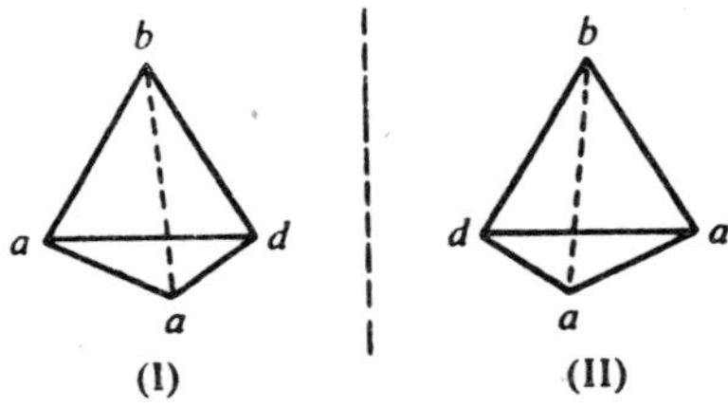

Fig. 19

When a molecule is not superimposable on its mirror image, then if only one enantiomer is present in the solution, the rotation produced by each individual molecule will (presumably) depend on the angle of incidence (with respect to any face), but there will be no compensating molecules (*i.e.*, mirror image molecules) present. Hence, in this case, there will be a net rotation that is not zero, the actual value being the statistical sum of the individual contributions (which are all in the same direction).

Thus, if we consider the behaviour of a compound in a solution (or as a pure liquid) as a whole, then the observed experimental results are always in accord with the statement that if the molecular structure of the compound is chiral, that compound will be optically active. Any compound composed of molecules possessing a plane but not a centre of symmetry is, considered as a whole, optically inactive, the net zero rotation being the result of 'external compensation'. This point is of great interest in connection with molecules that can exist in different conformations.

Let us consider meso-tartaric acid, a compound that is optically inactive by internal compensation. X-ray studies (Stern *et al.*, 1950) have shown that the staggered form of the molecule is the favoured one

(Fig. 20a). This has a centre of symmetry, and so molecules in this configuration are individually optically inactive. On the other hand, meso-tartaric acid is usually represented by the plane-diagram formula in Fig. 20(b). This corresponds to the eclipsed form, and has a plane of symmetry. In this conformation the individual molecules are optically active except when the direction of the light is perpendicular (or parallel) to the plane of symmetry; the net rotation is zero by 'external compensation'. It is possible, however, for the molecule to assume, at least theoretically, many conformations which have no elements of symmetry, *e.g,* Fig. 20(c).

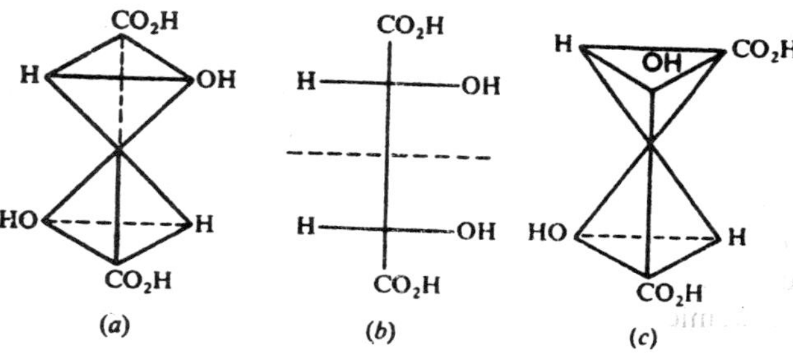

Fig. 20

All molecules in this conformation will contribute in the same direction to the net rotation. If the total number of molecules present were in this conformation, then meso-tartaric acid would have some definite rotation. On the theory of probability, however, for every molecule taking up the conformation in Fig. 20(c). there will also be present its mirror image molecule, thereby giving a net zero rotation due to 'external compensation'. As we have seen, meso-tartaric acid is optically inactive (as shown experimentally), and by common usage the inactivity is said to be due to internal compensation.

CORRELATIONS OF SIGN AND MAGNITUDE OF ROTATION WITH ABSOLUTE CONFIGURATION

Brewster (1959, 1961) has devised an empirical correlation (for rotation with the sodium D line) and has used a number of general rules for this purpose. These general rules are based on the following hypothesis: A centre of optical activity can usefully be described as an asymmetric screw pattern of polarisability. This screw pattern, however, may arise in one of two ways or as a combination of both of them:

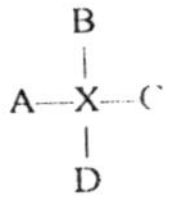

B
|
A—X—C
|
D

(I)

(i) **Atomic asymmetry :** If the tetrahedral system XABCD has the absolute configuration shown in (I), it is dextrorotatory when the order of polarisability of the groups is A > B > C > D. Provided the groups A, B, C, and D are atoms or simple groups (which do not introduce conformational problems), the rules for the prediction of the sign (and magnitude) of rotation are readily applied.

(ii) **Conformational asymmetry :** In this case, the polarisability is caused by the conformation of the groups in the molecule. When this is present, its contribution to the molecular rotation is usually larger than that due to atomic asymmetry. Application of the rules for the prediction of the sign (and magnitude) of rotation is far more difficult for conformational asymmetry than for atomic asymmetry.

Atoms and groups can be arranged in order of decreasing polarisability, the individual polarisabilities being derived from the atomic refractions of the atoms attached to the asymmetric carbon atom. In this way, the following order of polarisabilities has been obtained: I > Br > SH > Cl > Ph = CO2H > Me > NH2 > OH >H>D>F.

Let us now use a-phenylethyl chloride (II) to illustrate the application of the rules for atomic asymmetry. The absolute configuration of (R)-α-phenylethyl chloride has been shown to be (II). Reference to the order

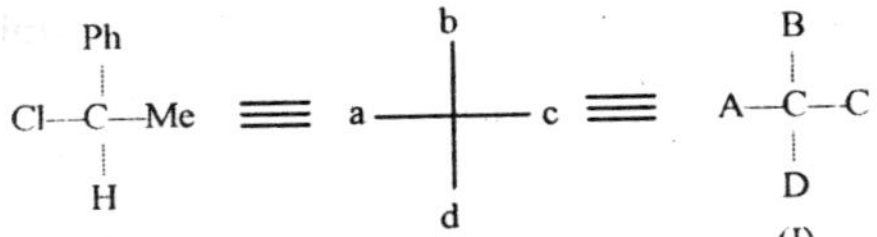

Ph
|
Cl—C—Me ≡ a—+—c ≡ A—C—C
|
H

(II) (b above, d below the cross) (I) (B above, D below)

of the polarisabilities of the groups gives the configuration (I). Therefore this enantiomer is predicted to be dextrorotatory. This is the case in practice (the (R)-form is the dextrorotatory enantiomer). If we referred to a table of values of polarisabilities, we would also find that the magnitudes of the predicted and observed rotations are in fair agreement.

Optical activity due to crystalline structure : There are many substances which are optically active in the solid state only, *e.g*, quartz, sodium chlorate, benzil, etc. Let us consider quartz, the first substance shown to be optically active (Arago, 1811). Quartz exists in two crystalline forms, one of which is dextrorotatory and the other laevorotatory. These two forms are mirror images and are not superimposable. Such pairs of crystals are said to be enantiomorphous (quartz crystals are actually hernihedral and are mirror images). X-ray analysis has shown that the quartz crystal lattice is built up of silicon and oxygen atoms arranged in left- and right-handed spirals. One is the mirror image of the other, and the two are not superimposable. When quartz crystals are fused, the optical activity is lost. Therefore the optical activity is entirely due to the asymmetry of the crystalline structure, since fusion brings about only a physical change. Thus we have a group of substances which are optically active only so long as they remain solid; fusion, vaporisation or solution in a solvent causes loss of optical activity.

Optical activity due to molecular structure : There are many compounds which are optically active in the solid, fused, gaseous or dissolved state, *e.g,* glucose, tartaric acid, etc. In this case the optical activity is entirely due to the asymmetry of the molecular structure. The original molecule and its non-superimposable mirror image are known as enantiomorphs (this name is taken from crystallography), enantiomers, or optical antipodes.

Properties of enantiomers : It appears that enantiomers are identical physically except in two respects:

(i) their manner of rotating polarised light; the rotations are equal but opposite.

(ii) the absorption coefficients for dextro- and laevocircularly polarised light are different; this difference is known as circular dichroism.

The crystal forms of enantiomers may be mirror images of each other, *i.e.,* the crystals themselves may be enantiomorphous, but this is unusual. Enantiomers are simila chemically, but their rates of reaction with other optically active compounds are usually different. They may also be different physiologically, *e.g,* (+)-histidine is sweet, (–)-tasteless; (–)-nicotine is more poisonous than (+). The mass spectra of enantiomers (and their corresponding racemates) are identical, and so are their NMR spectra.

2

Differences in Stability and Reactivity of Diastereoisomers

INTRODUCITON

The stabilities of diastereoisomers of compounds containing two asymmetric carbon atomsare generally different, but these differences are usually small. The meso-form is generally more stable than either of the active forms, and the erythro compounds are generally more stable than the threo compounds. This may be demonstrated by consideration of the molecule CSML–CSML, where S, M, and L represent the smallest, medium, and largest groups, respectively. First let us consider the meso-form and one active form, each being drawn in its most stable conformation (S = S, M = M, L = L; *e.g,* $HO_2CCHOHCHOHCO_2H$).

The basis of the argument is the general principle that crossed steric interactions between groups of different size are less than the sum of the steric interactions between groups of equal size. The sum of the skew interactions in the meso-form is: 2(L: M) + 2(M: S) + 2(L: S), whereas in the active form the sum is : 2(L: M) + (M: M) + 2(L: S) + (S: S). Therefore, since (M: M + S : S) > 2(M: S), the meso-form is under less steric strain than the active form, and consequently is the more stable diastereoisomer.

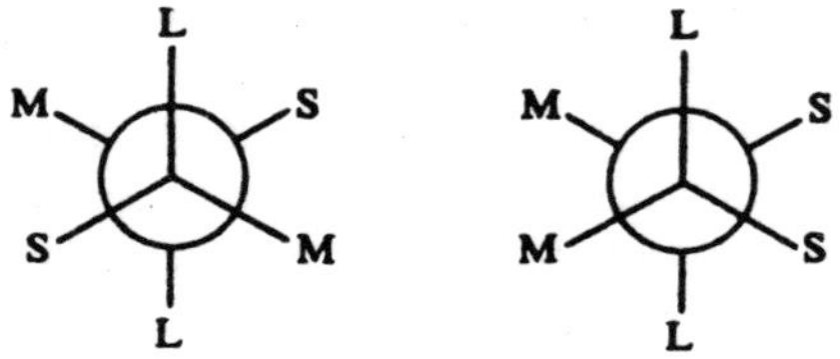

L M S S M L *meso*

L M S M S L active

Fig. 1

If any two of the groups are not the same, *e.g.* $HO_2CCHOHCHOH$ CHO, the compound will be the erythro- or threo-isomer, and using the same arguments as before it will be found that the former is more stable than the latter. This has been established experimentally for many pairs of meso- and racemic and of erythro- and threo-isomers.

When we consider the differences in reactivity between diastereoisomers, the main controlling factor is the height of the energy barrier leading to the transition state. This may be assessed in terms of two factors : the steric factor which is concerned with the conformational requirements of groups not involved in the reaction, and the stereoelectronic factor which is concerned with the spatial relationships that exist between electrons involved in bond formation and/or bond breaking in the transition state. Acyclic systems can usually adjust themselves to the stereoelectronic requirements of the transition state, about 4.2-8.4 kJ/mol being required in the process. With cyclic systems, because of their relative rigidity, adjustment in a similar fashion may be far more difficult.

The foregoing account of the problem of conformations of molecules has been mainly qualitative. It is also of interest to consider the problem from a semi-quantitative point of view.

The most convenient parameters for defining the spatial arrangement of the atoms in a molecule are bond lengths, bond angles, and torsional (dihedral) angles, and the changes in energy content in the molecule will depend on the changes in these parameters. These changes-bond stretching (and compression), bond angle bending, and bond torsion—are collectively called molecular deformations.

BOND STRETCHING AND COMPRESSION

If the potential energy of two particles in their equilibrium position, *i.e.*, when they are separated by the bond length, (r), is taken as zero, then the P.E., V_r, of the two particles when the bond length is changed by Δr is given by the expression

$$V_r = \frac{1}{2} k_r (\Delta r)^2$$

where k_r is the bond stretching force constant. For both C–C and C–H bonds, k_r is about 5×10^5 dynes/cm (5 N/cm) and the above expression (for these two bonds) reduces to

$$V_r = 350(\Delta r)^2 \text{ kcal/mol/Å}^2$$
$$= 1464(\Delta r)^2 \text{ kJ/mol/Å}^2$$

With the C–C bond length equal to 1.54Å, then a change of 2 per cent, *i.e.*, 0.031Å, is equal to a change in P.E. of ~ 1.42 kJ/mol.

Bond angle bending. If the C–C–C valency angle in saturated n-hydrocarbons is taken as the standard value (~112°), then any deviation from this value produces angle strain (also known as Baeyer strain or classical strain). If the angle deformation is $\Delta\theta$, then the angle strain, V_θ, is given by the expression

$$V_\theta = \frac{1}{2}k_\theta(\Delta\theta)^2$$

where k_θ is the bond bending force constant. Since k_θ has similar values for most C–C–C bond angles, the above expression may be reduced to

$$\begin{aligned} V_\theta &= 0.01(\Delta\theta)^2 \text{ kcal/mol/deg}^2 \\ &= 0.042(\Delta\theta)^2 \text{ kJ/mol/deg}^2 \end{aligned}$$

Thus, if $\Delta\theta = 6°$, then $V_\theta = 1.5$ kJ/mol. This is roughly the same value as V_r for a 2 per cent change in r (see above). If we consider acyclic compounds only, then the maximum value for $\Delta\theta$ is about 10–12°. Thus, an angle deformation of 6° is about 50 per cent of the maximum value. Hence angle deformation is more easily brought about than linear deformation.

Bond torsion : As we have seen, the P.E. of the system varies with the torsional angle, $\Delta\phi$. If V_0 is the torsional energy barrier, *i.e.*, the barrier height between a maximum and a minimum, then the variation in P.E., V_ϕ, (the torsional strain or Pitzer strain) is given by the expression

$$\begin{aligned} V_\theta &= \frac{1}{2}V_0(1 + n \cos \Delta\theta) \text{ kcal/mol} \\ &= 2.1\ V_0(1 + n \cos \Delta\phi) \text{ kJ/mol} \end{aligned}$$

where n is the number of P.E. minima that occur in the rotation through 360°. This equation may be applied to molecules such as ethane; n = 3.

Steric repulsion : This interaction between non-bonded atoms is also a function of the three parameters r, θ, and ϕ. These three may be replaced by a fourth parameter, ρ, the distance between non-bonded atoms, *i.e.*, V_ρ, the steric strain, is estimated in terms of ρ.

It can therefore be seen that molecular strain energy is the sum of the four contributing factors, V_r, V_θ, V_ϕ, and V_ρ. Furthermore, since a molecule will normally be in the state corresponding to its lowest P.E., strain energy is thus the increase in energy which is produced by deviations of the parameters from their most favourable values. Unfortunately, it is not easy to estimate the various contributions, but even so, it may be possible to obtain approximate values which may be used in judging the stabilities of various conformations. It also appears that, in general, molecules are subject to very little linear deformations. This, however, is not the case when the transition state is entered by the molecule during reaction.

CORRELATIONS BY USE OF MONOCHROMAT OPTICAL ROTATIONS

A consequence of the Distance Rule is that molecular rotations of higher members of homologous series containing one chiral centre tend to reach a limiting value or zero value, *e.g,* long-chain fatty acids containing an a-methyl group have molecular rotations which approach the value of -28°. As the methyl group shifts nearer to the centre of the chain, the molecular rotations get smaller and smaller. The application of the method of monochromatic optical rotations is based on the above generalisation. This may be restated as follows: If two compounds have the same absolute configuration, then if their structures differ only at some distance from their chiral centres, the molecular rotations have the same sign and have approximately the same magnitude, *e.g,* acyclic secondary alcohols with the formula shown in (I; $m > n$) are all dextrorotatory(for the sodium D-line).

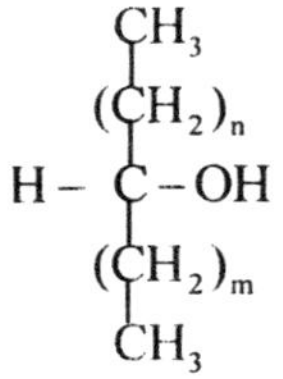

$$\begin{array}{c} CH_3 \\ | \\ (CH_2)_n \\ | \\ H-C-OH \\ | \\ (CH_2)_m \\ | \\ CH_3 \\ \text{(I)} \end{array} \qquad \begin{array}{c} CO_2H \\ | \\ H-C-OH \\ | \\ CH_2 \\ | \\ R \\ \text{(II)} \end{array}$$

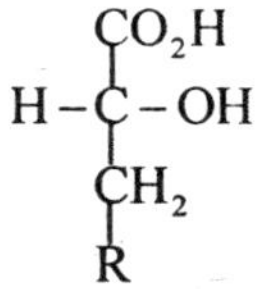

A very interesting point about this generalisation is that deviations are most likely to occur when the structural difference involves the introduction of a group that absorbs in the near ultraviolet. Thus, the hydroxy-acids represented by (II), in which R = H, Me, Et, Pr, etc., are all laevorotatory, whereas when R = Ph, the acid is dextrorotatory.

Specification of Absolute Configurations

Since the configuration of (+)-tartaric acid has been related to that of (+)-glyceraldehyde and since the absolute configuration of (+)-tartaric acid has been determined it is now possible to assign absolute configurations to many compounds whose relative configurations to (+)-glyceraldehyde are known. This raised the problem of using one system of specifying absolute configurations. Cahn *et al.*, (1956, 1964) have proposed such a system and this is now widely used. Let us first consider the procedure for a molecule containing one asymmetric carbon atom (one chiral centre).

(i) The four groups are first ordered according to the sequence rule. According to this rule, the groups are arranged in decreasing atomic number of the atoms by which they are bound to the asymmetric carbon atom. If two or more of these atoms have the same atomic number, then the relative priority of the groups is determined by a similar comparison of the atomic numbers of the next atoms in the groups (*i.e.*, the atoms joined to the atom joined to the asymmetric carbon atom). If this fails, then the next atoms of the group are considered. Thus one works outwards from the asymmetric carbon atom until a selection can be made for the sequence of the groups.

When multiple bonds or rings are present, the procedure for determining priority is as follows. Both atoms attached to the multiple bond are considered to be duplicated (for a double bond) or triplicated (for a triple bond), *e.g*,

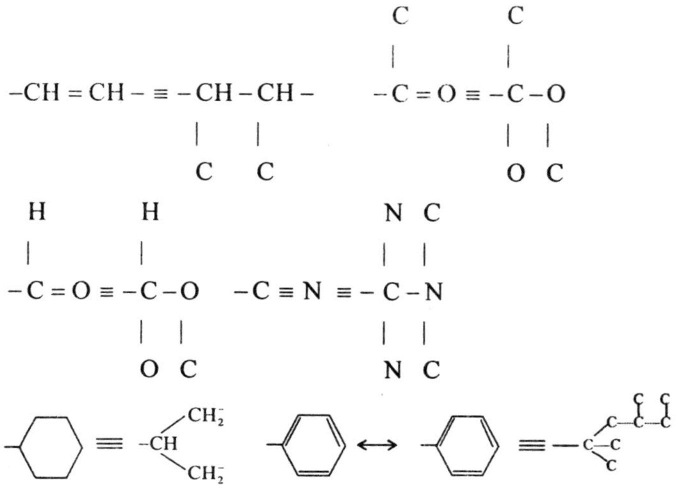

$$-CH=CH- \equiv -\underset{C}{\underset{|}{CH}}-\underset{C}{\underset{|}{CH}}- \qquad -\overset{C}{\overset{|}{C}}=O \equiv -\overset{C}{\overset{|}{\underset{O}{\underset{|}{C}}}}-\underset{C}{\underset{|}{O}}$$

$$-\overset{H}{\overset{|}{C}}=O \equiv -\overset{H}{\overset{|}{\underset{O}{\underset{|}{C}}}}-\underset{C}{\underset{|}{O}} \qquad -C\equiv N \equiv -\overset{N}{\overset{|}{\underset{N}{\underset{|}{C}}}}-\overset{C}{\overset{|}{\underset{C}{\underset{|}{N}}}}$$

$$\equiv -CH\begin{smallmatrix}CH_2\\ CH_2\end{smallmatrix} \qquad \longleftrightarrow \qquad \equiv -C \ldots$$

Fig. 2

The priority sequence is then determined by consideration of the duplicated or triplicated 'structure' in which there arephantom-atoms, *e.g,* –CHO is –CH(O)–O(C) and –CH$(OH)_2$ is –CHOH–OH (the phantom atoms in the former are those in parentheses). Both groups contain a carbon atom joined to two oxygen atoms, but since C precedes H, –CHO precedes –CH$(OH)_2$ in priority.

Ring systems are treated as branched chains, and if unsaturated, then duplication is used for a double bond (or triplication for a triple bond).

By using these rules, it can be shown that the order of priority sequence (for some of the common substituents) is: I, Br, Cl, SO_3H, SH, F, OCOR, OR, OH, NO_2, NR_2, NHR, NH_2, CO_2R, CO_2H, COR, CHO, CH_2OH, CN, Ph, CR_3, CHR_2, CH_2R, CH_3, D, H.

(ii) Next is determined whether the sequence describes a right- or left-handed pattern on the molecular model as viewed according to the conversion rule.

When the four groups in the molecule *Cabcd* have been ordered in the priority a, b, c, d, the conversion rule states that their spatial pattern shall be described as right- or left-handed according as the sequence a → b → c is clockwise or anticlockwise when viewed from an external point on the side remote from d (the group with the lowest priority), *e.g,* (I) in Fig. 3 shows a right-handed (*i.e.,* clockwise) arrangement.

(iii) Absolute configuration labels are then assigned. The asymmetry leading under the sequence and conversion rules to a right- and left-handed pattern is indicated by R and S respectively (R; rectus, right; S; sinister, left).

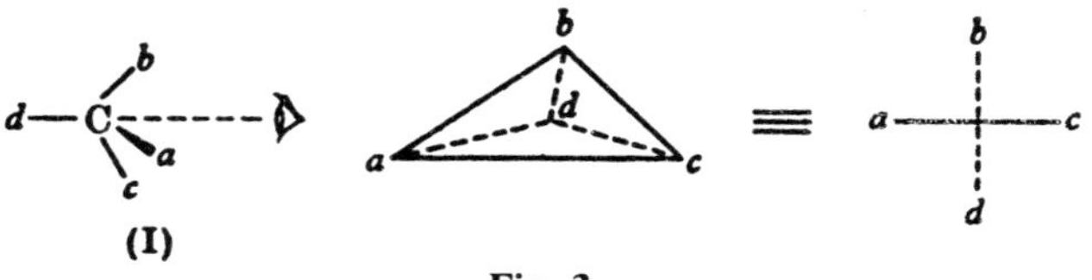

Fig. 3

Let us first consider bromochloroacetic acid (II). The priority of the groups according to the sequence rule is Br (a), Cl (b), CO_2H (c) and H (d). Hence by the conversion rule,, (II) is the (R)-form (a → b → c is clockwise).

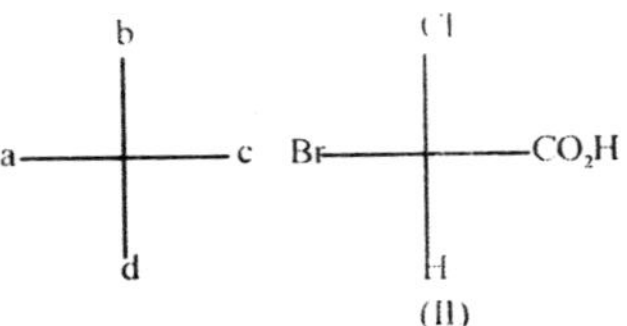

(II)

Now let us consider D(+)-glyceraldehyde. By convention it is drawn as (III) (this is also the absolute configuration). Reference to the sequence list gives the priority sequence : OH (a), CHO (b), CH_2OH (c), and H (d).

Since the interchanging of two groups inverts the configuration, the sequence (III) → (IV) → (V) gives the original configuration. Since (V) corresponds to (VI), it thus follows that D(+)-glyceraldehyde is (R)-glyceraldehyde.

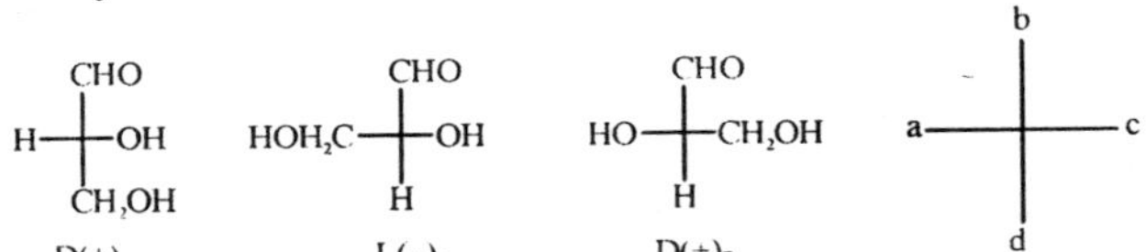

CHO, H—|—OH, CH_2OH — D(+)- (III)

CHO, HOH_2C—|—OH, H — L(–)- (IV)

CHO, HO—|—CH_2OH, H — D(+)- (V)

b, a—|—c, d — (VI)

This scheme can be applied to the deutero compound (VII), which is therefore the (R)-form.

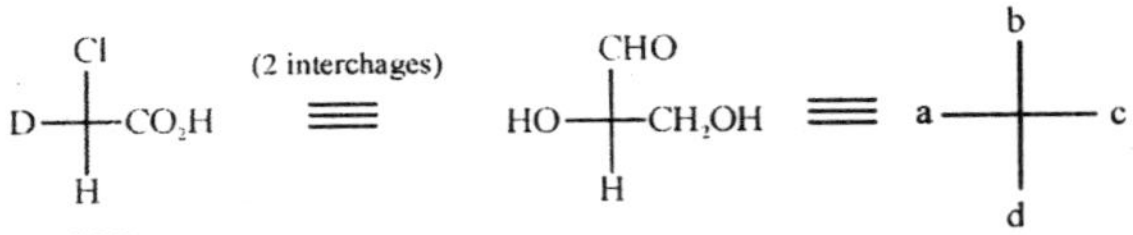

Cl, D—|—CO_2H, H (VII) ≡ (2 interchages) CHO, HO—|—CH_2OH, H ≡ b, a—|—c, d

On the other hand, (VIII) is the (S)-form, since a → b → c is anticlockwise.

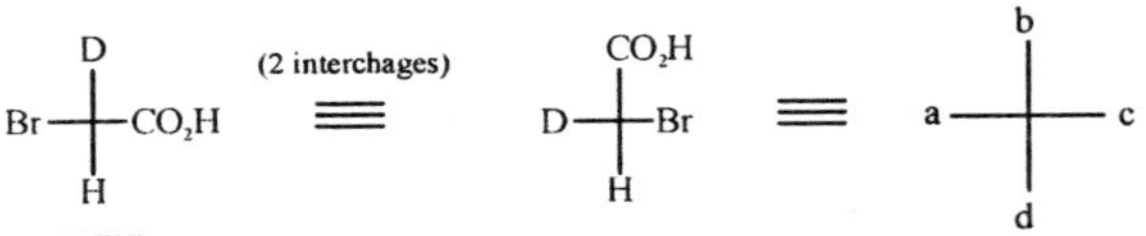

D, Br—|—CO_2H, H (VIII) ≡ (2 interchages) CO_2H, D—|—Br, H ≡ b, a—|—c, d

By reference to the sequence list above, it. can be seen that (IX) is the (S)-form.

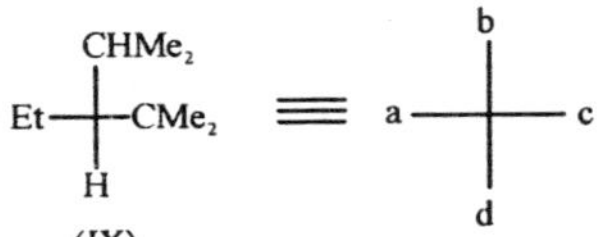

$CHMe_2$, Et—|—CMe_2, H (IX) ≡ b, a—|—c, d

Now let us consider some ring systems. As pointed out above, these systems are treated as branched chains, etc. Hence (X) is the (S)-form, since the CHOH group in the left-hand ring is reached before that in the right-hand ring.

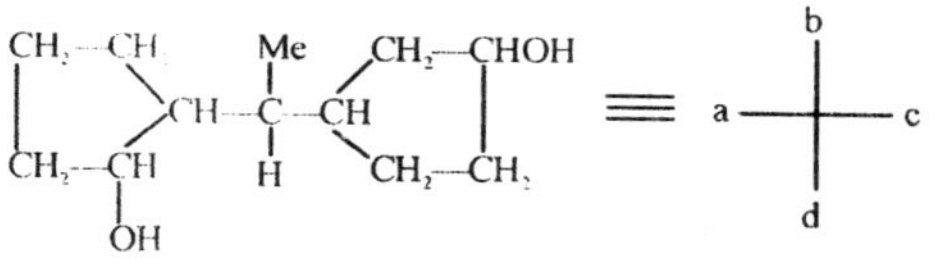

(X)

The same procedure is used when the asymmetric carbon atom is in the ring, *e.g*, (XI) is the (S)-form.

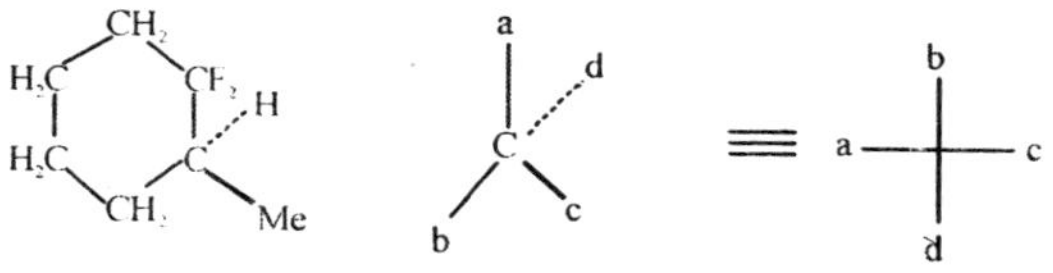

(XI)

When a molecule contains two or more chiral centres, each chiral centre is assigned a configuration according to the sequence and conversion rules and is then specified with R or S, *e.g*, (+)-tartaric acid. Thus the absolute configuration of (+)-tartaric acid is (RR)-tartaric acid:

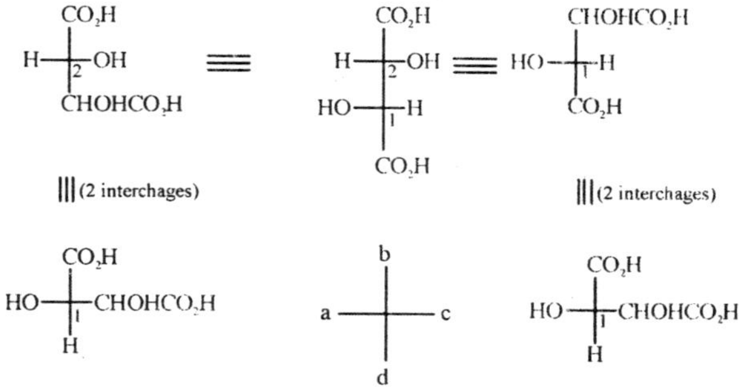

(2 interchages)

(2 interchages)

In a similar way it can be demonstrated that D(+)-glucose has the absolute configuration shown.

The sequence rule was designed to relate the symbols D and L with the symbols R and S. However, D and L are obtained by means of chemical transformations, whereas R and S are derived from geometrical models and are independent of correlations.

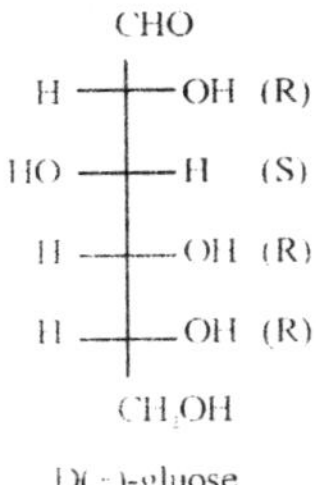

D(+)-gluose

Because of this, R and S must he applied only to compounds whose absolute stereochemistry has been determined; they do not necessarily correlate chemical families, *e.g,* (+)-tartaric acid, whether it be n or L (according to the method of correlation) has an absolute configuration specified by (RR). It should be noted that in the same way as n and L are not necessarily connected with the direction of rotation, nor are R and S.

The absolute configurations of chiral molecules which do not contain asymmetric carbon atoms may also be specified by an extension of the system described above. The, test of superimposing a formula (tetrahedral) on its mirror image definitely indicates whether the molecule is symmetric or not; it is asymmetric if the two forms are not superimposable. The most satisfactory way in which superimposability may be ascertained is to build up models of the molecule and its mirror image.

Usually this is not convenient, and so, in practice, one determines whether the molecule possesses (i) a plane of symmetry, (ii) a centre of symmetry or (iii) an alternating axis of symmetry. If the molecule contains at least one of these elements of symmetry, the molecule is symmetric; if none of these elements of symmetry is present, the molecule is asymmetric. It should be remembered that it is the Fischer projection formula that is normally used for inspection. As pointed out is necessary, when dealing with conformations, to ascertain whether at least one of them has one or more elements of symmetry. If such a conformation can be drawn, then the compound is not optically active.

(i) **A plane of symmetry** divides a molecule in such a way that points (atoms or groups of atoms) on the one side of the plane form mirror images of those on the other side. This test may be applied to both solid (tetrahedral) and plane-diagram formulae, *e.g,* the plane-formula of the meso-form of Cabd possesses a plane of symmetry; the other two, (+) and (–), do not.

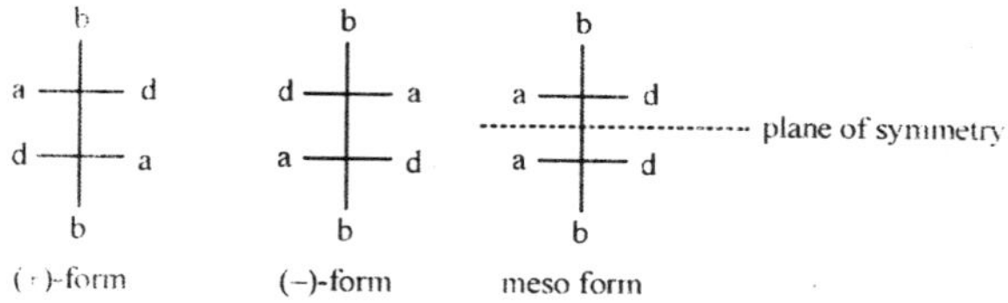

(+)-form (–)-form meso form

plane of symmetry

Fig. 4

(ii) **A centre of symmetry** is a point from which lines, when drawn on one side and produced an equal distance on the other side, will meet identical points in the molecule. This test can be satisfactorily applied only to three-dimensional formulae, particularly those offering systems, *e.g,* 2,4-dimethylcyclo-butane-l,3-dicarboxylic acid (Fig. 5). The form shown possesses a centre of symmetry which is the centre of the ring. This form is therefore optically inactive.

Another example we shall consider here is dimethyldiketo-piperazine; this molecule can exist in two geometrical isomeric forms, cis and trans. The cis-isomer has no elements of symmetry and can therefore exist in two enantiomeric forms; both are known. The trans-isomer has a centre of symmetry and is therefore optically inactive.

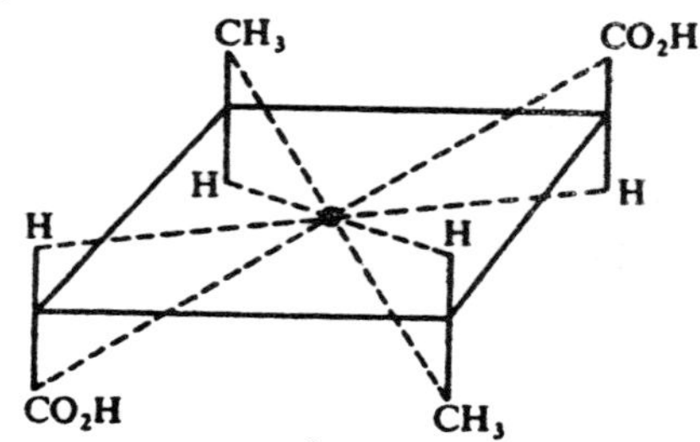

Fig. 5

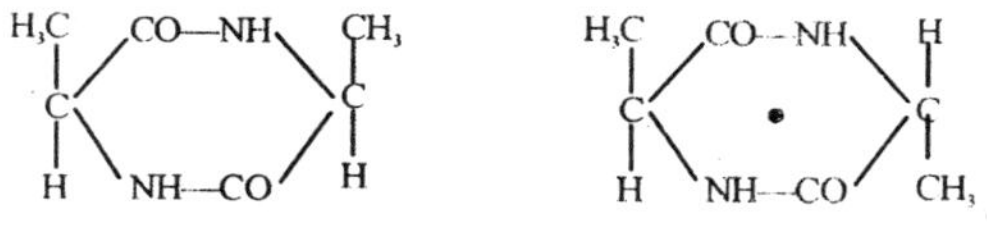

cis trans

It is important to note that only even-membered rings can possibly possess a centre of symmetry.

(iii) **Alternating axis of symmetry :** A molecule possesses an n-fold alternating axis of symmetry if, when rotated through an angle of 360°/n about this axis and then followed by reflection in a plane perpendicular to the axis, the molecule is indistinguishable from the original molecule.

Let us consider the molecule shown in Fig. 6(a) [1,2,3,4-tetramethylcyclobutane]. This contains a four-fold alternating axis of symmetry. Rotation of (a) through 90° about axis AB which passes through the centre of the ring perpendicular to its plane gives (b), and reflection of (b) in the plane of the ring gives (a).

It also happens that this molecule possesses two vertical planes of symmetry (through each diagonal of the ring), but if the methyl groups are replaced alternately by the chiral groups (+)—$CH(CH_3)C_2H_5$ and (–)—$CH(CH_3)C_2H_5$, represented by Z^+ and Z^- respectively, the resulting molecule (Fig. 6c) now has no planes of symmetry. Nevertheless, this molecule is not optically active since it does possess a four-fold alternating axis of symmetry [reflection of (d) (which is produced by rotation of (c) through 90° about the vertical axis) in the plane of the ring gives (c); it should be remembered that the reflection of a (+)-form is the (–)-form].

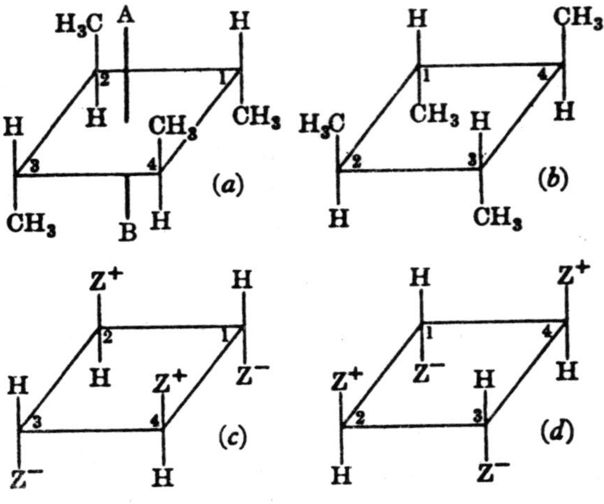

Fig. 6

The cyclobutane derivative (c) given above to illustrate the meaning of an alternating axis of symmetry is an imaginary molecule. No compound was known in which the optical inactivity was due to the existence of only an alternating axis until McCasland and Proskow (1956) prepared such a molecule for the first time. This is a spiro-type of molecule, viz., 3, 4, 3′, 4′-tetramethylspiro-(1, 1′) dipyrrolidinium p-toluenesulphonate, (I) (the p-toluenesulphonate ion has been omitted). This molecule is discussed in some detail but here we shall examine it for its alternating axis of symmetry. Molecule (I) is superimposable on its mirror image and hence is not optically active. It does not contain a plane or centre of symmetry, but it does contain a four-fold alternating axis of symmetry. To show the presence of this axis, if (I) rotated through 900 about the co-axis of both rings, (II) is obtained. Reflection of (II) through the central plane (*i.e.*, through the N atom) perpendicular to this axis gives a molecule identical and coincident with (I).

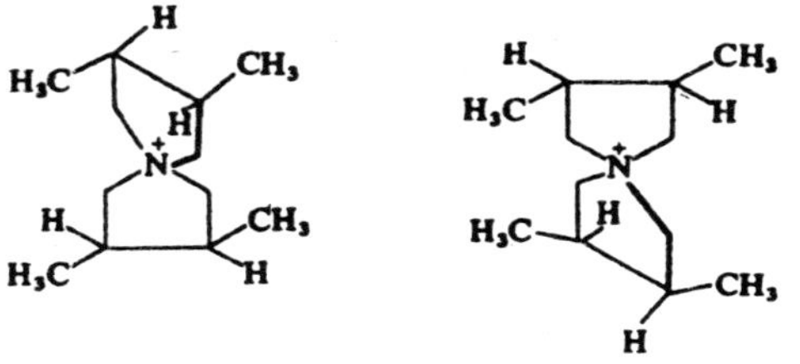

(I) (II)

Fig. 7

McCasland *et al.*, (1959) have now prepared a, second compound, a pentaerythritol ester, whose optical inactivity can be attributed only to the presence of a four fold alternating axis of symmetry:

(−)-ROCH$_2$COOCH$_2$ CH$_2$OCOCH$_2$OR(−)
C
(+)-ROCH$_2$COOCH$_2$ CH$_2$OCOCH$_2$OR(+)

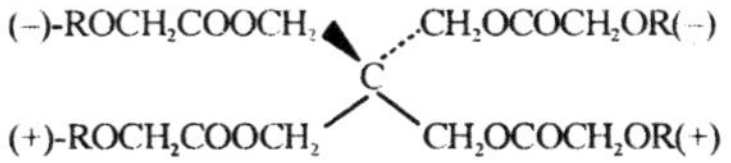

In practice one decides whether a molecule is symmetric or not by looking only for a plane or centre of symmetry, since no natural compound has yet been found to have an alternating axis of symmetry. The presence of two or more asymmetric carbon atoms will definitely give rise to optical isomerism, but nevertheless some isomers may not be optically active because these molecules as a whole are not asymmetric.

Molecular symmetry: In the above account, the test of determining whether a molecule is optically active has been to show the absence of the three elements of symmetry, (i), (ii), and (iii). We shall now consider this problem in some more detail. Symmetry problems are solved by mathematical methods known as Group Theory. A complete set of symmetry elements in any given molecule is known as a point group, *i.e.*, a point group describes the type of symmetry to which the molecule belongs. A point group is one example of groups which form the basis of group theory.

The symmetry of a molecule can be completely described in terms of symmetry elements, and the operations carried out to ascertain the presence of symmetry elements are known as symmetry operations. The basic operations are rotations and reflections. A symmetry operation may be defined as an operation which results in the conversion of the molecule into an equivalent configuration, *i.e.*, the molecule obtained after the operation is indistinguishable from the original molecule. As far as molecules are concerned, a symmetry element is a point, line or plane with respect to which one or more symmetry operations are performed. There are four basic kinds of symmetry elements, and each of these is designated by a symbol which is also used to represent the corresponding symmetry operation.

Axis of symmetry (C_n) : A C_n axis of symmetry is an axis about which the molecule can be rotated by 360°/n (2π/n rad) and thereby produce a molecule indistinguishable from the original molecule (rotation is usually taken as clockwise). The subscript n indicates the order of the axis, *i.e.*, the largest value of n for which the rotation through 360°/n produces an equivalent configuration. Some values of n (for the vertical axis) are as shown.

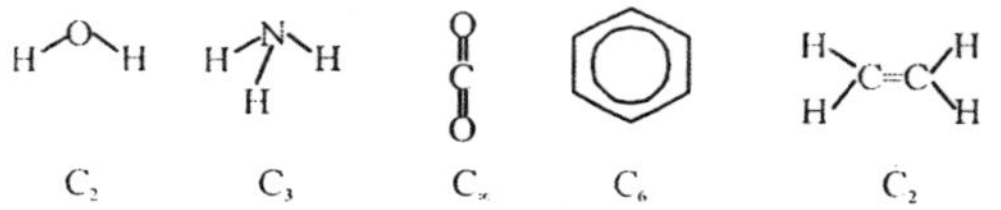

Fig. 8

All linear molecules have a C_∞ axis; an equivalent configuration is always obtained whatever is the angle of rotation. Benzene possesses a C_6 axis perpendicular to the plane of the ring; it also has six C_2 axes in the plane of the ring. Ethylene has three C_2 axes, one collinear with the C–C double bond, the second perpendicular to the plane of the molecule and passing through the centre of the double bond, and the third perpendicular to the first two and passing through the centre of the double bond.

When the value of n is one, the axis is a C_1 axis. The C_1 symmetry operation is carried out by rotating the molecule through 360°. The result is a molecule identical with the original molecule. The C_1 axis is said to be a trivial axis; all molecules possess a trivial axis. Also, every molecule possesses an identity of symmetry (E), which is observed by an identity operation (E). The operation can be carried out in various ways, *e.g*, by rotation through 360°, *i.e.*, one identity element of symmetry is C_1, a trivial axis. Since, *e.g*, C_3 represents the operation of rotation by 120° (2 × 2π/3 rad) about the C_3 axis, repetition of this operation effects the overall rotation of 240° (2 × 2π/3 rad), and another repetition effects the overall rotation of 360° (2π rad). Each result may be indicated by the symbol C_3, C_3^2, and C_3^3, respectively. At the end of the last operation, the molecule is identical with the original molecule. Hence $C_3^3 = E$.

The C_n axis is known as a proper axis; only one or more rotations about the axis are involved. When the molecule contains two or more axes of the same order, they are usually differentiated by superscript dashes, *e.g*, C_2, C'_2, and C''_2, etc. If two or more are equivalent, this is indicated by use of the same superscript dash, *e.g*, C_2, two C'_2, and two C''_2.

It is important to note that symmetry operations involving rotations are applied to the whole molecule. Since rotations of one part of the molecule with respect to another bring about changes in conformation strictly speaking the application of symmetry operations is to molecular conformations and not to 'molecules'. This may be illustrated with ethylene dichloride (see Fig. 2.11). The staggered form has a centre of symmetry, but not the fully eclipsed form. Hence, the study of molecular symmetry is the study of the molecule in a particular conformation.

Plane of symmetry (σ) : This has been defined above [see (i)]. If we suppose that the plane of symmetry is in the xy plane (of the Cartesian co-ordinates x, y, z) then, after changing the sign of the z co-ordinate for each atom from z to –z, the configuration of the molecule is equivalent to that of the original molecule. It also follows that repetition of the operation a results in the original molecule. Hence $\sigma^2 = E$. It might also be noted that the reflection plane contains a C_2 axis ($C_2^2 = E$).

When the C_n axis with the largest order (n is the greatest) is regarded as being vertical (coincident with the z-axis), planes of symmetry which are also vertical are indicated by the subscript v, *i.e.*, σ_v, If the reflection plane is in the plane of the paper, this plane of symmetry is indicated by σ_v, and if it is perpendicular to the plane of the paper, then by σ'_v. When the reflection plane is horizontal (*i.e.,* in the xy plane; C_n axis coincident with the z-axis), the plane is represented as σ_h. When a reflection plane is diagonal, *i.e.,* bisects the angles between two equivalent axes, this is indicated by σ_d.

Centre of symmetry (i) : This has been defined above [see (ii)]. If we use the Cartesian co-ordinate system, it can be seen that if the molecule has a centre of symmetry, then changing the co-ordinates (x, y, z) of every atom, with this centre as origin, to (–x, –y, –z) produces an equivalent configuration of the molecule. This operation, also denoted by i, is known as inversion and hence a centre of symmetry is also called a centre of inversion.

Alternating axis -of symmetry (S_n) : This has been defined above [see (iii)]. Since the operation S_n involves rotation followed by reflection, an alternating axis of symmetry is also called a rotationreflection axis of symmetry. This type of axis is called an improper axis; two steps are involved: rotation first (about an axis) followed by reflection (the order of operations may be reversed without affecting the result).

From what has been said above, it can be shown that $S_1 \equiv \sigma$ and $S_2 \equiv i$. The cyclobutane and spiro-N-compounds described have an S_4 axis. In this case, there is neither a plane nor a centre of symmetry present.

As we have seen, if a molecule is not superimposable on its mirror image, that molecule can exhibit optical activity. Such molecules are (a) asymmetric; these are completely devoid of any symmetry elements (except a trivial axis); or (b) dissymmetric; these have proper axes but no improper axis. It therefore follows that if a molecule has an S_n axis, that molecule is not optically active, whereas if it has no S_n axis, that molecule is optically active. Alternatively, if a C_n axis is the only symmetry element present in a molecule, that molecule is optically active.

Of all the possible point groups (see above), those of C_n and D_n contain only proper axes of rotation as their symmetry element. Hence, only molecules belonging to these groups are capable of exhibiting optical activity.

For our purpose, we may define a C_n point group as one which contains the symmetry element C_n. A D_n point group is one which contains the symmetry elements C_n and n C_2 axes. The C_2 axes are all perpendicular to the C_n axis and make equal angles with each other. The

D_n point group is also known as dihedral symmetry (n is the principal axis). All other point groups contain at least one of the symmetry elements S_n, σ, or i, *e.g.* an S_n point group contains the symmetry element S_n, a C_{nv} point group contains a C_n axis of symmetry and nσ planes of symmetry, all of which contain C_n.

ISOMERS IN OPTICALLY ACTIVE COMPOUNDS

The number of optical isomers that can theoretically be derived from a molecule containing one or more chiral centres is of fundamental importance in stereochemistry.

Compounds Containing One Chiral Centre

With the molecule Cabde only two optical isomers are possible, and these are related as object and mirror image, *i.e.*, there is one pair of enantiomers, *e.g*, D- and L-lactic acid. If we examine an equimolecular mixture of dextrorotatory and laevorotatory lactic acids, we shall find that the mixture is optically inactive. This is to be expected, since enantiomers have equal but opposite rotatory power. Such a mixture (of equimolecular amounts) is said to be optically inactive by external compensation, and is known as a racemic modification; it is designated as r-, (±)- or DL-, *e.g*, r-tartaric acid, (±)-limonene, DL-lactic acid.

Thus a compound containing one chiral centre can exist in three forms: (+), (–) and (±). Conversion of molecule Ca_2bd into Cabde. Let us consider as an example the bromination of propionic acid to give α-bromopropionic acid.

$$CH_3CH_2CO_2H \xrightarrow{Br_2/P} CH_3CHBrCO_2H$$

(II) and (III) (Fig. 9) are enantiomers, and since molecule (I) is symmetrical about its vertical axis, it can be anticipated from the theory of probability that either hydrogen atom should be replaced equally well to give (±)-α-bromopropionic acid. This actually does occur in practice.

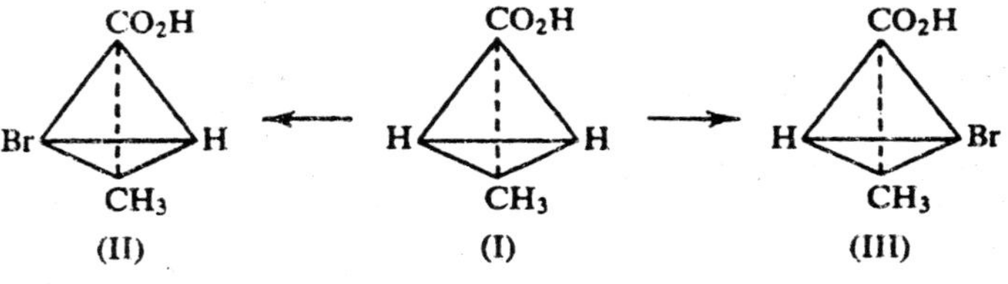

Fig. 9

From what has been said above, it would appear that the two hydrogen atoms in (I) are alike. This is certainly true for their behaviour towards bromine, but the point to note is that a pair of enantiomers was produced, *i.e.*, replacement of one or the other hydrogen does not produce the same molecule.

These two hydrogen atoms are therefore said to be enantiotopic. This term may be defined as follows: Two atoms or groups in a molecule are enantiotopic if replacement of each in turn by some other group leads to a pair of enantiomers.

The case of propionic acid is an example of enantiotopic groups by internal comparison, *i.e.*, the two groups are in the same molecule. There are also enantiotopic groups by external comparison.

Enantiotopic groups of this type are corresponding atoms or groups in a pair of enantiomers, *e.g*, the two methyl groups or bromine atoms in (II) and (III) are enantiotopic. If separate replacement produces the same molecule, the atoms or groups are said to be homotopic.

Prochirality. If a centre in a molecule bears enantiotopic groups, that centre is said to be prochiral. Alternatively, a molecule that contains enantiotopic groups is prochiral, and vice versa. Thus, the carbon atom in (I) is prochiral since replacement of one hydrogen atom by bromine produces a chiral centre.

Also, the prefix pro is used to designate a hydrogen (or any ligand) attached to a prochiral centre and the two enantiotopic hydrogens (or ligands) are distinguished by use of the symbols R and S. The symbol to be used is determined by the specification of the chiral molecule produced by replacing a hydrogen atom by deuterium, *e.g*,

(I): CO_2H (top); H—C—H; CH_2 (bottom) $\longrightarrow$ (Ia): CO_2H (top); D—C—H; CH_3 (bottom) $\equiv$ a (top); c—C—d; b (bottom) $\equiv$ (S): b (top); c—C—a; d (bottom)

(Ib): CO_2H (top); H_S—C—H_R; CH_3 (bottom)

(I) (Ia) (S) (Ib)

The hydrogen replaced in (I) to give (Ia) is therefore pro-S-hydrogen, and consequently the other prochiral hydrogen is pro-R-hydrogen. This may be indicated by writing the formula of (I) as (Ib).

Nuclei which experience equal magnetic shielding have identical chemical shifts; such nuclei are said to be isochronous. Thus, chemically equivalent protons are isochronous, but so are enantiotopic (prochiral) protons, since these also experience equal magnetic shielding, *i.e.,* the signals have the same chemical shifts. However, if dissolved in chiral solvents, then the chemical shifts of enantiotopic (prochiral) protons in a compound may be different.

As we have seen, when (I) reacts with bromine, the pair of enantiomers (II) and (III) are formed in equal amounts. This is due to the fact that when enantiotopic (prochiral) groups react with an achiral reagent, the transition states involved have equal energy contents. On the other hand, if the reagent is chiral, the transition states are diastereoisomeric.

Since these have different energy contents, the two rates of reaction are different, thereby resulting in the formation of a pair of enantiomers in unequal amounts. This may be illustrated by the oxidation of ethanol with the enzyme alcohol dehydrogenase; only one of the two enantiotopic (prochiral) hydrogen atoms is removed to form acetaldehyde. This may be formulated as shown, and it should be noted that the product in this case is not optically active.

Me | H—C—H | OH —enzyme→ Me | H—C=O

In addition to referring to groups as being enantiotopic or prochiral, faces of double bonds are also said to be enantiotopic or prochiral if stereoisomers are produced by addition reactions, *e.g,* the reaction between acetaldehyde and phenylmagnesium bromide (attack at 'front' and at 'back')

H(Me)C=O —PhMgBr→ Me—C(Ph)(H)—OH + Me—C(OH)(H)—Ph

'front' 'back'

For the purpose of naming the enantiotopic or prochiral faces, the sequence rule is used in two dimensions.

b(c)C=a c(b)C=a

re-face si-face

If the order of precedence is a > b > c, then, if the groups are in a clockwise arrangement, that face is called re (rectus), and if in an anticlockwise arrangement, si (sinister). This nomenclature may be extended to the ethylenic double bond, each end of the double bond being treated separately, *e.g,*

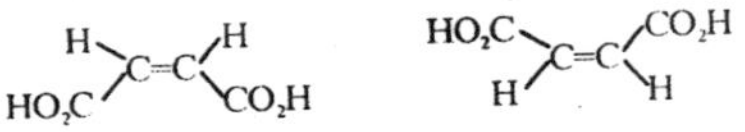

si-re re-si

maleic acid

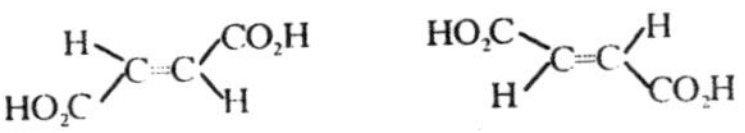

si-si re-re

fumaric acid

Compounds Containing two Different Chiral Centres

When we examine the molecule Cahd Cabe, *e.g*, α, β-dibromobutyric acid, $CH_3CHBrCHBrCO_2H$, we find that there are four possible spatial arrangements for this type of molecule (Fig. 10). (I) and (II) are enantiomers (the configurations of both asymmetric carbons are reversed), and an equimolecular mixture of them forms a racemic modification; similarly for (III) and (IV).

Thus, there are six forms in all for a compound of the type CabdCabe: two pairs of enantiomers and two racemic modifications.

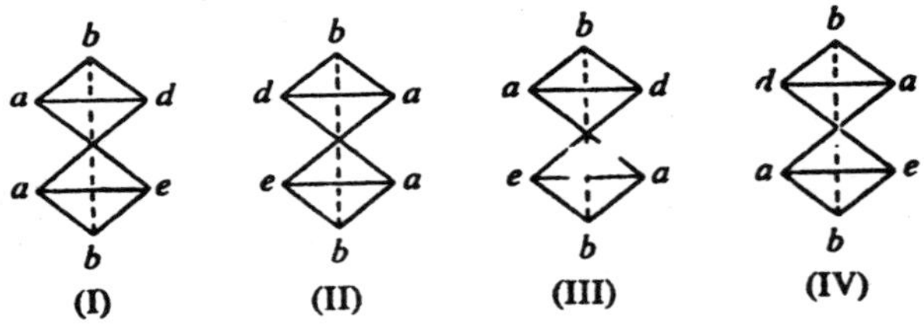

Fig. 10

(I) and (III) are not identical in configuration and are not mirror images (the configuration of one of the two asymmetric carbon atoms is reversed); they are known as diastereoisomers, *i.e.*, they are optical isomers but not enantiomers. Diastereoisomers differ in physical properties such as melting point, density, solubility, dielectric constant and specific rotation. Chemically they are similar, but their rates of reaction with other optically active compounds are different.

The mass spectra of diastereoisomers may exhibit differences. These differences are usually too small to be significant for acyclic diastereoisomers; this is believed to be due mainly to the fact that these molecules are capable of free rotation about the single bonds joined to the chiral centres. On the other hand, the mass spectra of alicyclic diastereoisomers may differ to such an extent that it may be possible to deduce the stereochemistry of each diastereoisomer from its mass spectrum.

The NMR spectra of diastereoisomers are also different.

The plane-diagrams of molecules (I-IV) will be (V-VIII), respectively, as shown. It should be remembered that groups joined to horizontal lines lie above the plane of the paper, and those joined to vertical lines lie below the plane of the paper.

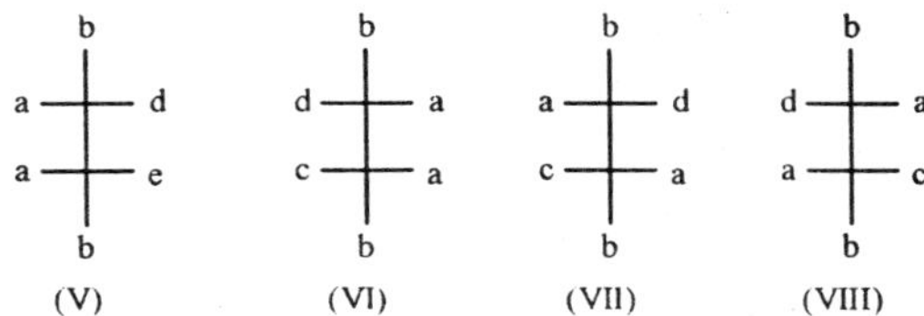

Instead of writing down all the possible configurations, the number of optical isomers for a compound of the type CabdCabe may be obtained by indicating the configuration of each asymmetric carbon atom by the symbol + or –, or by D or L; thus;

$$\underbrace{\begin{matrix}+ & -\\ + & -\end{matrix}}_{(\pm)}\quad \underbrace{\begin{matrix}+ & -\\ - & +\end{matrix}}_{(\pm)}\quad \text{or}\quad \underbrace{\begin{matrix}D_1 & L_1\\ D_2 & L_2\end{matrix}}_{DL}\quad \underbrace{\begin{matrix}D_1 & L_1\\ L_2 & D_2\end{matrix}}_{DL}$$

Pairs of enantiomers of the type CabdCabe are distinguished by the prefixes erythro and threo. The former is the one in which the identical groups can eclipse each other (a,a and b,b) in one conformation, whereas the latter is the one in which this cannot be done. These names are derived from erythrose and threose, the tetrose sugars. For the relative stabilities of the diastereoisomers.

Conversion of molecule Ca_2bCabe into CabdCabe. Let us consider the bromination of β-methylvaleric acid to give α-bromo-β-methylvaleric acid.

$$CH_3CH_2CH(CH_3)CH_2CO_2H \xrightarrow{Br_2/P} CH_3CH_2CH(CH_3)CHBrCO_2H$$

β-Methylvaleric acid contains one asymmetric carbon atom, but the bromine derivative contains two. Let us first consider the case where the configuration of the asymmetric carbon atom in the starting material is D_1 (IX). Bromination of this will produce molecules (X) and (XI); these are diastereoisomers and are produced in unequal amounts. This is to be anticipated; the two a-hydrogen atoms are not symmetrically placed with respect to the lower half of the molecule, and consequently different rates of substitution can be expected. In the same way, bromination of the starting material in which the configuration of the asymmetric carbon atom is L_1 (XII) leads to the formation of a mixture of diastereoisomers (XIII and XIV) in unequal amounts. One can expect, however, that the amount of (XIII) produced from (XII) would be the same as that of (X) from (IX) since, in both cases, the positions of the

CO_2H		CO_2H		CO_2H
H—D_2—Br	←	H——H	→	Br—L_2—H
H—D_1—CH_3		H—D_1—CH_3		H—D_1—CH_3
C_2H_5		C_2H_5		C_2H_5
(X)		(IX)		(XI)

bromine atoms with respect to the methyl group are the same. Similarly, the amount of (XIV) from (XII) will be the same as that of (XI) from (IX). Thus, bromination of (±)-β-methylvaleric acid will result in a mixture of four bromo derivatives which will consist of two racemic modifications in unequal amounts, and the mixture will be optically inactive.

CO_2H		CO_2H		CO_2H
Br—L_2—H	←	H——H	→	H—L_2—Br
CH_3—L_1—H		H_3C—L_1—H		H_3C—D_2—H
C_2H_5		C_2H_5		C_2H_5
(XII)		(XII)		(XIV)

As we have already seen enantiotopic groups react with achiral reagents at the same rates. In the molecules (IX) and (XII), however, the two hydrogen atoms react with bromine, an achiral reagent, at different rates to give a pair of diastereoisomers in unequal amounts. These two hydrogen atoms are therefore said to be diastereotopic. This term may be defined as follows:

Two atoms or groups in a molecule are diastereotopic if replacement of each in turn by some other group leads to a pair of diastereoisomers. The molecule we have discussed is an example of diastereotopic groups by internal comparison, *i.e.,* the two groups are in the same molecule. On the other hand, corresponding groups in a pair of diastereoisomers are said to be diastereotopic by external comparison.

The term diastereotopic faces may be used with respect to the faces of a double bond when one of the groups attached to the unsaturated carbon atom contains a chiral centre.

A centre in a molecule which bears diastereotopic groups is also said to be prochiral and the term 'heterotopic' has been used to describe atoms or groups which are not homotopic without differentiation being made whether the atoms or groups are enantiotopic or diastereotopic.

For enantiotopic groups, the transition states for reactions with achiral reagents have the same energy contents, but for reactions with chiral reagents, the transition states are diastereoisomeric and the energy contents are different.

For diastereotopic groups, the transition states with both achiral and chiral reagents are diastereoisomeric and consequently the diastereoisomeric products are formed in unequal amounts.

Protons in diastereotopic groups show different chemical shifts (since their environments are different). Such protons are said to be anisochronous, *i.e.,* the signals do not have the same chemical shifts. These different chemical shifts are exhibited whether the solvent is chiral or achiral.

Compounds Containing three Chiral Centres

A molecule of this type is CabdCabCabe, *e.g,* the pentoses, and the number of optical isomers possible is eight (four pairs of enantiomers):

D_1	L_1	D_1	L_1	D_1	L_1	L_1	D_1
D_2	L_2	D_2	L_2	L_2	D_2	D_2	L_2
D_3	L_3	L_3	D_3	D_3	L_3	D_3	L_3
DL		DL		DL		DL	

All the cases discussed so far are examples of a series of compounds which contain n structurally distinct carbon atoms, *i.e.,* they belong to the series $Cabd(Cab)_{n-2}Cabe$. In general, if there are n asymmetric carbon atoms in the molecule (of this series), then there will be 2^n optically active forms and 2^{n-1} resolvable forms (*i.e.,* 2^{n-1} pairs of enantiomers).

These formulae a also apply to monocyclic compounds containing n different asymmetric carbon atoms; they may or may not apply to fused ring systems since spatial factors may play a part in the possible existence of various configurations.

Compounds of the Type Cabd $(Cab)_x$ Cabd

In compounds of this type the two terminal asymmetric carbon atoms are similar, and the number of optically active forms possible depends on where x is odd or even.

(i) **Even series.** (a) CabdCahd, *e.g,* tartaric acid. In a compound of this type the rotatory power of each asymmetric carbon atom is the same. Now let us consider the number of optical isomers possible.

D	L	D	L
D	L	L	D
(I)	(II)	(III)	(IV)

In molecules (I) and (II), the upper and lower halves reinforce each other; hence (I), as a whole, has the dextro- and (II) the laevo-configuration, *i.e.,* (I) and (II) are optically active, and enantiomeric. On the other hand, in (III) the two halves are in opposition, and so the molecule, as a whole, will not show optical activity. It is also obvious that (III) and (IV) are identical, *i.e.,* there is only one optically inactive form of CabdCabd. Molecule (III) is said to be optically inactive by internal compensation, and is known as the meso-form, and is a diastereoisomer of the pair of enantiomers (I) and (II). The meso-form is also known as the inactive form and has been represented as the i-form; the meso form cannot be resolved.

CO_2H / HO—H / H—OH / CO_2H — L-

CO_2H / H—OH / HO—H / CO_2H — D-

DL-

CO_2H / H—OH / ------ plane of symmetry / H—OH / CO_2H — meso-(i-)

Inspection of these formulae shows that the D- and L- forms do not possess any elements of symmetry; the meso-form, however, possesses a plane of symmetry.

(b) CabdCabCabCabd, *e.g,* saccharic acid,

$$HO_2CCHOHCHOHCHOHCHOHCO_2H$$

The rotatory powers of the two terminal asymmetric carbon atoms are the same, and so are those of the middle two (the rotatory powers of the latter are almost certainly different from those of the former;

equality would be fortuitous). The possible optical isomers are as follows (V-XIV):

D_1	L_1	D_1	L_1	D_1	L_1	D_1	L_1	D_1	D_1
D_2	L_2	L_2	D_2	D_2	L_2	D_2	L_2	D_2	L_2
D_2	L_2	L_2	D_2	D_2	L_2	D_2	L_2	L_2	D_2
D_1	L_1	D_1	L_1	L_1	D_1	D_1	L_1	L_1	D_1
(V)	(VI)	(VII)	(VIII)	(IX)	(X)	(XI)	(XII)	(XIII)	(XIV)
DL		DL		DL		DL		meso-forms	

Molecules (V) and (VI) are optically active (enantiomeric) and are not 'internally compensated'; (VII) and (VIII) are optically active (enantiomeric) and are not 'internally compensated'; (IX) and (X) are optically active (enantiomeric) but are 'internally compensated at the ends'; (XI) and (XII) are optically active (enantiomeric) but are 'internally compensated in the middle'; (XIII) and (XIV) are meso-forms and are optically inactive by (complete) internal compensation. Thus, there are eight optically active forms (four pairs of enantiomers), and two meso-forms. In general, in the series of the type $Cabd(Cab)_{n-2}Cabd$, if n is the number of asymmetric carbon atoms and n is even, then there will be 2^{n-1} optically active forms, and $2^{(n-2)\backslash 2}$ meso-forms.

(ii) *Odd series :* (a) CabdCabCabd, *e.g*, trihydroxyglutaric acid. If the two terminal asymmetric carbon atoms have the same configuration, then the central carbon atom has two identical groups joined to it and hence cannot be asymmetric. If the two terminal configurations are opposite, then the central carbon atom has apparently four different groups attached to it (the two ends are mirror images and not superimposable).

Thus, the central carbon atom becomes asymmetric, but at the same time the two terminal atoms 'compensate internally' to make the molecule as a whole symmetric (there is now a plane of symmetry), and consequently the compound is not optically active. In this molecule the central carbon atom is said to be pseudo-asymmetric, and is designated 'D' and 'L' (or $\oplus$ and $\ominus$ if the + and – convention is used).

There will, however, be two meso-forms since the pseudo-asymmetric carbon atom can have two different configurations (see XV-XVIII). Thus there are five forms in all: two optically active forms (enantiomers), one racemic

Cabd	D	L	D	D
Cab			----'D'----------	'L'---- plane of symmetry
Cabd	D	L	L	L
	(XV)	(XVI)	(XVII)	(XVIII)
	DL		meso	meso

modification, and two meso-forms. The following are the corresponding trihydroxyglutaric acids, all of which are known.

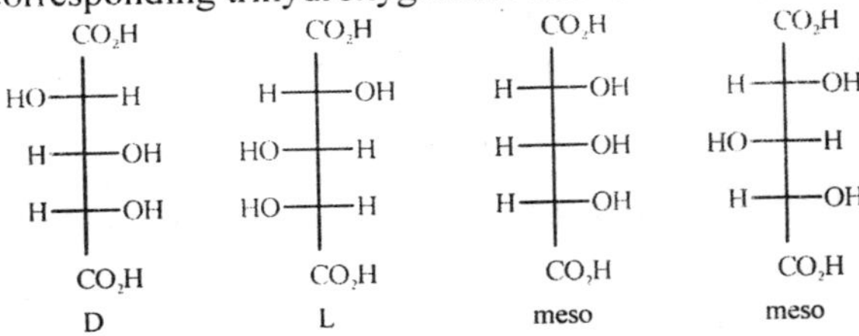

Inspection of the structure of the trihydroxyglutaric acid shows that the four groups attached to the pseudo-asymmetric carbon atoms are of two types in pairs: one pair consists of two different groups which are achiral (H and OH), and the other pair consists of two groups which are enantiomeric (–$CHOHCO_2H$). These are the characteristics of a pseudo-asymmetric carbon atom, *i.e.*, the molecule is of the type $Cabd_Dd_L$.

(b) *CabdCabCabCabCabd* : In this molecule the central carbon atom is pseudo-asymmetric when the left-hand side of the molecule has the opposite configuration to that of the right-hand side; the central carbon atom is symmetrical when both sides have the same configuration. In all other cases the central carbon atom is asymmetric, the molecule now containing five asymmetric carbon atoms. The following table shows that there are sixteen optical isomers possible, of which twelve are optically active (six pairs of enantiomers), and four are meso-forms.

Ends with Opposite Configurations

D_1	D_1	D_1	D_1
D_2	D_2	L_2	L_2
'D'	'L'	'D'	'L'
L_2	L_2	D_2	D_2
L_1	L_1	L_1	L_1
meso	meso	meso	meso

Note the characteristics of the pseudo-asymmetric carbon atom (the central one): two different achiral groups (a and b) and two enantiomeric groups in pairs (D_1D_2, L_1L_2; D_1L_2, L_1D_2).

Ends with the Same Configurations

$$
\begin{array}{cc}
\begin{array}{cc} D_1 & L_1 \\ D_2 & L_2 \\ & \\ D_2 & L_2 \\ D_1 & L_1 \end{array} & \begin{array}{cc} D_1 & L_1 \\ L_2 & D_2 \\ & \\ L_2 & D_2 \\ D_1 & L_1 \end{array} \\
\underbrace{\qquad\quad}_{DL} & \underbrace{\qquad\quad}_{DL}
\end{array}
$$

Molecule with Five Asymmetric Carbon Atoms

$$
\begin{array}{cccc}
\begin{array}{cc} D_1 & L_1 \\ D_2 & L_2 \\ D & L \\ D_2 & L_2 \\ L_1 & D_1 \end{array} & \begin{array}{cc} D_1 & L_1 \\ D_2 & L_2 \\ L & D \\ D_2 & L_2 \\ L_1 & D_1 \end{array} & \begin{array}{cc} D_1 & L_1 \\ D_2 & L_2 \\ D & L \\ L_2 & D_2 \\ D_1 & L_1 \end{array} & \begin{array}{cc} D_1 & L_1 \\ D_2 & L_2 \\ L & D \\ L_2 & D_2 \\ D_1 & L_1 \end{array} \\
\underbrace{\qquad\quad}_{DL} & \underbrace{\qquad\quad}_{DL} & \underbrace{\qquad\quad}_{DL} & \underbrace{\qquad\quad}_{DL}
\end{array}
$$

In general, in the series of the type $Cabd(Cab)_{n-2}Cabd$, if n is the number of 'asymmetric' carbon atoms and n is odd, then there will be 2^{n-1} optical isomers, of which $2^{(n-1)/2}$ are meso-forms and the remainder optically active forms.

THE RACEMIC MODIFICATION

The racemic modification is an equimolecular mixture of a pair of enantiomers, and it may be prepared in several ways.

(i) Mixing of equimolecular proportions of enantiomers produces the racemic modification.

(ii) Synthesis of chiral compounds from achiral compounds always results in the formation of the racemic modification. This statement is true only if the reaction is carried out in the absence of other optically active compounds or circularly polarised light.

(iii) **Racemisation :** The process of converting an optically active compound into the racemic modification is known as racemisation. The (+)- and (–)-forms of most compounds are capable of racemisation under the influence of heat, light or chemical reagents. Which agent is used depends on the nature of the compound, and at the same time the ease of racemisation also depends on the nature of the compound, *e.g,*

(a) Some compounds racemise so easily that they cannot be isolated in the optically active forms.

(b) A number of compounds racemise spontaneously when isolated in optically active forms.

(c) The majority of compounds racemise with various degrees of ease under the influence of different reagents.

(d) A relatively small number of compounds cannot be racemised at all.

When a molecule contains two or more asymmetric carbon atoms and the configuration of only one of these is inverted by some reaction, the process is then called epimerisation.

Many theories have been proposed to explain racemisation, but owing to the diverse nature of the structures of the various optically active compounds, one cannot expect to find one theory which would explain the racemisation of all types of optically active compounds. Thus we find that a number of mechanisms have been suggested, each one explaining the racemisation of a particular type of compound.

A number of compounds which are easily racemisable are those in which the asymmetric carbon atom is joined to a hydrogen atom and can undergo tautomeric change. Let us consider the case of keto-enol tautomerism: In the keto-form (I) the carbon joined to the hydrogen atom and the oxo group is asymmetric; in the enol-form, (II), this carbon atom has lost its asymmetry. When the enol-form reverts to the keto-form, it can do so to produce the original keto molecule (I), but owing to its symmetry, the enol-form can produce equally well the keto-form (III) in which the configuration of the asymmetric carbon atom is opposite

$$\underset{(I)}{-\overset{\overset{H}{|}}{\underset{|}{C}}-C{=}O} \rightleftharpoons \underset{(II)}{-\underset{|}{C}{=}\underset{|}{C}-OH} \rightleftharpoons \underset{(III)}{-\overset{|}{\underset{\underset{H}{|}}{C}}-C{=}O}$$

to that in (I). Thus racemisation, according to this scheme, occurs via the enol-form, *e.g,* (-)-lactic acid is racemised in aqueous sodium hydroxide, and this change may be formulated:

$$H_3C-\underset{H}{\overset{OH}{C}}-C\overset{O}{\underset{O^-}{\lessgtr}} \underset{}{\overset{-H^+}{\rightleftharpoons}} H_3C-\overset{OH}{\underset{-}{C}}-C\overset{O}{\underset{O^-}{\lessgtr}} \longleftrightarrow \overset{HO}{\underset{H_3C}{>}}C=C\overset{O}{\underset{O}{\lessgtr}} \overset{-H^+}{\rightleftharpoons} H_3C-\underset{OH}{\overset{O}{C}}-C\overset{O}{\underset{O^-}{\lessgtr}}$$

(–)

There is a great deal of evidence to support this tautomeric mechanism. When the hydrogen atom joined to the asymmetric carbon atom is replaced by some group that prevents tautomerism (enolisation) then racemisation is also prevented (at least under the same conditions as the original compound), *e.g,* mandelic acid, $C_6H_5CHOHCO_2H$, is readily racemised by warming with aqueous sodium hydroxide.

On the other hand, atrolactic acid, $C_6H_5C(CH_3)(OH)CO_2H$, is not racemised under the same conditions; in this case keto-enol tautomerism is no longer possible (*i.e.,* formation of the intermediate carbanion is not possible).

Racemisation of compounds capable of exhibiting keto-enol tautomerism is catalysed by acids and bases. Since keto-enol tautomerism is also catalysed by acids and bases, then if racemisation proceeds via enolisation, the rates of racemisation and enolisation should be the same.

This relationship has been established by means of kinetic studies, *e.g,* Bartlett *et al.,* (1935) found that the rate of acid-catalysed iodination of 2-butyl phenyl ketone was the same as that of racemisation in acid solution.

This is in keeping with both reactions involving the rate-controlling formation of the enol.

$$\underset{(+)}{PhCOCHMeEt} \underset{fast}{\overset{slow}{\rightleftharpoons}} Ph\overset{OH}{C}=CMeEt \underset{slow}{\overset{fast}{\rightleftharpoons}} \underset{(-)}{PhCOCHMeEt}$$

$$Ph\overset{OH}{C}=CMeEt \xrightarrow[\text{fast}]{I_2} PhCOCIMeEt$$

On the other hand, on the basis that the rate-determining step in base-catalysed enolisation and racemisation is the formation of the enolate ion, then the two processes will also occur at the same rate.

$$B + RCOCHR_2 \underset{fast}{\overset{slow}{\rightleftharpoons}} BH^+ + R\overset{O^-}{C} = CR_2 \underset{slow}{\overset{fast}{\rightleftharpoons}} B + R\overset{OH}{C} = CR_2$$

Hsu *et al.,* (1936) found that the rates of bromination and racemisation (in the presence of acetate ions) of 2-o-carboxybenzyl-l-indanone were identical.

$$\text{(indanone ring: } C_6H_4\text{-}CH_2\text{-}CH(\text{-}CO\text{-})\text{)}\text{—}CH_2\text{—}C_6H_4\text{—}CO_2H$$

Further support for this mechanism is the work of Ingold *et al.*, (1938) who showed that the rate of racemisalon of (+)-2-butyl phenyl ketone in dioxan-deuterium oxide solution in the presence of NaOD is the same as the rate of deuterium exchange. This is in keeping with the formation of the enolate ion (or carbanion), which is common to both reactions. There are many compounds containing an asymmetric carbon atom which can be racemised under suitable conditions although there is no possibility of tautomerism. A number of different types of compounds fall into this group, and the mechanism proposed for racemisation depends

$$PhC(O^-)=CMeEt$$

$$(+)\text{-}PhCOCHMeEt + OD^- \underset{\text{fast}}{\overset{\text{slow}}{\rightleftharpoons}} HOD + PhCOC^-MeEt$$

$$HOD + PhCOC^-MeEt \xrightarrow[\text{slow}]{D_2O;\ \text{fast}} PhCOCDMeEt$$

$$HOD + PhCOC^-MeEt \underset{\text{slow}}{\overset{\text{fast}}{\rightleftharpoons}} (-)\text{-}PhCOCHMeEt + OD^-$$

on the type of compound under consideration. In the case of compounds of the type of (–)-limonene which is racemised by strong heating, the mechanisms proposed are highly speculative, *e.g*, according to Kincaid *et al.*, (1940), molecules of the type Cabde can only be racemised by the breaking of bonds. A number of optically active secondary alcohols can be racemised by heating with a sodium alkoxide. This has been explained by a reversible dehydrogenation (Huckel, 1931) and there is some evidence to support this mechanism (Doering *et al.*, 1947, 1949). It has also been found that the presence of a trace of carbonyl compound (generally formed by atmospheric oxidation) is necessary for this reaction.

$$\underset{(+)\text{-}}{R^1\text{—}C(R^2)(H)\text{—}OH} \underset{}{\overset{-2H}{\rightleftharpoons}} R^1\text{—}C(R^2)=O \overset{+2H}{\rightleftharpoons} \underset{(-)\text{-}}{R^1\text{—}C(H)(R^2)\text{—}OH}$$

Another different type of compound which can be readily racemised is that represented by a-chloroethylbenzene. When the (+)- or (-)-form is dissolved in liquid sulphur dioxide, spontaneous racemisation occurs. This has been explained by assuming ionisation into a carbonium ion (Polanyi *et al.,* 1933).

$$\underset{(+)\text{-}}{C_6H_2CHClCH_3} \rightleftharpoons C_6H_5\overset{+}{C}HCH_3 + Cl^- \rightleftharpoons \underset{(-)\text{-}}{C_6H_5CHClCH_3}$$

The carbonium ion is planar (the positively charged carbon atom is in a state of trigonal hybridisation) and consequently symmetric; recombination with the chlorine ion can occur equally well to form the (+)- and (–)-forms, *i.e.,* racemisation occurs. The basis of this mechanism is that alkyl halides in liquid sulphur dioxide exhibit an electrical conductivity, which has been taken as indicating ionisation. Hughes, Ingold *et al.,* (1936), however, found that pure a-chloroethylbenzene in pure liquid sulphur dioxide does not conduct, but when there is conduction, then styrene and hydrogen chloride are present. These authors showed that under the conditions of purity, the addition of bromine leads to a quantitative yield of styrene dibromide, and so suggested that the rate of racemisation is accounted for by the rate of formation of hydrogen chloride; thus:

$$C_6H_5CHClCH_3 \xrightarrow{\text{slow}} C_6H_5\overset{+}{C}HCH_3$$

$$C_6H_5\overset{+}{C}HCH_3 \xrightarrow{\text{fast}} C_6H_5CH = CH_2 + H^+$$

It is the recombination of the styrene with the hydrogen chloride that produces the racemised product; this may be written as follows

$$\underset{(+)\text{-}}{C_6H_5CHClCH_3} \rightleftharpoons C_6H_5CH = CH_2 + HCl \rightleftharpoons \underset{(-)\text{-}}{C_6H_5CHClCH_3}$$

α-Chloroethylbenzene can also be readily racemised by means of Lewis acids, *e.g,* SbC15, $HgCl_2$, etc. In this case, the mechanism is believed to be similar to that proposed by Polanyi (see above). Thus:

$$\underset{(+)\text{-}}{C_6H_5CHClCH_3} + SbCl_5 \rightleftharpoons C_6H_5\overset{+}{C}HCH_3 + SbCl_6^- \rightleftharpoons C_6H_5CHClCH_3 + SbCl_5$$

The racemisation of optically active hydrocarbons containing a tertiary hydrogen atom is very interesting. It has been shown that such hydrocarbons undergo hydrogen exchange when dissolved in concentrated sulphuric acid (Ingold *et al.,* 1936), and the mechanism is believed to occur via a carbonium ion (Burwell *et al.,* 1948).

$$R_3CH + 2H_2SO_4 \rightarrow R_3C^+ + HSO_4^- + SO_2 + 2H_2O$$

$$R_3C^+ + R_3CH \rightarrow R_3CH + R_3C^+, \text{ etc.}$$

This reaction is very useful for racemising optically active hydrocarbons, *e.g,* Burwell *et al.*, (1948) racemised optically active 3-methylheptane in concentrated sulphuric acid (the carbonium ion is flat):

$$\underset{(+)-}{C_2H_5-\overset{\overset{CH_3}{|}}{CH}-C_4H_9} + C_2H_5-\overset{\overset{CH_3}{|}}{C^+}-C_4H_9 \longrightarrow$$

$$C_2H_5-\overset{\overset{CH_3}{|}}{C^+}-C_4H_9 + \underset{(\pm)-}{C_2H_5-\overset{\overset{CH_3}{|}}{CH}-C_4H_9}$$

Optically active hydrocarbons can also be racemised by means of aluminium chloride, the mechanism again probably being via the formation of a carbonium ion, *e.g,* 2-phenylbutane

(I) $\underset{(+)-}{C_6H_5CH(CH_3)C_2H_5} + AlCl_2 \longrightarrow C_6H_5\overset{+}{C}(CH_3)C_2H_5 + HAlCl_2^-$

(II) $\underset{(+)-}{C_6H_5CH(CH_3)C_2H_5} + C_6H_5\overset{+}{C}(CH_3)C_2H_5 \longrightarrow$

$$C_6H_5\overset{+}{C}(CH_3)C_2H_5 + \underset{(\pm)-}{C_6H_5CH(CH_3)C_2H_5}$$

The racemisation of other types of optically active compounds is described later.

PROPERTIES OF THE RACEMIC MODIFICATION

The racemic modification may exist in three different forms in the solid state.

(i) **Racemic mixture :** This is also .known as a (±)-conglomerate, and is a mechanical mixture of two types of crystals, the (+)- and (–)-forms; there are two phases present. The physical properties of the racemic mixture are mainly the same as those of its constituent enantiomers. The most important difference is the m.p.

(ii) **Racemic compound :** This consists of a pair of enantiomers in combination as a molecular compound; only one solid phase is present. The physical properties of a racemic compound are different from those of the constituent enantiomers, but in solution racemic compounds dissociate into the (+)- and (–)-forms.

(iii) **Racemic solid solution :** This is also known as a pseudo-racemic compound, and is a solid solution (one phase system) formed by a pair of enantiomers crystallising together due to their being isomorphous. The properties of the racemic solid solution are mainly the same as those of its constituent enantiomers; the m.p.s may differ.

Methods for Determining the Nature of a Racemic Modification

One simple method of examination is to estimate the amounts of water of crystallisation in the enantiomers (only one need be examined) and in the racemic modification; if these are different, then the racemic modification is a racemic compound. Another simple method is to measure the densities of the enantiomers and the racemic modification; again, if these are different, the racemic modification is a racemic compound; *e.g*, tartaric acids.

Table 1

	D-Tartaric acid	*L.-Tartaric acid*	*Racemic tartaric acid*
Melting point	170°C	170°C	206°C
Water of crystallisation	None	None	1H_2O
Density	1.7598	1.7598	1.697
Solubility in H_2O (at 20°C)	139 g/100 ml	139 g/100 ml	20.6 g/100 ml

There are, however, two main methods for determining the nature of a racemic modification: a study of the freezing-point curves and a study of the solubility curves (Roozeboom, 1899; Andriani, 1900).

Freezing Point Curves

These are obtained by measuring the melting points of mixtures containing different amounts of the racemic modification and its corresponding enantiomers. Various types of curves are possible according to the nature of the racemic modification. In Fig. 11(a) the melting points of all mixtures are higher than that of the racemic modification alone.

In this case the racemic modification is a racemic mixture (a eutectic mixture is formed at the point of 50 per cent composition of each enantiomer), and so addition of either enantiomer to a racemic mixture raises the melting point of the latter; (±)-pinene is an example of this type. In Fig. 11(b) and (c) the melting points of the mixtures are lower than the melting point of the racemic modification which, therefore, is a racemic compound.

The melting point of the racemic compound may be above, that of each enantiomer (Fig. 11b) or below (Fig. 11c); in either case the melting point is lowered when the racemic compound is mixed with an enantiomer; an example of Fig. 11(h) is methyl tartrate, and one of Fig. 11(c) is mandelic acid.

When the racemic modification is a racemic solid solution, three types of curves are possible (Fig. 12). In Fig. 12(a) the freezing-point curve is a horizontal straight line, all possible compositions having the same melting point, *e.g,* (+)- and (–)-camphor. In Fig. 12(b) the freezing-point curve shows a maximum, *e.g,* (+)- and (–)-carvoxime; and in Fig. 12(c) the freezing-point curve shows a minimum, *e.g,* (+)- and(–) isopentyl (isoamyl) carbamate.

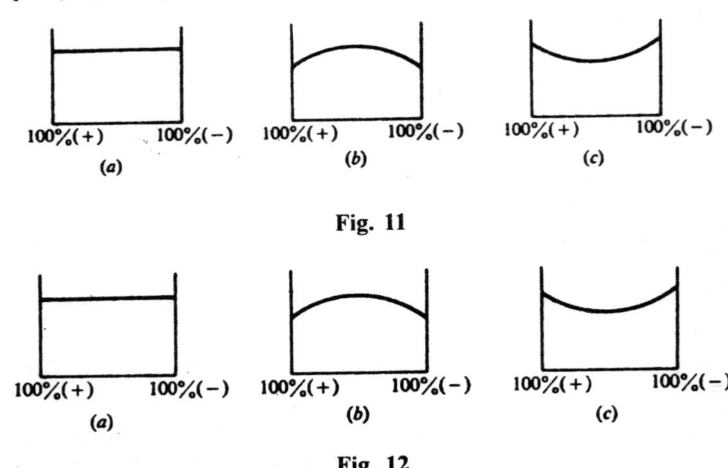

Fig. 11

Fig. 12

In a number of cases there is a transition temperature at which one form of the racemic modification changes into another form, *e.g,* (±)-camphoroxime crystallises as the racemic solid solution above 103°C, whereas below this temperature it is the racemic compound that is obtained.

Correlation of Configurations by Means of Quasi-racemic Compounds

Fredga (1944) has introduced the study of quasi-racemic compounds as a means of correlating configurations their formation being detected by studying the melting-point curves of the two components.

The curves obtained are similar to those of the racemic modification shown in Fig. 3.3a and b, and 3.4a, but with the quasi-racemic compounds these curves are unsymmetrical (since the m.p.s of the components will be different). An unsymmetrical curve 3.3a indicates a eutectic mixture, an unsymmetrical 3.4a a solid solution, and an unsymmetrical 3.3b a quasi-racemic compound. Curves for quasi-racemic compounds are given only by compounds (containing one asymmetric carbon atom) which have closely similar structures but opposite configurations, *e.g,*

```
   a          a
   |          |
b—C—e      e—C—f
   |          |
   d          d
  (I)        (II)
```

(I) and (II). On the other hand, curves of the other two types are given by compounds of like configuration (but some cases are known where the configurations have been opposite). Various examples of this method of correlating configurations have now been described; *e.g,* Fredga (1941) showed (partly by chemical methods and partly by using the quasi-racemate method) that (+)-malic acid (III) and (–)-mercaptosuccinic acid (IV) had opposite configurations. He then showed (1942) that

```
     CO2H            CO2H            CO2H
  H——+——OH      HS——+——H        H——+——Me
     CH2CO2H         CH2CO2H         CH2CO2H

     (III)           (IV)            (V)
```

compound with (+)-methylsuccinic acid (V). Therefore (IV) and (V) have opposite configurations and consequently (+)-malic acid and (+)-methylsuccinic acid have the same configuration. It is of interest to note that McPhail *et al.,* (1966) have confirmed, by X-ray analysis, the absolute configuration of methylsuccinic acid established by Fredga.

Mislow *et al.,* (1956) have applied the m.p. curves in a somewhat different manner. They worked with 3-mercapto-octanedioic acid (VI) and 3-methyl-octanedioic acid (VII). These authors found that compounds

(–)-VI and (+)-(VII) gave solid solutions for all mixtures (unsymmetrical 2.20a), whereas (+)-(VI) and (+)-(VII) give a diagram with a single

$$H-\underset{(CH_2)_4CO_2H}{\overset{CO_2H}{C}}-SH \qquad H-\underset{(CH_2)_4CO_2H}{\overset{CH_2CO_2H}{C}}-Me$$

(–)-form (VI)   (+)-form (VII)

eutectic. These results indicate that (–)-(VI) and (+)-(VIII) have the same absolute configuration, whereas (+)-(VI) and (+)-(VII) have opposite configurations.

Solubility curves : The interpretation of solubility curves is difficult, but in practice the following simple scheme based on solubility may be used. A small amount of one of the enantiomers is added to a saturated solution of the racemic modification, and the resulting solution is then examined in a polarimeter. If the solution exhibits a rotation, then the racemic modification is a compound, but if the solution has a zero rotation, then the racemic modification is a mixture or a solid solution. The reasons for this behaviour are as follows.

If the racemic modification is a mixture or a solid solution, then the solution (in some solvent) is saturated with respect to each enantiomer and consequently cannot dissolve any of the added enantiomer. If, however, the racemic modification is a compound, then the solution (in a solvent) is saturated with respect to the compound form but not with respect to either enantiomer; hence the latter will dissolve when added and thereby produce a rotation. It should be noted that this simple method does not permit a differentiation to be made between a racemic mixture and a racemic solid solution.

Infrared spectroscopy is also being used to distinguish a racemic compound from a racemic mixture or a racemic solid solution. In the latter the spectra are identical, but are different in the former. These observations are also true for X-ray powder diagrams, and so X-ray analysis in the solid state may also be used.

RESOLUTION OF RACEMIC MODIFICATIONS

Resolution is the process whereby a racemic modification is separated into its two enantiomers. In practice the separation may be far from quantitative, and in some cases only one form may be obtained. Furthermore; the form isolated need not be optically pure, *i.e.,* it may consist of the (+)and (–)-forms in unequal amounts, but in this case the process is usually referred to as partial resolution. A large variety of methods for resolution have now been developed, and the method used in a particular case depends largely on the chemical nature of the compound under consideration.

(i) **Mechanical separation :** This method is also known as spontaneous resolution by crystallisation, and was introduced by Pasteur (1848). It depends on the crystallisation of the two forms separately, which are then separated by hand. The method is applicable only for racemic mixtures where the crystal forms of the enantiomers are themselves enantiomorphous. Pasteur separated sodium ammonium racemate in this way. The transition temperature of sodium ammonium racemate is 28°C; above this temperature the racemic compound crystallises out, and below this temperature the racemic mixture.

Now Pasteur crystallised his sodium ammonium racemate from a concentrated solution at room temperature, which must have been below 28°C, since, had the temperature been above this, he would have obtained the racemic compound, which cannot be separated mechanically. Actually, Staedel (1878) failed to repeat Pasteur's separation since he worked at a temperature above 28°C.

(ii) **Preferential crystallisation by inoculation :** A supersaturated solution of the racemic modification is treated with a crystal of one enantiomer (or an isomorphous substance), whereupon this form is precipitated. The resolution of glutamic acid by inoculation has been perfected for industrial use (Ogawa *et al.*, 1957; Oeda, 1961). Harada *et al.*, (1962) have also resolved the copper complex of DL-aspartic acid by inoculation.

Except for the two amino-acids mentioned above, this method of resolution has been found impractical or resulted in partial resolution only. Harada (1965), however, has now obtained total optical resolution of free a-amino-acids by the inoculation method. Resolution was effected by seeding the supersaturated aqueous solutions with pure crystals of L- or D-isomer of the amino-acid.

(iii) **Biochemical separation (Pasteur, 1858) :** Certain bacteria and moulds, when they grow in a dilute solution of a racemic modification, destroy one enantiomer more rapidly than the other, *e.g,* Penicillium glaucum (a mould), when grown in a solution of racemic ammonium tartrate, attacks the (+)-form and leaves the (–).

This biochemical method of separation has some disadvantages

(a) Dilute solutions must be used, and so the amounts obtained will be small.

(b) One form is always destroyed and the other form is not always obtained in 50 per cent yield since some of this may also be destroyed.

(c) It is necessary to find a micro-organism which will attack only one of the enantiomers.

(iv) **Conversion into diastereoisomers (Pasteur, 1858) :** This method, which is the best of all the methods of resolution, consists in converting the enantiomers of a racemic modification into diastereoisomers the racemic modification is treated with an optically active substance and the diastereoisomers thereby produced are separated by fractional crystallisation. Thus, racemic acids may be separated by optically active bases, and vice versa, *e.g,*

$$(D_{acid} + L^{acid}) + 2D_{base} \rightarrow (D_{acid}D_{base}) + (L_{acid}D_{base})$$

These two diastereoisomers may then be separated by fractional crystallisation and the acids (enantiomers) regenerated by hydrolysis with inorganic acids or with alkalis. In practice it is usually easy to obtain the less-soluble isomer in a pure state, but it may be very difficult to obtain the moresoluble isomer.

In a number of cases this second (more-soluble) isomer may be obtained by preparing it in the form of another diastereoisomer which is less soluble than that of its enantiomer. On the other hand, separation and purification of the diastereoisomers may be successfully achieved by chromatography.

Resolution by means of diastereoisomer formation may be used for a variety of compounds, *e.g,*

(a) **Acids :** The optically active bases used are mainly alkaloids: brucine, quinine, strychnine, cinchonine, cinchonidine and morphine. Synthetic optically active bases are also used, *e.g,* benzimidazoles, menthylamine, α-phenylethylamine.

(b) **Bases :** Many optically active acids have been used, *e.g,* tartaric acid, camphor-β-sulphonic acid and particularly α-bromocamphor-π-sulphonic acid.

(c) **Alcohols :** These are converted into the acid ester derivative using either succinic or phthalic anhydride (Pickard and Kenyon, 1912). The acid ester, consisting of equimolecular amounts of

$$C_6H_4(CO)_2O + ROH \rightarrow C_6H_4(CO_2R)(CO_2H)$$

the (+)- and (-)-forms, may now be resolved as for acids. Racemic alcohols may also be resolved by diastereoisomer formation with optically active acyl chlorides (to form esters) or with optically active isocyanates (to form urethans)

$$R^1OCH_2COCl + R^2OH \rightarrow R^1OCH_2CO_2R^2 + HCl$$

$$R^1NCO + R^2OH \rightarrow R^2NHCO_2R^2$$

In these equations R^1 is the (–)-menthyl group; recently N-(–)-menthyl-p-sulphamylbenzoyl chloride, (I), has been used (Mills *et al.,* 1950).

$$C_{10}H_{19}NHSO_2{-}C_6H_4{-}COCl$$

β-Acetoxy-D^5-etienic acid has been found very useful for resolving (±)-alcohols (inter alia, Djerassi *et al.,* 1961).

(d) **Aldehydes and Ketones :** These have been resolved by means of optically active hydrazines, *e.g,* (–)-menthylhydrazine. Sugars have been resolved with (+)-isopentanethiol. Nerdel *et al.,* (1952) have resolved oxo compounds with D-tartramide acid hydrazide,

$$NH_2COCHOHCHOHCONHNH_2;$$

this forms diastereoisomeric tartramazones. On the other hand, Shillington *et al.,* (1958) have converted oxo compounds into their 4-carboxyphenylsemicarbazones by means of a 4-carboxyphenylsemicarbazide,

$$HO_2C{-}C_6H_4{-}NHCONHNH_2$$

Since these derivatives contain a carboxyl group, they can be resolved like acids, *e.g,* with brucine, and finally hydrolysed to liberate the optically active oxo compounds.

Another method of resolution is reduction of the oxo compound to the corresponding alcohol, which is then resolved and the separated enantiomers re-oxidised.

Adams *et al.,* (1966) have resolved ketones via enamine formation. The enamine, produced by condensation of the ketone with pyrrolidine in the presence of a trace of p-toluenesulphonic acid is converted into iminium salts containing optically active anions, *e.g,* (+)-camphor10-sulphonate anion (represented as Z^- in the equation):

$$R^1CH_2COR^2 + HN\langle\text{pyrrolidine}\rangle \xrightarrow{TsOH} R^1CH{=}CR^2{-}N\langle\text{pyrrolidine}\rangle \xrightarrow{HZ} \left\{R^1CH{=}CR^2{-}\overset{+}{N}H\langle\text{pyrrolidine}\rangle\right\}Z^-$$

iminium salt

Recrystallisation from suitable solvents gives the (+)- and (–)-forms.

(e) **Amino-compounds :** These may be resolved by conversion into diastereoisomeric anils by means of optically active aldehydes. Amines have also been resolved via their salts using, *e.g,* (+)-tartaric acid. a-Amino-acids have been resolved by preparing (he acyl derivative with an optically active acyl chloride, *e.g,* (–)-menthoxyacetyl chloride (cf. alcohols). Another method of resolving DL-amino-acids is asymmetric enzymic synthesis. The racemic amino-acid is converted into the acyl derivative which is then allowed to react with aniline in the presence of the enzyme papain at the proper pH (Albertson, 1951). Under these conditions only the L-amino-acid derivative reacts to form an insoluble anilide; the D-acid does not react but remains in the solution.

$$\underset{\text{DL-acid}}{R^1CH(NHCOR^2)CO_2H} + C_6H_5NH_2 \xrightarrow{\text{papain}} \underset{\text{L-acid}}{R^1CH(NHCOR^2)CONHC_6H_5} + \underset{\text{D-acid}}{R^1CH(NHCOR^2)CO_2H}$$

Amino-acids have also been resolved by other means.

Although amino-acids contain both an amino-group and a carboxyl group, they usually cannot be resolved in the straightforward way as amines or acids. This is due to the fact that amino-acids behave as dipolar ions.

Asymmetric Transformation

Resolution of racemic modifications by means of salt formation (the diastereoisomers are salts; cf. acids and bases) may be complicated by the phenomenon of asymmetric transformation. This is exhibited by compounds that are optically unstable, *i.e.*, the enantiomers are readily interconvertible

$$(+)\text{–C} \rightleftharpoons (-)\text{–C}$$

Suppose we have an optically stable (+)-base (one equivalent) dissolved in some solvent, and this is then treated with one equivalent of an optically unstable (±)-acid. At the moment of mixing, the solution will contain equal amounts of [(+)-Base .(+)-Acid] and [(+)-Base.(–)-Acid]; but since the acid is optically unstable, the two diastereoisomers will be present in unequal amounts when equilibrium is attained.

$$[(+)\text{-Baseo}(+)\text{-Acid}] \rightleftharpoons [(+)\text{-Base.}(-)\text{-Acid}]$$

According to Jamison and Turner, first-order asymmetric transformation is the establishment of equilibrium in solution between the two diastereoisomers which must have a real existence. In secondorder asymmetric transformation it is necessary that one salt should crystallise from solution; the two diastereoisomers need not have a real existence in solution. In second-order asymmetric transformation it is possible to get a complete conversion of the acid into the form that crystallises; the form may be the (+)- or (–)-, and which one it is depends on the nature of the base and the solvent.

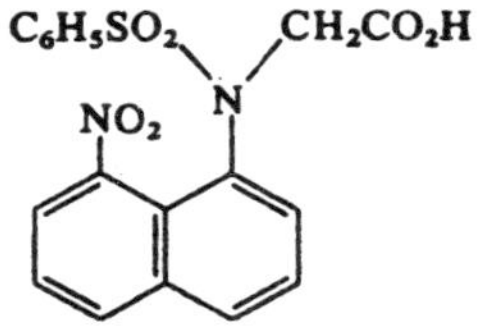

$C_6H_5SO_2$ CH_2CO_2H N NO_2

(II)

Many examples of first- and second-order asymmetric transformation are known, and a large number of these compounds are those which owe their chirality to restricted rotation about a single bond, *e.g*, Mills and Elliott (1928) tried to resolve N-benzenesulphonyl-8-nitro-1-naphthylglycine, (II), by means of the brucine salt. These authors found that either diastereoisomer could be obtained in approximately 100 percent yield by crystallisation from methanol and acetone, respectively.

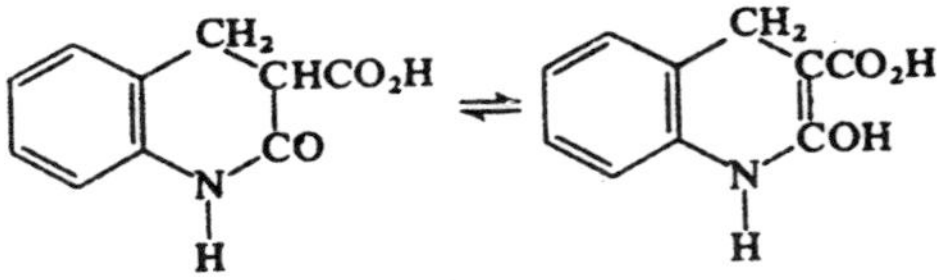

Another example of second-order asymmetric transformation is hydrocarbostyril-3carboxylic acid. This compound contains an asymmetric carbon atom, and Leuchs (1921), attempting to resolve it with quinidine, isolated approximately 90 per cent of the (+)-form. Optical instability in this case is due to keto-enol tautomerism.

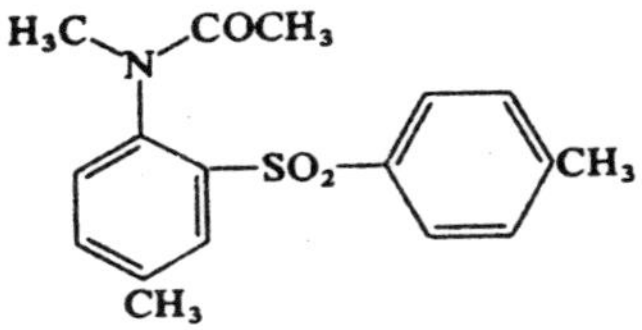

(III)

A very interesting example of second-order asymmetric transformation is 2-acetomethylamido4',5-dimethylphenylsulphone, (III). When this compound was crystallised from a supersaturated solution in ethyl (+)-tartrate, the crystals obtained had a rotation of +02°; evaporation of the mother liquor gave crystals with a rotation of -0 15° (Buchanan *et al.,* 1950).

(v) Another method of resolution that has been tried is the conversion of the enantiomers into volatile diastereoisomers, which are then separated by fractional distillation. So far, the method does not appear to be very successful, only a partial resolution being the result; *e.g,* Bailey and Hass (1941) converted (±)-pentan-2-ol into its diastereoisomers with L(+)-lactic acid, and then partially separated them by fractional distillation.

(vi) **Chromatography.** Optically active substances may be selectively adsorbed by some optically active adsorbent, *e.g,* Henderson and Rule (1939) partially resolved p-phenylenebisiminocamphor on lactose as adsorbent; Bradley and Easty (1951) have found that wool and casein selectively adsorb (+)-mandelic acid from an aqueous solution of (±)-mandelic acid. A particularly important case of resolution by chromatography is that of Troger's base.

Jamison and Turner (1942) have carried out a chromatographic separation without using an optically active adsorbent; they partially resolved the diastereoisomers of (–)-menthyl (±)mandelate by preferential adsorption on alumina. It is also interesting to note that the resolution of a racemic acid by salt formation with an optically active base is made more effective by the application of chromatography. More recently, enzymic and chromatographic methods have been developed for the direct separation of enantiomers (inter alia, Rogozhin, 1971; Gil-Av, 1972).

GSC and GLC have been used with great success for resolving racemic modifications, *e.g,* s-butanol and s-butyl bromide have been separated into two overlapping fractions using a column of starch or ethyl tartrate as the stationary phase (Karagounis *et al.,* 1959). On the other hand, Casanova *et al.,* (1961) have resolved the diastereoisomeric ketals from (±)-camphor by GLC, and Halpern *et al.,* (1965) have resolved in the same way DL-amino-acids via their (–)-menthyl ester derivatives and via their acyl derivatives. Amines may also be resolved via acylation with optically active acid chlorides. Halpern *et al.,* (1966) have used N-trifluoroacetyl-L-prolyl chloride, and separated the diastereoisomers by GLC, *e.g,* 2-aminobutane and 2-aminopentane have been obtained optically pure.

Beckett *et al.,* (1957) have introduced a novel method for correlating and determining configurations. These authors have prepared 'stereoselective adsorbents'. These are adsorbents prepared in the presence of a suitable reference compound of known configuration, *e.g,* silica gel in the presence of quinine. Such an adsorbent exhibits higher adsorptive power for isomers related to the reference compound than for their stereoisomers, provided that their structures are not too dissimilar from that of the reference compound. Thus, silica gel prepared in the presence of quinine adsorbs quinine more readily than its stereoisomer quinidine; cinchonidine (configurationally related to quinine) is adsorbed more readily than its stereoisomer cinchonine (configurationally related to quinidine).

(vii) **Kinetic method of resolution.** Marckwald and McKenzie (1899) found that (–)-menthol reacts more slowly with (–)-mandelic acid than with the (+)-acid. Hence, if insufficient (–)-menthol is used to completely esterify (±)-mandelic acid, the resulting mixture of diastereoisomers will contain more (–)-menthyl (+)-mandelate than (–)-menthyl (–)-mandelate. Consequently there will be more (–)-mandelic acid than (+)-mandelic acid in the unchanged acid, *i.e.*, a partial resolution of (±)-mandelic acid has been effected.

(viii) Ferreira (1953) has partially, resolved (±)-narcotine and (±)-laudanosine (1–2.5 per cent resolution) without the use of optically active reagents. He dissolved the racemic alkaloid in hydrochloric acid and then slowly added pyridine; the alkaloid was precipitated, and it was found to be optically active. The explanation offered for this partial resolution is as follows (Ferreira). When a crystalline racemic substance is precipitated from solution, a crystallisation nucleus is first developed. Since this nucleus contains a relatively small number of molecules, there is more than an even chance that it will contain an excess of one enantiomer or other. If it be assumed that the forces acting on the growth of crystals are the same kind as those responsible for adsorption [cf. (vi)], the nucleus will grow preferentially, collecting one enantiomer rather than the other. Crystallisation, when carried out in the usual manner, results in the formation of crystals containing more or less equivalent numbers of both enantiomers.

(ix) **Channel complex formation** has also been used to resolve racemic modifications. This also offers a means of carrying out a resolution without chiral reagents, *e.g,* Schlenk (1952) added (±)-2-chloro-octane to a solution of urea and obtained, on fractional crystallisation, the two urea inclusion complexes urea/(+)-2-chloro-octane and urea/(–)-2-chloro-octane.

Baker *et al.,* (1952) have prepared tri-o-thymotide, and found that it formed clathrates with ethanol, n-hexane, etc. Powell *et al.,* (1952) have shown that tri-o-thymotide crystallises as a racemate, but that resolution takes place when it forms clathrates with n-hexane, benzene or chloroform. By means of seeding and slow growth of a single crystal, it is possible to obtain the (+)- or (–)-form depending on the nature of the seed. Furthermore, crystallisation of tri-o-thymotide (dl) from a solvent which is itself a racemic modification (d′l′) and which forms a clathrate, produces crystals of the types dd′ and ll′. Thus such (solvent) racemic modifications can be resolved, *e.g,* s-butyl bromide has been resolved in this way.

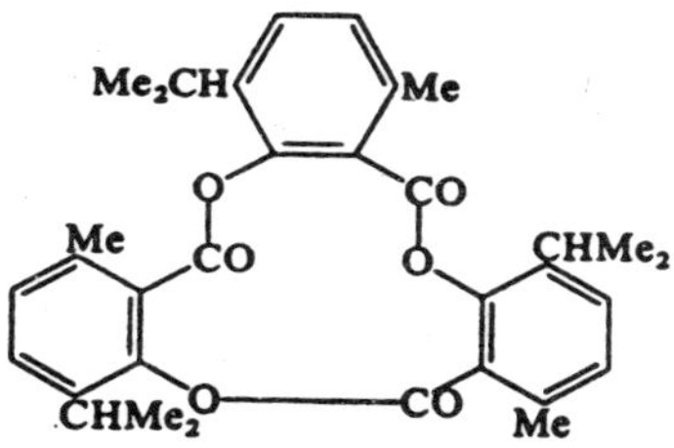

tri-*o*-thymotide

Fig. 13

Optical Purity

An optically pure compound is one which has been prepared in 100 per cent purity, *i.e.,* optical purity is expressed as a percentage, *e.g,* if the (maximum) specific rotation of compound A is + 50° and an impure sample has a rotation of + 30°, this sample is 60 per cent optically pure. For a racemic modification, the optical purity is zero. The difficult problem with respect to optical purity is to be able to ascertain whether a specimen of an enantiomer is 100 per cent optically pure. Several criteria may be used.

The simplest criterion is that which considers a crystalline compound to be optically pure if, after repeated crystallisation, the melting point and rotation remain unchanged. This, however, may not be a correct conclusion, *e.g,* the resolution of a racemic solid solution may lead to the isolation of a partially resolved enantiomer which, after repeated crystallisation, does not change its rotation. The conclusion, however, that the compound is optically pure is strengthened if both of its enantiomers can be prepared and their rotations are equal and opposite. There are several methods which may be used for ascertaining optical purity and are reliable within certain experimental limits. One method uses isotopic dilution. A known weight of the enantiomer being examined is mixed, in solution, with a known weight of its racemic modification which has been labelled with an isotope.

After recrystallisation, the isotope content of the racemic modification is then determined. Suppose the enantiomer under consideration is the (+)-form. In this case, the recovered racemic modification will contain unlabelled (+)-form as well as labelled (+)-form, and it is therefore possible to calculate the dilution factor (since known weights of both were used). If, however, the (+)-enantiomer is not optically pure, some unlabelled (–)-form will also be present in the recovered racemic modification. In this case, the isotope dilution factor will be less than the predicted one. Other methods make use of enzymes or conversion into other compounds of known optical purity. NMR spectroscopy may also be used to determine optical purity.

$$\underset{(+)\text{-(I)}}{\begin{array}{c} R^2 \\ | \\ R^1 - C - NH_2 \\ | \\ R^3 \end{array}} + \underset{(-)\text{-(I)}}{\begin{array}{c} R^3 \\ | \\ R^1 - C - NH_2 \\ | \\ R^2 \end{array}} \xrightarrow{(+)\text{-ChabCOCl}}$$

$$\underset{\text{(IIa)}}{\begin{array}{ccccccccc} & & R^2 & & & & O & & H \\ & & | & & & & \| & & | \\ R^1 & - & C & - & NH & - & C & - & C - a \\ & & | & & & & & & | \\ & & R^3 & & & & & & N \end{array}} + \underset{\text{(IIb)}}{\begin{array}{ccccccccc} & & R^3 & & & & O & & H \\ & & | & & & & \| & & | \\ R^1 & - & C & - & NH & - & C & - & C - a \\ & & | & & & & & & | \\ & & R^2 & & & & & & b \end{array}}$$

It has already been pointed out that the NMR spectra of a pair of enantiomers are identical. Now suppose that a racemic modification is completely converted into a pair of diastereoisomers, *e.g,* In the enantiomers (+)-(I) and (–)-(I), corresponding pairs of groups are enantiotopic and their chemical shifts are identical. This is no longer the case for corresponding groups in (IIa) and (IIb). Corresponding pairs are now in diastereoisomeric environments. Thus, the protons of the group CH are diastereotopic, and if the acid chloride is optically pure, the two different proton signals will have the same intensities in their NMR spectra. If reaction between the optically pure acid chloride (in excess) is carried out with a resolved specimen of (I), then only one signal for the CH proton will be observed if (I) is optically pure. If (I) is not optically pure, then (IIa) and (IIb) will be formed in unequal amounts and two signals will be observed for the CH proton. It is then possible to calculate the optical purity of (1) from the ratio of the intensities of the two signals.

THE TETRAHEDRAL CARBON ATOM

In 1874, van't Hoff and Le Bel, independently, gave the solution to the problem of optical isomerism in organic compounds. van't Hoff proposed the theory that if the four valencies of the carbon atom are arranged tetrahedrally (not necessarily regular) with the carbon atom at the centre, then all the cases of isomerism known are accounted for. Le Bel's theory was substantially the same as van't Hoff's, but differed in that whereas van't Hoff believed that the valency distribution was definitely tetrahedral and fixed as such, Le Bel believed that the valency directions were not rigidly fixed, and did not specify the tetrahedral arrangement, but thought that whatever the spatial arrangement, the molecule Cabde would be asymmetric. Later work has shown that van't Hoff's theory is more in keeping with the facts.

Both van't Hoff's and Le Bel's theories were based on the assumption that the four hydrogen atoms in methane are equivalent; this assumption has been shown to be correct by means of chemical and physico-chemical methods. Before the tetrahedral arrangement was proposed, it was believed that the four carbon valencies were planar, with the carbon atom at the centre of a square (Kekule, 1858).

Pasteur (1848) stated that all substances fell into two groups, those which were superimposable on their mirror images, and those which were not. In substances such as quartz, optical activity is due to the dissymmetry of the crystal structure, but in compounds like sucrose the optical activity is due to molecular dissymmetry. Since it is impossible to have molecular dissymmetry if the molecule is flat, Pasteur's work is based on the idea that molecules are three-dimensional and arranged dissymmetrically.

A further interesting point in this connection is that Pasteur quoted an irregular tetrahedron as one example of a dissymmetric structure. Also, Paterno (1869) had proposed tetrahedral models for the structure of the isomeric compounds $C_2H_4Cl_2$ (at that time it was thought that there were three isomers with this formula; one ethylidene dichloride and two ethylene dichlorides).

Evidence for the Tetrahedral Carbon Atom

The molecule CX_4 constitutes a five-point system, and since the four valencies of carbon are equivalent, their disposition in space may be assumed to be symmetrical. Thus there are three symmetrical arrangements possible for the molecule CX_4, one planar and two solid-pyramidal and tetrahedral. By comparing the number of isomers that have been prepared for a given compound with the number predicted by the above three spatial arrangements, it is possible to decide which one is correct.

Compounds of the types Ca_2b_2 and Ca_2bd. Both of these are similar, and so we shall only discuss molecule Ca_2b_2.

(i) If the molecule is planar, then two forms are possible (Fig. 14c). This planar configuration can be either square or rectangular; in each case there are two forms only.

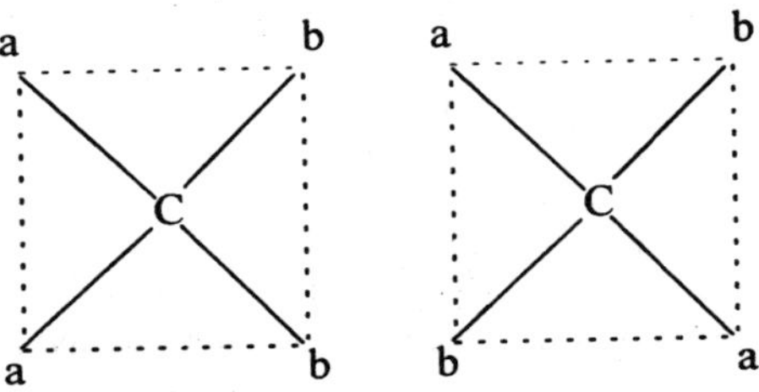

Fig. 14

(ii) If the molecule is pyramidal, then two forms are possible (Fig. 15). There are only two forms, whether the base is square or rectangular.

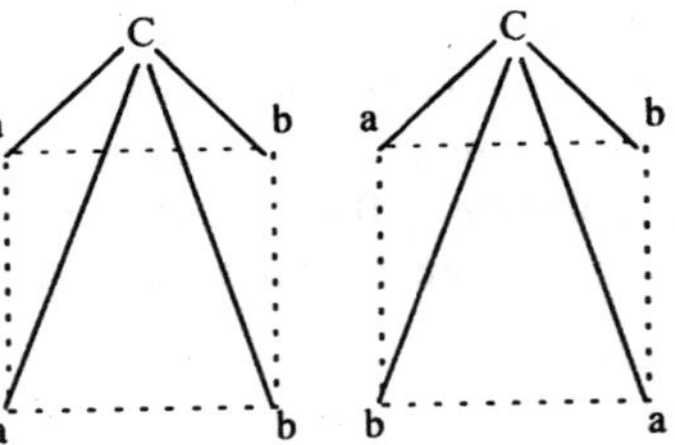

Fig. 15

(iii) If the molecule is tetrahedral, then only one form is possible (Fig. 16 the carbon atom is at the centre of the tetrahedron).

In practice, only one form is known for each of the compounds of the types Ca_2b_2 and Ca_2bd; this agrees with the tetrahedral configuration.

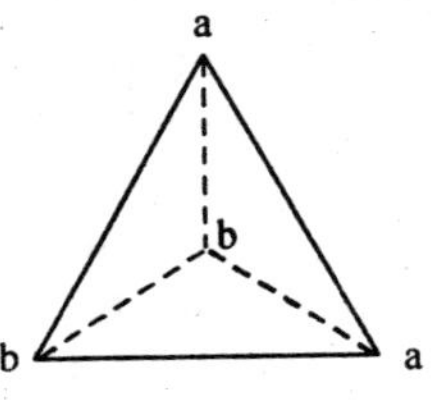

Fig. 16

Compounds of the type Cabde.

(i) If the molecule is planar, then three forms are possible (Fig. 17).

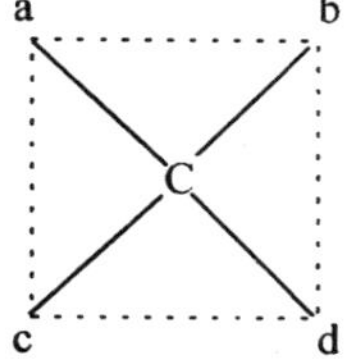

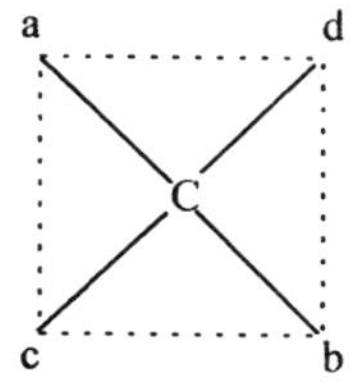

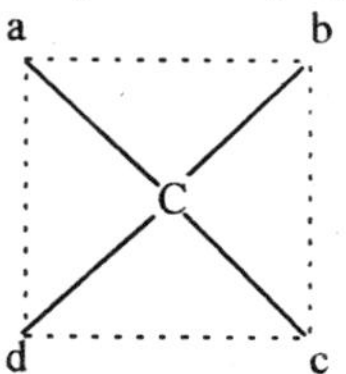

Fig. 17

(ii) If the molecule is pyramidal, then six forms are possible; there are three pairs of enantiomers. Each of the forms in Fig. 2.4, drawn as a pyramid, is not superimposable on its mirror image, *e.g*, Fig. 18 shows one pair of enantiomers.

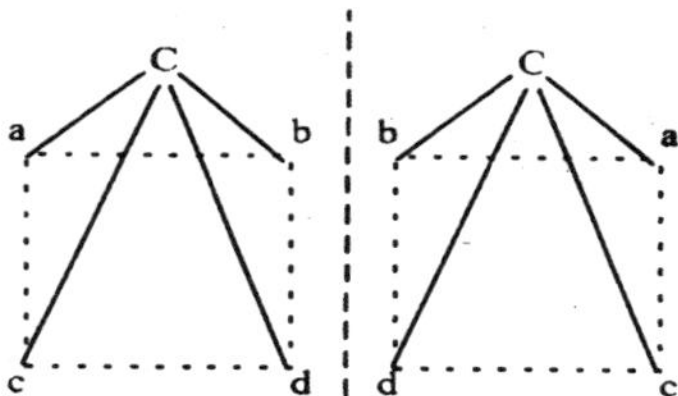

Fig. 18

(iii) If the molecule is tetrahedral, there are two forms possible, one related to the other as object and mirror image, which are not superimposable, *i.e.*, the tetrahedral configuration gives rise to one pair of enantiomers (Fig. 19).

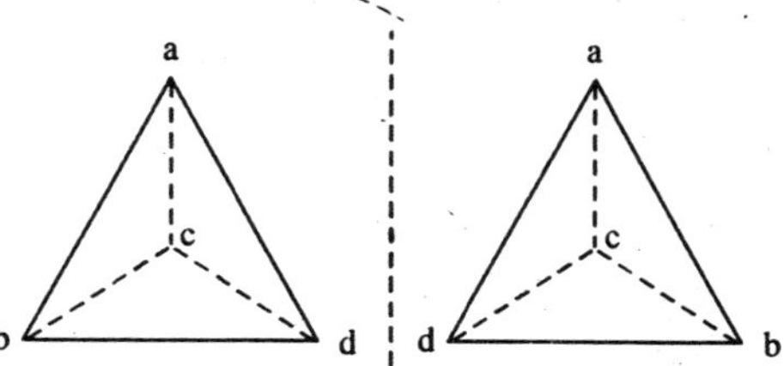

Fig. 19

In practice, compounds of the type Cabde give rise to only one pair of enantiomers; this agrees with the tetrahedral configuration.

When a compound contains four different groups attached to a carbon atom, that carbon atom is said to be asymmetric (actually, of course, it is the group which is asymmetric; a carbon atom cannot be asymmetric). The majority of optically active compounds (organic) contain one or more asymmetric carbon atoms. It should be remembered, however, that the essential requirement for optical activity is the asymmetry of the molecule. A molecule may contain two or more asymmetric carbon atoms and still not be optically active.

A most interesting case of an optically active compound containing one asymmetric carbon atom is the resolution of s-butylmercuric bromide, EtMeCHHgBr (Hughes, Ingold *et al.,* 1958). This appears to be the first example of the resolution of a simple organometallic compound where the asymmetry depends only on the carbon atom attached to the metal.

Chirality

As we have seen, the structures of enantiomers differ only in 'handedness', one being left-handed and the other being right-handed. Any molecule which is not superimposable on its mirror image is said to possess chirality (from the Greek kheir, hand). Thus, the term chirality means 'having handedness'. This term was first introduced by Kelvin (1884), and has been used by Cahn *et al.,* in their system for the specification of absolute configuration.

Chirality expresses the necessary and sufficient condition for the existence of enantiomers. The adjective chiral is equivalent to being left- or right-handed, and so a chiral centre is one which can be left- or right-handed.

On the other hand, when a molecule is superimposable on its mirror image, that molecule does not possess 'handedness' (*i.e.,* it is not asymmetric) and is said to be achiral. The commonest cause of optical activity is the presence of one or more chiral centres which, in organic chemistry, are usually asymmetric carbon atoms.

From what has been said above, it can be seen that chiral molecules are those which can exist as enantiomers, and that enantiomers have opposite chirality. Furthermore, since chirality expresses the necessary and sufficient condition for the existence of enantiomers, chirality is therefore, strictly speaking, equivalent to dissymmetry. Achiral molecules are symmetric or non-dissymmetric (the latter term is preferred by many authors).

Isotopic Asymmetry

In the optically active compound Cabde, the groups a, b, d and e (which may or may not contain carbon) are all different, but two or more may be structural isomers, *e.g*, propylisopropylmethanol is optically active. The substitution of hydrogen by deuterium has also been investigated in recent years to ascertain whether these two atoms are sufficiently different to give rise to optical isomerism.

The earlier work gave conflicting results, but later work, however, is definitely conclusive in favour of optical activity, *e.g*, Eliel (1949) prepared optically active phenylmethyldeuteromethane, $CH_3CHDC_6H_5$, by reducing optically active phenylmethylmethyl chloride (x-phenylethyl chloride), $CH_3CHClC_6H_5$, with lithium aluminium deuteride; Ross *et al.*, (1956) have prepared (–)-2-deuterobutane by reduction of (–)-2-chlorobutane with lithium aluminium deuteride; and Alexander *et al.*, (1949) reduced trans-2-p-menthene with deuterium (Raney nickel catalyst) and obtained a 2,3-dideutero-trans-p-menthane (I) that was slightly laevorotatory. Alexander (1950) also reduced (–)-menthyl toluene-p-sulphonate and obtained an optically active 3deutero-trans-p-menthane (II).

```
H3C     CH3            H3C     CH3
   \   /                  \   /
    CH                     CH
    |                      |
    CH                     CH
   /  \                   /  \
H2C    CHD             H2C    CHD
 |      |               |      |
H2C    CHD             H2C    CH2
   \   /                  \   /
    CH                     CH
    |                      |
    CH3                    CH3
    (I)                    (II)
```

Some other optically active compounds with deuterium asymmetry are, *e.g*, (III; Streitwieser, 1955) and (IV; Levy *et al.*, 1957):

$CH_3CH_2CH_2CHDOH$ (III) CH_3CHDOH (IV)

A point of interest here is that almost all optically active deuterium compounds have been prepared from optically active precursors. Exceptions are (V) and (VI), which have been resolved by Pocker (1961).

$C_6H_5CHOHC_6D_5$ (V) $C_6H_5CDOHC_6D_5$ (VI)

It might be noted that chirality produced by replacement of hydrogen by deuterium is of two types:

(i) deuterium is directly attached to the chiral centre (asymmetric carbon atom), *e.g,* (III) and (IV);

(ii) deuterium is not directly attached to the chiral centre, *e.g,* (V).

Further evidence for the tetrahedral carbon atom

(i) Conversion of the two forms (enantiomers) of the molecule Cabde into Ca_2bd results in the formation of one compound only (and disappearance of optical activity), *e.g,* both dextro- and laevorotatory lactic acid may be reduced to the sane propionic acid, which is not optically active. These results are possible only with a tetrahedral arrangement (Fig. 20).

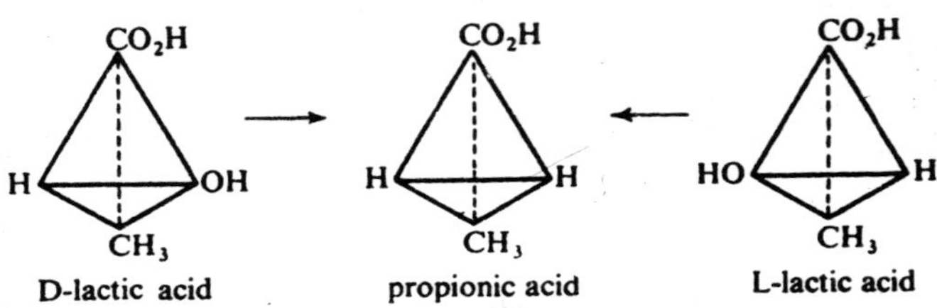

Fig. 20

(ii) If the configuration is tetrahedral, then interchanging any two groups in the molecule Cabde will produce the enantiomer, *e.g,* b and e (Fig. 21). Fischer and Brauns (1914), starting with (+)-isopropylmalonamic acid, carried out a series of reactions whereby the carboxyl and the carbonamide groups were interchanged: the product was (–)-isopropylmalonamic acid.

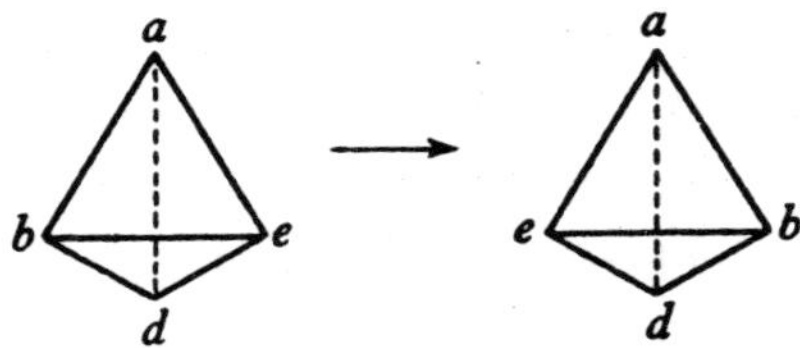

Fig. 21

It is most important to note that in this series of reactions no bond connected to the asymmetric carbon atom was ever broken. This change from one enantiomer into the other is in agreement with the tetrahedral theory. At the same time, this series of reactions shows that optical isomers have identical structures, and so the difference must be due to the spatial arrangement.

$$\underset{(+)-\text{acid}}{\begin{array}{c}CONH_2\\|\\H-C-CH(CH_3)_2\\|\\CO_2H\end{array}} \xrightarrow{CH_3N_2} \underset{(+)}{\begin{array}{c}CONH_2\\|\\H-C-CH(CH_3)_2\\|\\CO_2H_2\end{array}} \xrightarrow{HNO_2}$$

$$\underset{(-)}{\begin{array}{c}CO_2H_2\\|\\H-C-CH(CH_3)_2\\|\\CO_2H_3\end{array}} \xrightarrow{N_3H_4} \underset{(-)}{\begin{array}{c}CO_2H\\|\\H-C-CH(CH_3)_2\\|\\CONHNH_2\end{array}} \xrightarrow{HNO_2}$$

$$\underset{(-)}{\begin{array}{c}CO_2H\\|\\H-C-CH(CH_3)_2\\|\\CON_3\end{array}} \xrightarrow{NH_3} \underset{(-)-\text{acid}}{\begin{array}{c}CO_2H\\|\\H-C-CH(CH_3)_2\\|\\CONH_3\end{array}}$$

(iii) X-ray crystallography, dipole moment measurements, absorption spectra and electron diffraction studies show that the four valencies of carbon are arranged tetrahedrally with the carbon atom inside the tetrahedron.

It should be noted in passing that the tetrahedra are not regular unless four identical groups are attached to the central carbon atom; only in this case are the four bond lengths equal. In all other cases the bond lengths will be different, the actual values depending on the nature of the atoms joined to the carbon atom.

TWO POSTULATES UNDERLIE THE TETRAHEDRAL THEORY

(i) **The principle of constancy of the valency angle :** Mathematical calculation of the angle subtended by each side of a regular tetrahedron at the central carbon atom (Fig. 22) gives a value of 109° 28′. Originally, it was postulated (van't Hoff) that the valency angle was fixed at this value. It is now known, however, that the valency angle may deviate from this value. The four valencies of carbon are formed by hybridisation of the $2s^2$ and $2p^2$ orbitals, *i.e.*, there are four spa bonds. Quantum mechanical calculations show that the four carbon valencies in the molecule Ca_4 are equivalent and directed towards the four corners of a regular tetrahedron.

Fig. 22

Furthermore, quantum-mechanical calcula-tions require the carbon bond angles to be close to the tetrahedral value, since change from this value is associated with loss in bond strength and consequently decrease in stability. According to Coulson *et al.*, (1949), calculation has shown that the smallest valency angle that one can reasonably expect to find is 104°. It is this value which is found in the cyclopropane and cyclobutane rings, these molecules being relatively unstable because of the 'bent' bonds.

(ii) **The principle of free rotation about a single bond.** Originally, it was believed that internal rotation about a single bond was completely free. Let us consider the ethane molecule, CH_3—CH_3, and let us imagine that one methyl group is rotated about the C–C bond as axis with the other group at rest. Suppose we use, as the starting point, the position in which two C–H bonds are parallel, *i.e.*, these four atoms lie in a plane. In this position, the dihedral angle (angle of rotation or angle of torsion) is zero, and if the rotation is free, then as the dihedral angle changes, the energy content of the molecule will remain constant (the plot of energy content against the dihedral angle will be a horizontal line). In this situation, the two 'halves' can assume, with complete freedom, an infinite number of positions relative to each other. Thus, the entropy of the molecule will be a maximum. Pitzer *et al.*, (1936) calculated the entropy of ethane based on the assumption that the internal rotation was free, and found that calculated value was greater than the observed value.

This means that there is less freedom to assume all possible dihedral angles than was expected on the principle of free rotation. Pitzer therefore assumed that the internal rotation is hindered by a potential energy barrier, and calculated the change in entropy with increasing barrier height. When he assumed a barrier of 12.55 kJ mol^{-1} (3 kcal), the calculated and observed entropies were brought into good agreement. The potential energy curve obtained for ethane is shown in Fig. 23a and, by convention, the potential energy is measured relative to the energy of the most stable form. Because of the existence of potential energy barriers, the result is mutual oscillation (libration) about the conformations with minimum potential energy, thereby producing 'more order' (more restriction) than had there been no barriers.

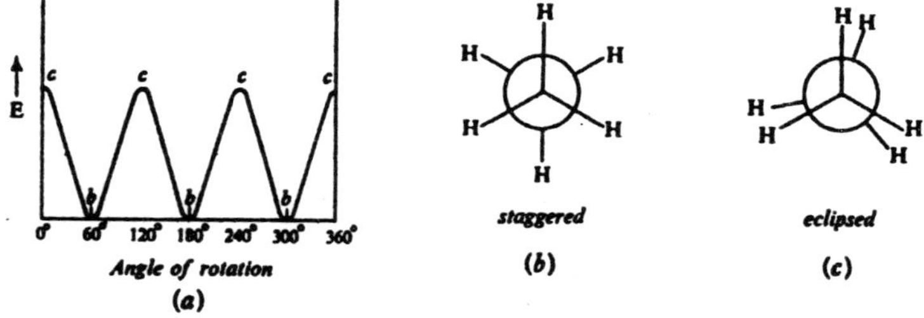

Fig. 23

Fig. 23b is the Newman projection formula. This is obtained by viewing the molecule along the bonding line of the two carbon atoms, with the carbon atom nearer to the eye being designated by equally spaced radii, and the carbon atom further from the eye by a circle with three equally spaced radial extensions. Figure 23b represents the staggered (or transoid) conformation (in which the hydrogen atoms are as far apart as possible), and Fig. 23c the eclipsed (or cisoid) conformation in which the hydrogen atoms are as close together as possible). It can be seen from Fig. 23a that the energy of the eclipsed conformation is greater than that of the staggered conformation. The potential energy barrier (11.92 kJ mol[-1]) is much too small for either conformation to remain stable, *i.e.*, the eclipsed and staggered forms are readily interconvertible and hence neither can be isolated as such. However, the staggered conformation is the preferred form, *i.e.*, its population is greater than that of the eclipsed form.

Now let us consider the case of ethylene dichloride. According to Bernstein (1949), the potential energy of ethylene dichloride undergoes the changes shown in Fig. 24 when one CH_2Cl group is rotated about the C–C bond with the other CH_2Cl at rest.

There are two positions of minimum energy, one corresponding to the staggered (transoid or anti) form and the other to the gauche (skew) form, the latter possessing approximately 4.6 kJ more than the former. The fully eclipsed (cisoid) form possesses about 18.83 kJ more energy than the staggered form and thus the latter is the preferred form, *i.e.*, the molecule is largely in this form.

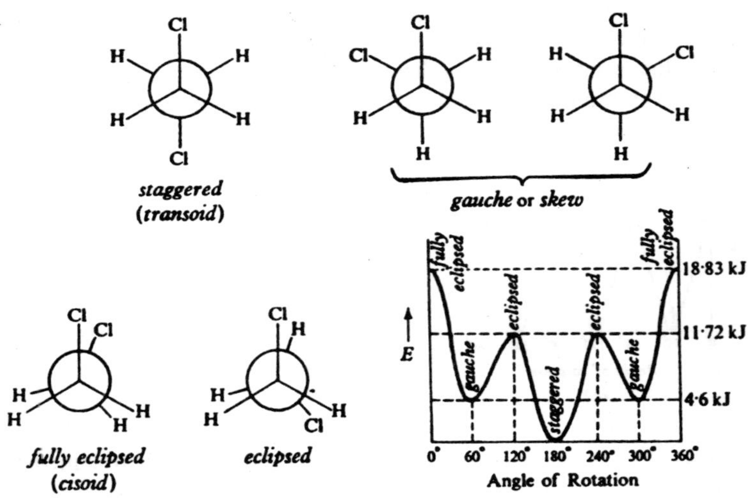

Fig. 24

Dipole moment studies show that this is so in practice, and also show (as do Raman spectra studies) that the ratio of the two forms varies with the temperature. Furthermore, infrared, Raman spectra and electron diffraction studies have shown that the gauche form is also present. According to Mizushima *et al.*, (1938), only the staggered form is present at low temperatures.

The problem of internal rotation about the central C–C bond in n-butane is interesting, since the values of the potential energies of the various forms have been used in the study of cyclic compounds. The various forms are shown in Fig. 25, and if the energy content of the staggered form is taken as zero, then the other forms have the energy contents shown (Pitzer, 1951).

From the foregoing account it can be seen that in theory, there is no free rotation about a single bond. In practice, however, it may occur if the potential barriers of the various forms do not differ by more than about 40 kJ mol^{-1}. Free rotation about a single bond is generally accepted in simple molecules. Restricted rotation, however, may occur when the molecule contains groups large enough to impede free rotation, *e.g,* in ortho-substituted biphenyls. In some cases resonance can give rise to restricted rotation about a 'single' bond.

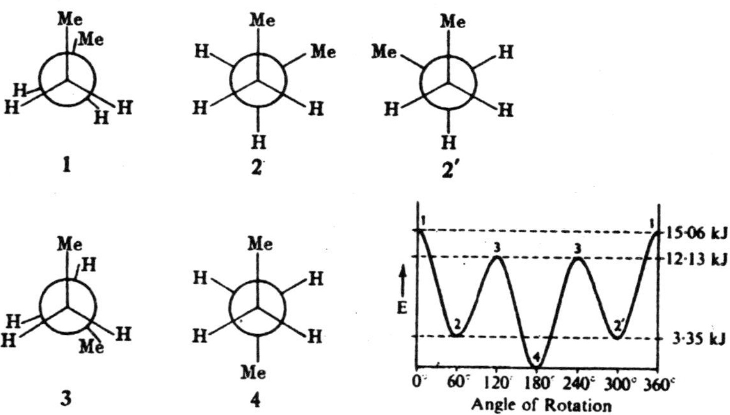

Fig. 25

In addition to the nomenclature of conformers described above-staggered (anti), skew (gauche), etc., conformers are also described in terms of the dihedral angle between specified groups. Thus, the dihedral angle for the skew form of n-butane (2 and 2′, Fig. 25) is 60°; for the staggered (anti) form (4, Fig. 25) the dihedral angle is 180°.

In many cases, however, the exact dihedral angle is not known, and to describe these cases, the following nomenclature in terms of approximate dihedral angles has been proposed (Klyne and Prelog, 1960).

Table 2

Dihedral angle	*Designation*	*Symbol*
– 30° to + 30°	± syn-periplanar	± SP
+ 30° to + 90°	+ syn-clinal	+sc
+ 90° to + 150°	+ anti-clinal	+ac
+ 150° to – 150°	± anti-periplanar	± ap
– 30° to – 90°	– syn-clinal	– sc
– 90° to – 150°	– anti-clinal	– ac

The terms syn and anti are used to indicate respectively a dihedral angle smaller or greater than 90°. The terms 'periplanar' and 'clinal' respectively describe approximately planar (0° ± 30° and 180° ± 30°) and inclined positions (all other angles).

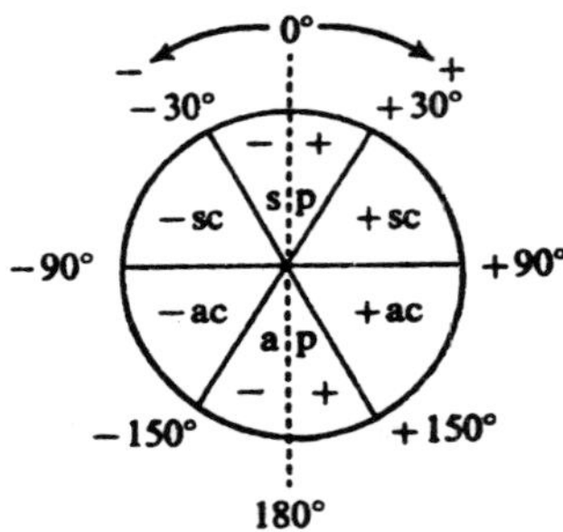

Conformational prefixes

Fig. 26

This method of nomenclature is summarised in the diagram.

Conformational Analysis

Isomers which are formed by rotation about single bonds are called different conformations or conformers. The terms rotational isomers and constellations have also been used in the same sense as conformations.

Various definitions have been given to the term conformation (which was originally introduced by W. N. Haworth, 1929). In its widest sense, conformation has been used to describe different spatial arrangements of a molecule which are not superimposable. This means, in effect, that the terms conformation and configuration are equivalent. The definition of configuration, in the classical sense, does not include the problem of the internal forces acting on the molecule. The term conformation, however, is the spatial arrangement of the molecule when all the internal forces acting on the molecule are taken into account. In this more restricted sense, the term conformation is used to designate different spatial arrangements arising by twisting or rotation of bonds of a given configuration (used in the classical sense).

The existence of potential energy barriers between the various conformations shows that there are internal forces acting on the molecule. The nature of these interactions that prevent free rotation about single bonds, however, is not completely clear. According to one theory, the hindering of internal rotation is due to dipole-dipole forces. Calculation of the dipole moment of ethylene dichloride on the assumption of free rotation gave a value not in agreement with the experimental value. Thus free rotation cannot be assumed, but on the assumption that there is interaction between the two groups through dipole-dipole attractive or repulsive forces, there will be preferred conformations, *i.e.*, the internal rotation is not completely free.

This restricted rotation is shown by the fact that the dipole moment of ethylene dichloride increases with temperature; in the staggered form the dipole moment is zero, but as energy is absorbed by the molecule, rotation occurs to produce finally the eclipsed form in which the dipole moment is a maximum. Further work, however, has shown that factors other than dipole-dipole interactions must also be operating in opposing the rotation. One of these factors is steric repulsion, *i.e.*, repulsion between the non-bonded atoms (of the rotating groups) when they are brought into close proximity.

The existence of steric repulsion may be illustrated by the fact that although the bond moment of C–Cl is greater than that of C–Br, the energy difference between the eclipsed and staggered conformations of ethylene dichloride is less than that of ethylene dibromide. Furthermore, if steric repulsion does affect internal rotation, then in the ethylene dihalides, steric repulsion between the hydrogen and halogen atoms, if sufficiently large, will give rise to two other potential energy minima.

Other factors also affect stability of the various conformations. Staggered and skew forms always exist in molecules of the type CH_2Y–CH_2Z (where Y and Z are Cl, Br, I, CH_3, etc.), and usually the staggered form is more stable than the skew. In a molecule such as ethylene chlorohydrin or ethylene glycol, however, intramolecular hydrogen bonding is possible in the skew form but not the staggered. This would stabilise the molecule by about 20-29 kJ mol^{-1}, and this is great enough to make the skew form more stable than the staggered. Infrared spectroscopy has shown that the skew form predominates.

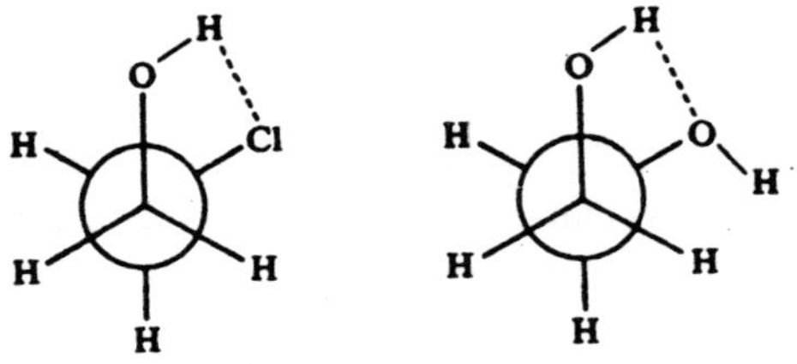

ethylene chlorohydrin **ethylene glycol**

Fig. 27

In addition to the factors already mentioned, there appear to be other factors that cause the absence of complete free rotation about a single bond, *e.g*, the energy barrier in ethane is too great to be accounted for by steric repulsion only.

Several explanations have been offered; *e.g*, Pauling (1958) has proposed that the energy barrier in ethane (and in similar molecules) results from repulsions between adjacent bonding pairs of electrons, *i.e.*, the bonding pairs of the C–H bonds on one carbon atom repel those on the other carbon atom. Thus the preferred conformation will be the staggered. It is still possible, however, that steric repulsion is also present, and this raises the barrier height.

When the stability of a molecule is decreased by internal forces produced by interaction between constituent parts, that molecule is said to be under strain. In view of the foregoing discussion, it can be seen that there are four contributing sources to strain:

(i) steric strain,

(ii) dipole-dipole interactions,

(iii) bond angle strain, and

(iv) bond opposition strain.

Which of these plays the predominant part depends on the nature of the molecule in question. This study of the existence of preferred conformations in molecules, and the relating of physical and chemical properties of a molecule to its preferred conformation, is known as conformational analysis. The energy differences between the various conformations determine which one is the most stable, and the ease of transformation depends on the potential energy barriers that exist between these conformations.

It should be noted that the molecule, in its unexcited state, will exist largely in the conformation of lowest energy content. If, however, the energy differences between the various conformations are small, then when excited, the molecule can take up a less favoured conformation, *e.g*, during the course of reaction with other molecules.

So far, we have considered the conformations of saturated compounds. Conformational studies of unsaturated compounds and compounds containing the oxo group have led to some unexpected results, *e.g*, microwave spectroscopy has shown that the preferred conformations of propene (Herschbach *et al.*, 1958) and acetaldehyde (Kilb *et al.*, 1957) are the eclipsed forms, and NMR spectroscopy has shown that the predominant conformation of propionaldehyde is the one in which the methyl group and oxygen atom are eclipsed (Pople *et al.*, 1960). The reason for these observations is uncertain.

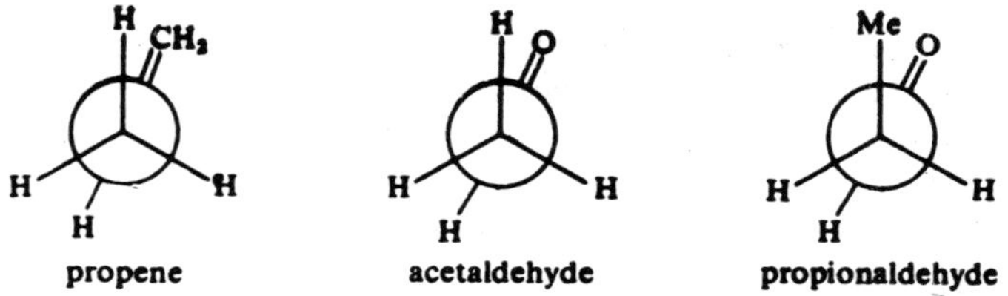

H CH_2 H H H propene

H O H H H acetaldehyde

Me O H H H propionaldehyde

Fig. 28

Many methods are now used to investigate the conformations of molecules: thermodynamic calculations, dipole moments, X-ray and electron diffraction, infrared and ultraviolet spectroscopy, chemical methods, etc. A particularly useful method for studying the conformations of molecules is NMR spectroscopy. However, before dealing with this, let us first consider the chemical shifts of the protons of CH_3, CH_2, and the CH group in acyclic compounds and the chemical shift of a proton attached to O, N, etc. As we have seen the more electronegative Z is in the groups –CH–Z and –Z–H, the more is the proton deshielded and consequently the lower is the r-value.

Let us now consider ethyl chloride, and since the most stable conformation is the staggered one, the molecule may be represented as (I), (II), and (III). In (I), the environments of H_a and H_c are identical, but differ from that of H_b. This can be seen to be the case by replacing each proton, one at a time, by Z. This procedure for H_a and H_c produces mirror images, and so the two protons are chemically equivalent, but for Hb the substituted product is different, and therefore H_b is not chemically equivalent to H_a and H_c. The coupling constants of H_a with H_d and H_e are different because of the different dihedral angles, and similarly for H_c.

Thus, H_a and H_c are chemically but not magnetically equivalent. By using similar arguments, it can be shown that H_d and H_c are also chemically but not magnetically equivalent. Hence, if the population of ethyl chloride conformation were completely represented by (I), protons Ha and. He would give one signal and Hb another signal provided that the chemical shifts were sufficiently different (0.1–0.2 p.p.m.), *i.e.*, the methyl group would give two signals.

In practice, the methyl group gives only one signal (a triplet), and therefore protons H_a, H_b, and H_c must be equivalent. This is explained on the basis that there is free rotation about the carbon-carbon single bond. When the methyl group rotates (with respect to the CH_2CI group), it can take up the other two stable staggered conformations, (II) and (III), the result being that each proton has an average environment. Since these average environments are identical, rotation results in the equivalence of H_a, H_b, and H_c.

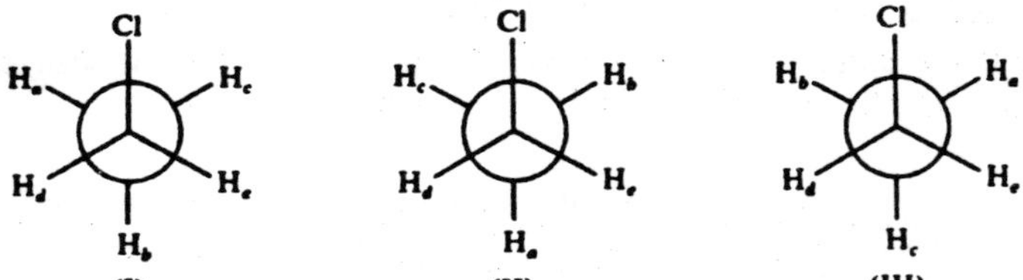

(I) (II) (III)

Fig. 29

This equivalence may be demonstrated as follows. Let P_I, P_{II}, P_{III} be the populations in conformations (I), (II), and (III), respectively, where $P_I, P_{II}, P_{III} = 1$.

Also, let δ (the chemical shift) of any proton trans (anti) to another proton be x, and y if trans to Cl. Now, the observed δ of a given proton is the sum of the δ-contributions in each conformation, and so it follows that:

$$\delta(H_a) = P_Ix + P_{II}y + P_{III}x$$
$$\delta(H_b) = P_Iy + P_{II}x + P_{III}x$$
$$\delta(H_c) = P_Ix + P_{II}x + P_{III}y$$

Since all three populations are equal (all three are indistinguishable because all protons, as such, are identical), *i.e.*, $P_I = P_{II} = P_{III} = 1/3$, therefore

$$\delta(H_a) = \delta(H_b) = \delta(H_c) = 2/3x + 1/3y$$

Hence, all three protons have the same (average) chemical shift (in the freely rotating state) and so there is no observable splitting for the methyl group.

Application of this method to protons H_d and H_e gives:

$$\delta(H_d) = P_Ix + P_{II}x + P_{III}x = x$$
$$\delta(H_e) = P_Ix + P_{II}x + P_{III}x = x$$

From the above discussion, it can be seen that if ethyl chloride were undergoing slow rotation, it would be an ABB′CC′ spin system, but when undergoing fast rotation, it is an A_3B_2 spin system.

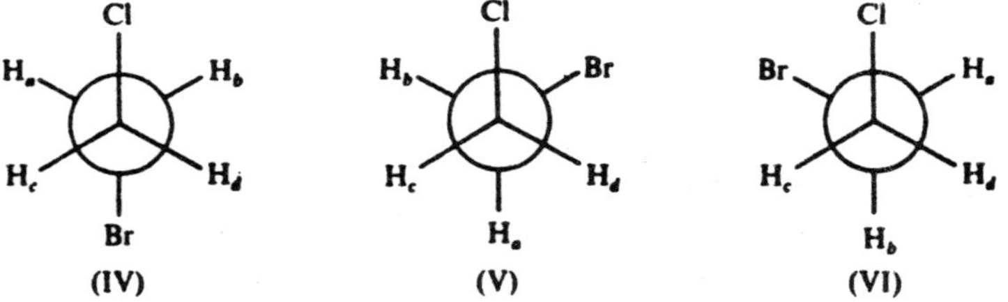

Fig. 30

Now let us consider the case of 1-bromo-2-chloroethane, and if we use the same arguments as before, then in (IV), H_a and H_b are chemically but not magnetically equivalent, and this is also true for H_c and H_d. Also, if δ for a proton trans to a proton is x, trans to Cl is y, and trans to Br is z, then

$$\delta(H_a) = P_{IV}x + P_Vy + P_{VI}x$$
$$\delta(H_b) = P_{IV}x + P_Vx + P_{VI}Y$$

In this molecule, although $P_V = P_{VI} \neq P_{IV}$, it still follows that $\delta(H_a) = \delta(H_b)$. In the same way, it can be shown that $\delta(H_c) = \delta(H_d)$.

We can extend the argument by also considering the possible coupling constants. If J_t and J_s represent, respectively, the vicinal trans and skew coupling, then:

$$J_{ad} = P_{IV}J_t + P_VJ_s + P_{VI}J_s$$

$$J_{bd} = P_{IV}J_s + P_VJ_t + P_{VI}J_s$$

Since $P_V = P_{VI} \neq P_{IV}$, therefore $J_{ad} \neq J_{bd}$. Hence, H_a and H_b (and also H_c and H_d) are chemically but not magnetically equivalent. Application of this method will show that the three protons in the methyl group in ethyl chloride are magnetically equivalent, as are also the two protons in the CH_2Cl group, for the freely rotating molecule. Finally, let us consider a molecule of the type $R^1R^2CHCHR^3R^4$. Each carbon atom is asymmetric, and the environments of H_a are different in all three conformations, and this is also true for H_b.

Furthermore, because of the different steric effects, the three populations will be different (*i.e.*, $P_{VII} \neq P_{VIII} \neq P_{IX}$). Hence, the chemical shifts of H_a (and those of H_b) are different in each conformation. If rotation were slow, the NMR spectrum would be a composite of all three spectra (three AB spizz systems for each pair of enantiomers). If the temperature is raised, the rate of rotation is increased, and the result is an average of chemical shifts and coupling constants to give one AB spin system (for each pair of enantiomers). These averages, however, will be weighted in favour of the most stable conformation, *i.e.*, the one with the highest population.

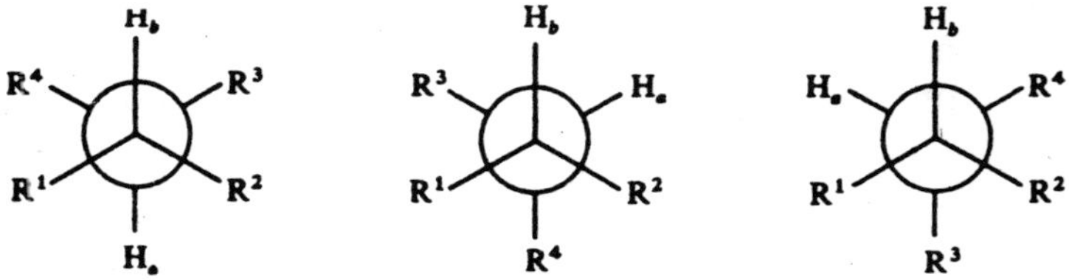

H_b, R^4, R^3, R^1, R^2, H_a (VII) — H_b, R^3, H_a, R^1, R^2, R^4 (VIII) — H_b, H_a, R^4, R^1, R^2, R^3 (IX)

Fig. 31

Now let us consider the case of amides. Infrared spectroscopic evidence has indicated that amides are resonance hybrids, and this' is supported by NMR studies, *e.g*, the two methyl groups in dimethylformamide give two separate signals at room temperature. This can be explained on the basis that, because of the partial double bond character of the C–N bond, the two methyl groups have different

Me Me
N
|
C
H O

⟶

Me Me
N
||
C
H O

environments, one methyl group being cis and the other trans to the hydrogen atom. At room temperature the rate of rotation about the C–N bond is slow enough for the two methyl groups to be non-equivalent. As the temperature is raised, the signals broaden and finally collapse to a single line.

At these higher temperatures, the molecule has absorbed sufficient energy to overcome the energy barrier to rotation to make the average environment of the two methyl groups the same, *i.e.*, the two methyl groups are now equivalent.

CORRELATION OF CONFIGURATIONS WITHOUT DISPLACEMENT AT THE CHIRAL CENTRE CONCERNED

Since no bond joined to the chiral centre is ever broken, this method is an extremely valuable method of correlation. Before discussing examples, the following point is worth noting. For amino-acids, natural (–)-serine, $HOCH_2CH(NH_2)CO_2H$, was chosen as the arbitrary standard. Thus correlation with glyceraldehyde was indicated by D_g or L_g, and with serine by D_s or L_s. These two standards have now been correlated, and it has been shown that L_g – L_s, *i.e.*, natural (–)-serine belongs to the L-series.

The following examples illustrate this method of correlation.

(i) CHO / HO—C—H / CH_2OH (L(–)-glyceraldehyde) $\xrightarrow{HgO}$ CO_2H / HO—C—H / CH_2OH (L(+)-glyceric acid) $\xleftarrow{HNO_2}$ CO_2H / HO—C—H / CH_2NH_2 (L(–)-isoserie)

$\xrightarrow{NOBr}$ CO_2H / HO—C—H / CH_2Br (L-) $\xrightarrow{Na/Hg}$ CO_2H / HO—C—H / CH_3 (L(+)-lactic acid)

It can be seen from this example that change in the sign of rotation does not necessarily indicate a change in configuration.

(ii) 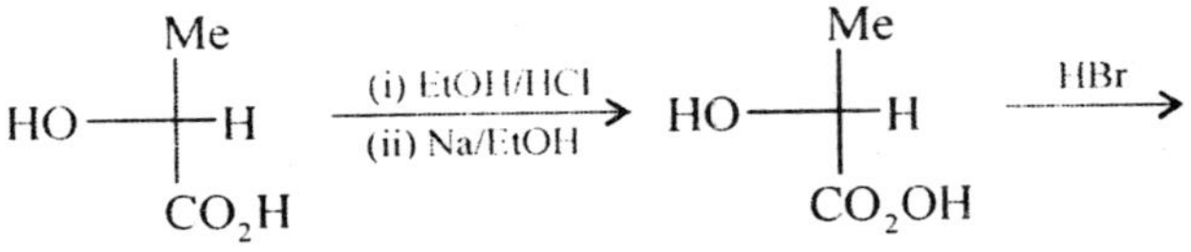 Fischer projections (Me top; HO left, H right; group at bottom):

D(+)-lactic acid (bottom CO_2H) $\xrightarrow[\text{(ii) Na/EtOH}]{\text{(i) EtOH/HCl}}$ D- (bottom CO_2OH) $\xrightarrow{\text{HBr}}$

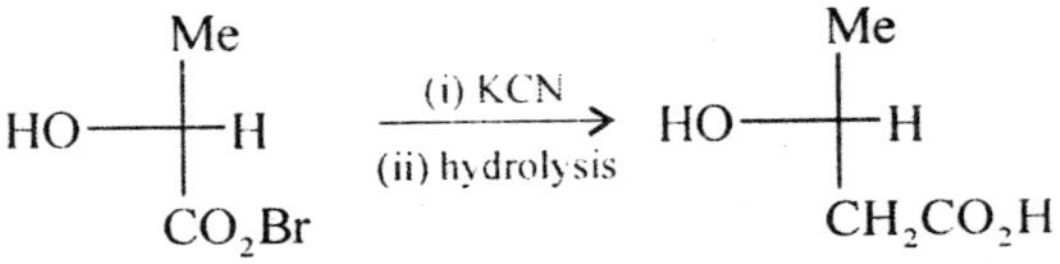

D- (bottom CO_2Br) $\xrightarrow[\text{(ii) hydrolysis}]{\text{(i) KCN}}$ D(−)-β-hydroxybutyric acid (bottom CH_2CO_2H)

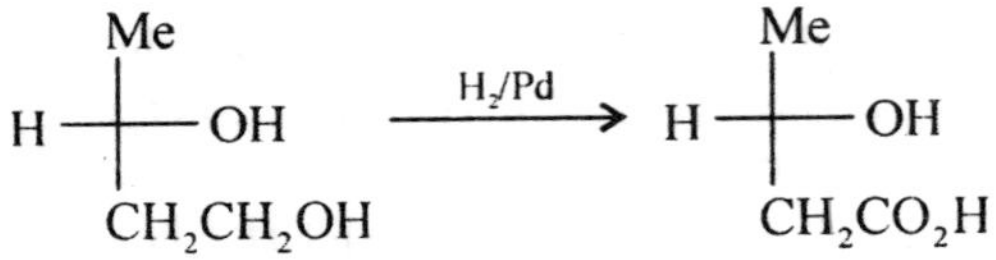

(iii) 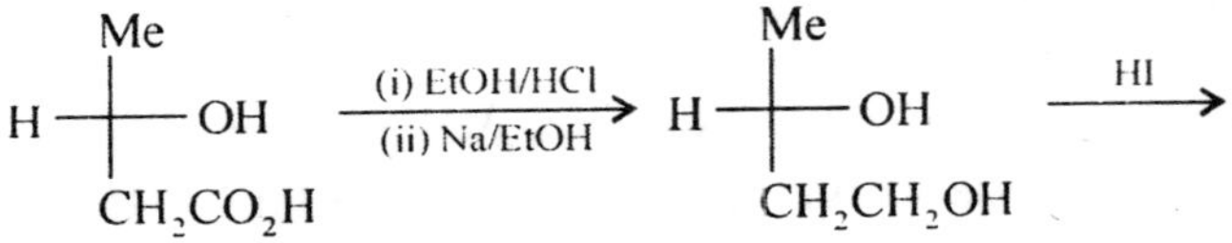 Fischer projections (Me top; H left, OH right; group at bottom):

L(+)-β-hydroxybutyric acid (bottom CH_2CO_2H) $\xrightarrow[\text{(ii) Na/EtOH}]{\text{(i) EtOH/HCl}}$ L- (bottom CH_2CH_2OH) $\xrightarrow{\text{HI}}$

L- (bottom CH_2CH_2OH) $\xrightarrow{H_2/Pd}$ L(+)-butan-2-ol (bottom CH_2CO_2H)

(iv) Another example is that in the terpene series.

(v) All the previous examples are acyclic or alicyclic compounds.

When, however, the compound contains a phenyl group attached to the asymmetric carbon atom, correlation with glyceraldehyde is carried out either by breaking down the phenyl group, leaving only C_1 (usually as CO_2H) attached to the asymmetric.

The carbon atom, or by building up a cyclohexane ring from C_1 (usually present as CO_2H), and preparing this compound by reduction of the original phenyl compound.

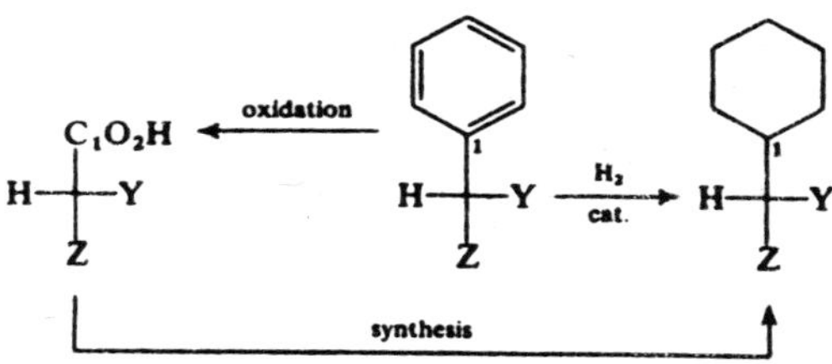

Fig. 32

An example of this method is the correlation of (+)-lactic acid with (–)-mandelic acid. Since the former is L(+), the latter is therefore L(-). At the same time, (–)-phenylmethylmethanol is also L(–).

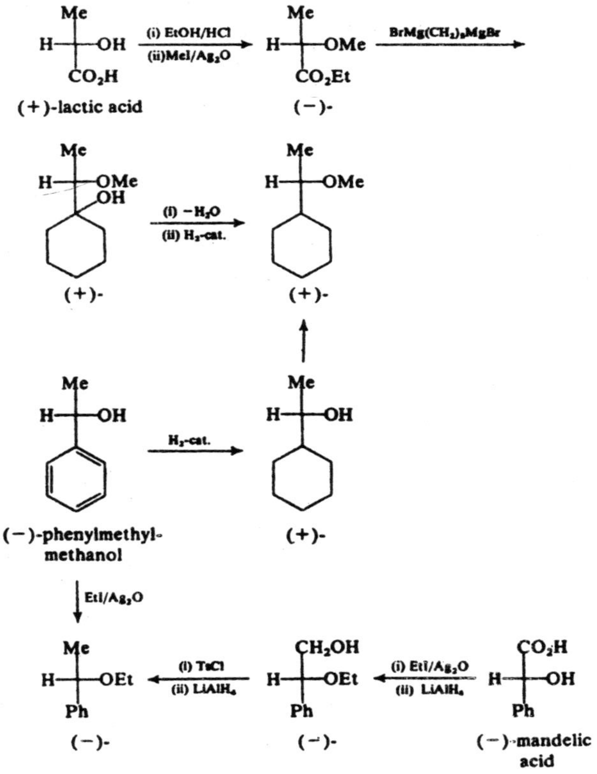

(i) EtOH/HCl (ii) MeI/Ag_2O

$BrMg(CH_2)_5MgBr$

(+)-lactic acid

(–)-

(i) $-H_2O$ (ii) H_2-cat.

(+)-

(+)-

H_2-cat.

(–)-phenylmethyl-methanol

(+)-

EtI/Ag_2O

(i) TsCl (ii) $LiAlH_4$

(i) EtI/Ag_2O (ii) $LiAlH_4$

(–)-

(–)-

(–)-mandelic acid

Fig. 33

CONVENTIONS USED IN STEREOCHEMISTRY

The original method of indicating enantiomers was to prefix each one by d or l according as it was dextrorotatory or laevorotatory. van't Hoff (1874) introduced a + and – notation for designating the configuration of an asymmetric carbon atom. He used mechanical models (built of tetrahedra), and the + and – signs were given by observing the tetrahedra of the mechanical model from the centre of the model.

Thus a molecule of the type CabdCabd may be designated + +, – –, and + –. E. Fischer (1891) pointed out that this + and – notation can lead to wrong interpretations when applied to molecules containing more than two asymmetric carbon atoms (the signs given to each asymmetric carbon atom depend on the point of observation in the molecule). Fischer therefore proposed the use of plane projection diagrams of the mechanical models instead of the + and – system. It is important to note here that the Fischer projection formulae are always those of the eclipsed conformations.

Fischer, working on the configurations of the sugar\$, obtained the plane formulae (I) and (II) for the enantiomers of saccharic acid, and arbitrarily chose (I) for dextrorotatory saccharic acid, and called it d-saccharic acid. He then, from this, deduced formula (III) for d-glucose. Furthermore, Fischer thought it was more important to indicate stereochemical relationships than merely to indicate the actual direction of rotation.

He therefore proposed that the prefixes d and l should refer to stereochemical relationships and not to the direction of rotation of the compound. For this scheme to be self-consistent (among the sugars) it is necessary to choose one sugar as standard and then refer all the others to it. Fischer apparently intended to use the scheme whereby the compounds derived from a given aldehyde sugar should be designated according to the direction of rotation of the parent aldose.

$$\begin{array}{rcl@{\qquad}rcl@{\qquad}rcl}
 & CO_2H & & & CO_2H & & & CHO & \\
H- & C & -OH & HO- & C & -H & H- & C & -OH \\
HO- & C & -H & H- & C & -OH & HO- & C & -H \\
H- & C & -OH & HO- & C & -H & H- & C & -OH \\
H- & C & -OH & HO- & C & -H & H- & C & -OH \\
 & CO_2H & & & CO_2H & & & CO_2H & \\
 & \text{(I)} & & & \text{(II)} & & & \text{(III)} &
\end{array}$$

Natural mannose is dextrorotatory. Hence natural mannose will be d-mannose, and all derivatives of d-mannose, *e.g*, mannonic acid, mannitol, mannose phenylhydrazone, etc., will thus belong to the d-series. Natural glucose is dextrorotatory. Hence natural glucose will be d-glucose, and all its derivatives will belong to the d-series. Furthermore, Fischer (1890) converted natural mannose into natural glucose as follows:

d-mannose → d-mannonic acid → d-mannolactone → d-glucose

Since natural glucose is d-glucose (according to Fischer's scheme), the prefix d for natural glucose happens to agree with its dextrorotation (with d-mannose as standard). Natural fructose can also be prepared from natural mannose (or natural glucose), and so will be d-fructose.

Natural fructose, however, is laevorotatory, and so is written as d(–)-fructose, the symbol d indicating its stereochemical relationship to the parent aldose glucose, and the symbol - placed in parentheses before the name indicating the actual direction of rotation.

More recently the symbols d and l have been replaced by D and L for configurational relationships, *e.g*, L(+)-lactic acid. Also, when dealing with compounds that cannot be referred to an arbitrarily chosen standard, (+)- and (–)- are used to indicate the sign of the rotation. The prefixes dextro and laevo (with hyphens) are also used.

Fischer's proposal to use each aldose as the arbitrary standard for its derivatives leads to some difficulties, *e.g*, natural arabinose is dextrorotatory, and so is to be designated D-arabinose. Now natural arabinose (D-arabinose) can be converted into mannonic acid which, if D-arabinose is taken as the parent aldose, will therefore be D-mannonic acid. This same acid, however, can also be obtained from L-mannose, and so should be designated as L-mannonic acid.

Thus, in cases such as this the use of the symbol D or L will depend on the historical order in which the stereochemical relationships were established. This, obviously, is an unsatisfactory position, which was realised by Rosanoff (1906), who showed that if the enantiomers of glyceraldehyde (a molecule which contains only one asymmetric carbon atom) are chosen as the (arbitrary) standard, then a satisfactory system for correlating stereochemical relationships can be developed.

He also proposed that the formula of dextrorotatory glyceraldehyde should be written, in order that the arrangement of its asymmetric carbon atom should agree with the arrangement of C_5 in Fischer's projection formula for natural glucose (see formula (III) above). It is of great interest to note in this connection that in 1906 the active forms of glyceraldehyde had not been isolated, but in 1914 Wohl and Momber separated DL-glyceraldehyde into its enantiomers, and in 1917 they showed that dextrorotatory glyceraldehyde was stereochemically related to natural glucose, *i.e.*, with D(+)-glyceraldehyde as arbitrary standard, natural glucose is D(+)-glucose

The accepted convention for drawing D(+)-glyceraldehyde-the agreed (arbitrary) standardis shown in Fig. 34(a). The tetrahedron is drawn so that three corners are imagined to be above the plane of the paper, and the fourth below the plane of the paper.

Furthermore, the spatial arrangement of the four groups joined to the central carbon atom must be placed as shown in Fig. 34(a), *i.e.*, the accepted convention for drawing D(+)-glyceraldehyde places the hydrogen atom at the left and the hydroxyl group at the right, with the aldehyde group at the top corner.

Now imagine the tetrahedron to rotate about the horizontal line joining H and OH until it takes up the position. This is the conventional position for a tetrahedron, groups joined to full horizontal lines being above the plane of the paper, and those joined to broken vertical lines being below the plane of the paper.

The conventional plane-diagram is obtained by drawing the full horizontal and broken vertical lines of Fig. 34(b) as full lines, placing the groups as they appear in Fig. 34(b), and taking the asymmetric carbon atom to be at the point where the lines cross. Although, Fig. 34(c) is a plane-diagram, it is most important to remember that horizontal lines represent groups

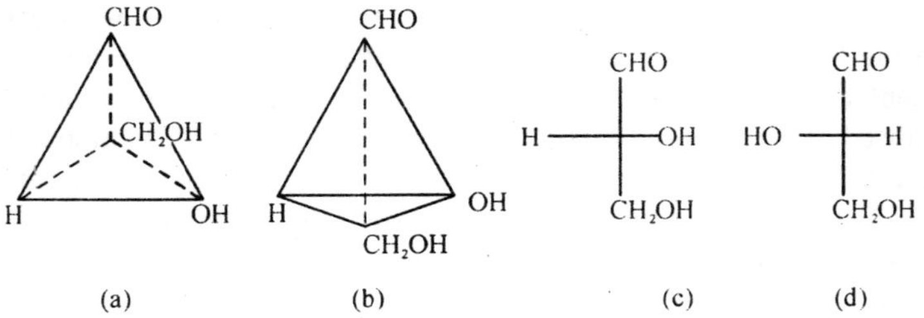

CHO, CH_2OH, H, OH (a)

CHO, H, OH, CH_2OH (b)

CHO, H—OH, CH_2OH (c)

CHO, HO—H, CH_2OH (d)

Fig. 34

above the plane, and vertical lines groups below the plane of the paper. Many authors prefer to draw Fig. 34(c) [and Fig. 34(d)] with a broken vertical line. Fig. 34(d) represents the plane-diagram formula of L(–)-glyceraldehyde; here the hydrogen atom is to the right and the hydroxyl group to the left.

Thus, any compound that can be prepared from, or converted into, D(+)-glyceraldehyde will belong to the D-series. Similarly, any compound that can be prepared from, or converted into, L(–)-glyceraldehyde will belong to the L-series. When representing relative configurational relationships of molecules containing more than one asymmetric carbon atom, the asymmetric carbon atom of glyceraldehyde is always drawn at the bottom, the rest of the molecule being built up from this unit (but see below).

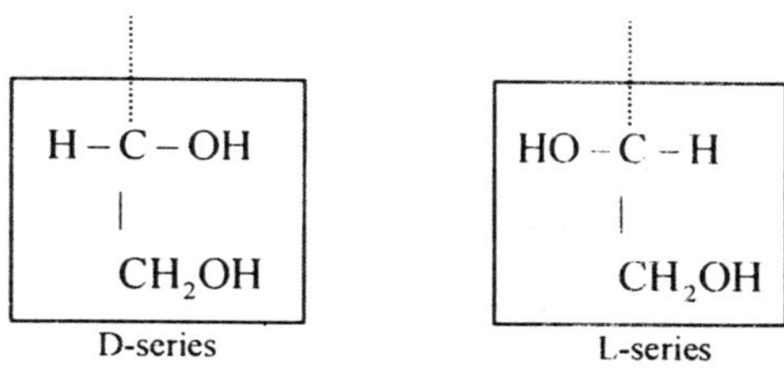

Fig. 35

Thus, we have a scheme of classification of relative configurations based on D(+)-glyceraldehyde as arbitrary standard. Even on this basis confusion is still possible in relating configurations to the standard (see later). Until recently there was no way of determining, with certainty, the absolute configuration of molecules.

Arbitrary choice makes the configuration of D(+)-glyceraldehyde have the hydrogen to the left and the hydroxyl to the right. Bijvoet *et al.*, (1951), however, have shown by X-ray analysis of sodium rubidium tartrate that it is possible to differentiate between the two optically active forms, *i.e.*, it is possible to determine the absolute configuration of these two enantiomers.

These authors showed that natural dextrorotatory tartaric acid has the configuration assigned to it by Fischer (who correlated its configuration with that of the saccharic acids). The configurations of the tartaric acids, however, are a troublesome problem. Fischer wrote the configuration of natural dextrorotatory tartaric acid as (IV). If we use the convention of writing the glyceraldehyde unit at the bottom, then (IV) is L(+)-tartaric

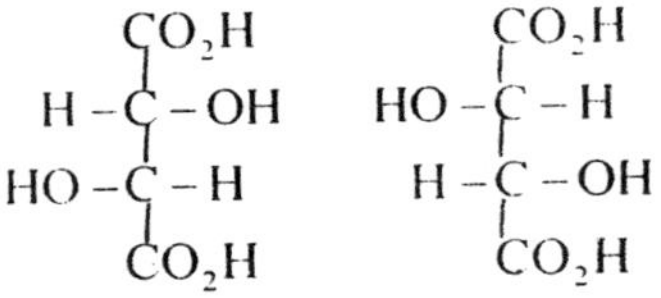

$$\begin{array}{c} CO_2H \\ H-C-OH \\ HO-C-H \\ CO_2H \\ (IV) \end{array} \qquad \begin{array}{c} CO_2H \\ HO-C-H \\ H-C-OH \\ CO_2H \\ (V) \end{array}$$

acid and (V) is D(–)-tartaric acid. This relationship (to glyceraldehyde) is confirmed by the conversion of D(+)-glyceraldehyde into laevorotatory tartaric acid via the Kiliani reaction.

Thus (–)-tartaric acid is D(–)-tartaric acid (V). On the other hand, (+)-tartaric acid can be converted into D(–)-glyceric acid, and so (+)-tartaric acid is D(+)-tartaric acid (IV). In this reduction of (+)-tartaric acid to (+)-malic acid (by hydriodic acid), it has been assumed that it is C_1 which has been

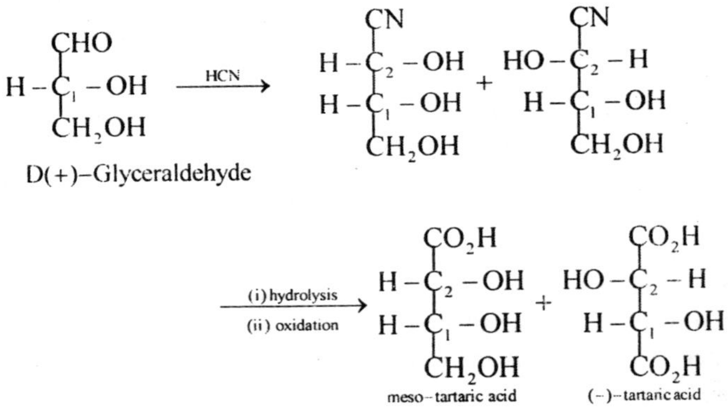

$$\underset{\text{D(+)-Glyceraldehyde}}{\begin{array}{c} CHO \\ H-C_1-OH \\ CH_2OH \end{array}} \xrightarrow{HCN} \begin{array}{c} CN \\ H-C_2-OH \\ H-C_1-OH \\ CH_2OH \end{array} + \begin{array}{c} CN \\ HO-C_2-H \\ H-C_1-OH \\ CH_2OH \end{array}$$

$$\xrightarrow[\text{(ii) oxidation}]{\text{(i) hydrolysis}} \underset{\text{meso-tartaric acid}}{\begin{array}{c} CO_2H \\ H-C_2-OH \\ H-C_1-OH \\ CH_2OH \end{array}} + \underset{\text{(–)-tartaric acid}}{\begin{array}{c} CO_2H \\ HO-C_2-H \\ H-C_1-OH \\ CO_2H \end{array}}$$

reduced, *i.e.*, in this case the configuration of C_2 has been correlated with glyceraldehyde and not that of C_1 as in the previous set of reactions. Had, however, C_2 been reduced, then the final result would have been

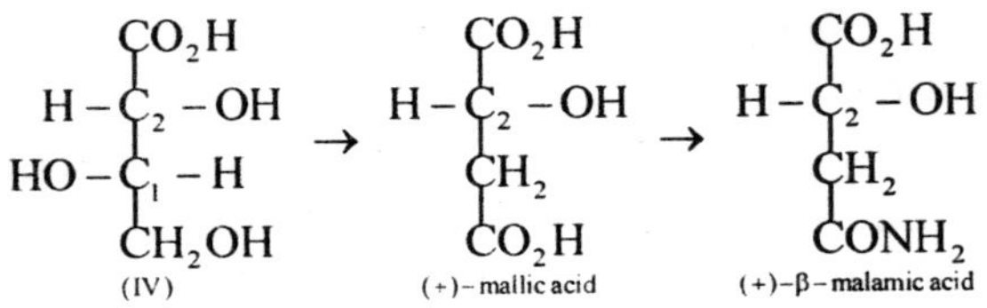

$$\underset{(IV)}{\begin{array}{c} CO_2H \\ H-C_2-OH \\ HO-C_1-H \\ CH_2OH \end{array}} \rightarrow \underset{\text{(+)-mallic acid}}{\begin{array}{c} CO_2H \\ H-C_2-OH \\ CH_2 \\ CO_2H \end{array}} \rightarrow \underset{\text{(+)-}\beta\text{-malamic acid}}{\begin{array}{c} CO_2H \\ H-C_2-OH \\ CH_2 \\ CONH_2 \end{array}}$$

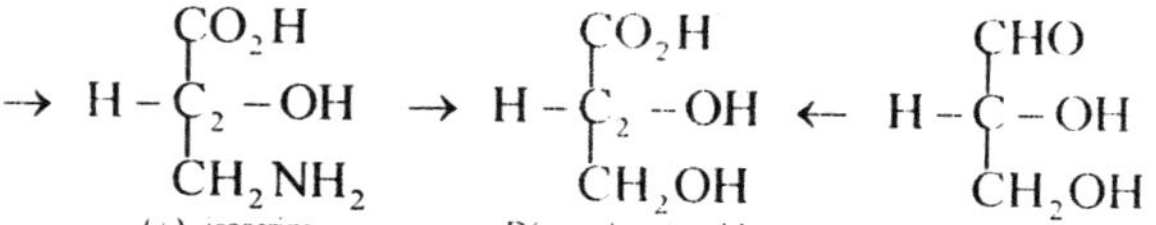

$$\rightarrow \begin{array}{c} CO_2H \\ H-C_2-OH \\ CH_2NH_2 \\ \text{(+)-isoserine} \end{array} \rightarrow \begin{array}{c} CO_2H \\ H-C_2-OH \\ CH_2OH \\ \text{D(-)-glyceric acid} \end{array} \leftarrow \begin{array}{c} CHO \\ H-C-OH \\ CH_2OH \\ \text{D(+)-glyceraldehyde} \end{array}$$

(+)-tartaric acid still through the intermediate, (+)-malic acid (two exchanges of groups give the same malic acid as before). Since (+)-malic acid has been correlated with (+)-glyceraldehyde, (+)-tartaric acid should be designated D(+)-tartaric acid.

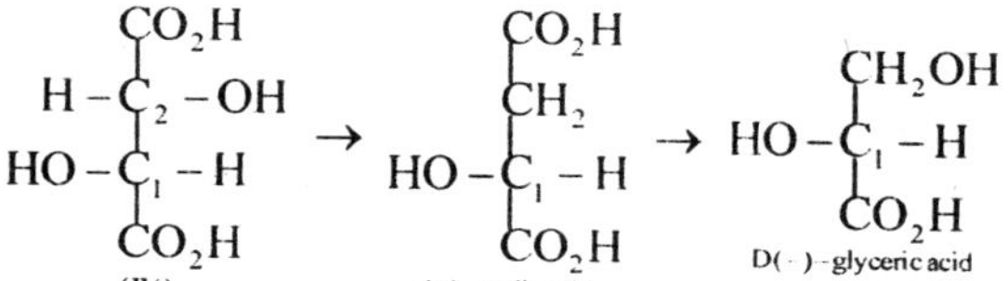

$$\begin{array}{c} CO_2H \\ H-C_2-OH \\ HO-C_1-H \\ CO_2H \\ \text{(IV)} \end{array} \rightarrow \begin{array}{c} CO_2H \\ CH_2 \\ HO-C_1-H \\ CO_2H \\ \text{(+)-malic acid} \end{array} \rightarrow \begin{array}{c} CH_2OH \\ HO-C_1-H \\ CO_2H \\ \text{D(-)-glyceric acid} \end{array}$$

The designation L(+)-tartaric acid is used by those chemists who regard this acid as a carbohydrate derivative.

CORRELATION OF CONFIGURATIONS

As we have seen since the relative configurations of (+)-tartaric acid and (+)-glyceraldehyde have been established, it is now possible to assign absolute configurations to many compounds whose relative configurations to (+)-glyceraldehyde are known, since the configurations assigned to them are actually the absolute configurations.

The methods used for correlating configurations are:

(i) Chemical reactions without displacement at the chiral centre concerned.

(ii) Chemical reactions with displacement at the chiral centre concerned.

(iii) X-ray analysis.

(iv) Asymmetric inductive correlation.

(v) Optical rotations:

  (a) Monochromatic rotations.

  (b) Rotatory dispersion.

(vi) The study of quasi-racemic compounds.

(vii) Enzyme studies.

The above methods normally apply to compounds containing one chiral centre. When several chiral centres are present, the usual procedure is to establish the stereochemistry of the centres relative to each other and then to correlate one of the centres with glyceraldehyde.

It can also be seen that it is better still if it is possible to correlate more than one centre with glyceraldehyde.

Knowledge of absolute configurations is very valuable in the study of optical activity and the mechanisms of organic reactions.

Also, it is considered necessary to know absolute configurations in order to obtain a complete understanding of biochemical processes.

3

Structural Formulae of Organic Compounds

INTRODUCTION

In 1857, Kekule postulated the *constant* quadrivalency (tetravalency) of carbon. From 1900 onwards, however, compounds containing *tervalent* carbon have been prepared, and their number is increasing rapidly. These compounds are free radicals, carbonium ions and carbanions, and are usually reactive intermediates in many reactions. More recently, compounds containing bivalent carbon (*carbenes*) are believed to be formed as intermediates during certain reactions. Hence, unless there is definite evidence to the contrary, carbon is always assumed to be quadrivalent. If 'valency units' or 'valency bonds' are represented by lines, then the number of lines drawn from the symbol shows the valency of that atom, *e.g.*,

$$\begin{array}{c} | \\ -\text{C}- \\ | \end{array};\ -\text{O}-;\ \begin{array}{c} | \\ -\text{N}- \\ \end{array};\ \text{H}-$$

The molecular formula shows the number of each kind of atom present in the molecule, but does not indicate their arrangement. In organic chemistry there are many cases where a given formula represents two or more compounds that differ in physical and chemical properties, *e.g.*, there are at least seven compounds having the same molecular formula $C_4H_{10}O$. Such compounds, having the same molecular formula, but differing in physical and chemical properties, are known as *isomers* or *isomerides*, the phenomenon itself being known as isomerism. The existence of *isomerism* may be explained by assuming that the atoms are arranged in a definite manner in a molecule, and that there is a different arrangement in each isomer, *i.e.*, the isomers differ in structure or *constitution*. This type of isomerism is known as *structural isomerism.*

Obviously, then, from what has been said above, it is always desirable to show the arrangement (if known) of the atoms in the molecule, and this is done by means of *structural formulae* or *bond-diagrams*; *e.g.*, the molecular formula of ethanol is C_2H_6O; its structural formula.

```
     H   H
     |   |
  H—C—C—O—H
     |   |
     H   H
```

So far nothing has been said about the spatial disposition of the four valencies of carbon. Later it will be shown that when carbon is joined to four univalent atoms or groups, the four valencies are directed towards the four corners of a tetrahedron. Thus, the above planestructural formula does not show the disposition of the atoms in space; a three-dimensional formula is necessary for this. Usually the plane-formula is satisfactory. Since the actual spatial arrangements of a given structure may differ, this gives rise to isomerism of the type known as *stereoisomerism. Stereoisomers* have different *configurations*, *i.e.*, the spatial arrangements are different but not their structures.

A structural formula is really a short-hand description of the properties of the compound. Hence the study of organic chemistry is facilitated by mastering the structural formula of every compound the reader meets. An organic molecule, however, is only completely described when the following facts are known: *structure* or *constitution* (this includes a knowledge of the electron distribution;.

ORGANIC COMPOUNDS ANALYSIS

The following is an outline of the methods used in the study of organic compounds.

Purification : Before the properties and structure of an organic compound can be completely investigated, the compound must be pure. Common methods of purification are:

(i) Recrystallisation from suitable solvents.

(ii) *Distillation* : (a) at atmospheric pressure; (b) under reduced pressure or in vacua; (c) under increased pressure.

(iii) Steam distillation.

(iv) Sublimation.

(v) *Chromatography* : This offers a means of concentrating a product that occurs naturally in great dilution, and is an extremely valuable method for the separation, isolation, purification and identification of the constituents of a mixture.

Chromatography depends on the distribution of the mixture between two phases, one fixed and the other moving. The mixture is dissolved in the moving phase and passed over a fixed phase which may be a column or a paper strip. The moving phase may be a liquid or a gas. When the fixed phase is a solid, *e.g.*, alumina, silica gel, etc., the process is usually called *adsorption chromatography*, and the solutes are adsorbed on *different* parts of the column.

The adsorbed compounds are then eluted by the passage of a series of *eluants* (solvents). *Thin-layer chromatography* (TLC) uses a chromatoplate as the column, *i.e.*, a glass strip coated with the adsorbent. Alternatively, the fixed phase may consist of a liquid strongly adsorbed on a solid (as support). The solutes now become distributed between the 'fixed' liquid and the moving liquid (solvent). This technique is known as *partition Chromatography. Paper chromatography* is a special case of partition Chromatography; a paper strip is now the adsorbent column. When the moving phase is a mixture of gases, the usual method is to use a stationary solid phase (*gas-solid Chromatography,* GSC) or a solid coated with a non-volatile liquid (*gas-liquid chromatography*, GLC; this was formerly called *vapour-phase chromatography*, VPC). Furthermore, in GSC and GLC, a 'carrier' gas such as nitrogen replaces the solvent in column chromatography. In *ion-exchange chromatography*, the solid phase consists of a synthetic resin. Two types of exchange resins are used, cation and anion exchangers. In this way, mixtures of bases or mixtures of acids may be separated.

(vi) *Zone melting* : This is a most recent method of purification of organic compounds. A solid in a horizontal tube is slowly moved through a heating apparatus, or alternatively, the heater is moved with the tube fixed. In this way, a relatively narrow zone of solid is liquefied. When this zone is moved out of the heater, the liquid solidifies to a purer solid and the impurity is now in solution (in the heater portion). The basic principle is the redistribution of the components in a mixture during a phase change. A well-known example is freezing a salt solution, whereupon pure ice separates first.

There are various criteria for purity. The most common one for solids is m.p.; for liquids, b.p., density and refractive index are used; and more recently, the examination of the inferred spectrum of a compound (whatever its normal physical state) has been used as a test for purity. In all cases, the process of purification is repeated until the physical constant or spectrum remains unchanged.

Before a natural product can be purified, it first has to be extracted from its source. Some of the earlier methods were solvent extraction, distillation and steam distillation. The methods used now are much superior, *e.g.*, counter-current separation (this has replaced solvent extraction), electrophoresis, dialysis, diffusion, molecular distillation, ultracentrifugation, etc.

The most important method nowadays is chromatography in its various forms (see (v) above). Zone melting ((vi) above) is now also used for separating constituents in a mixture.

QUALITATIVE ANALYSIS

The elements commonly found in organic substances are: carbon (always: by definition), hydrogen, oxygen, nitrogen, halogens, sulphur, phosphorus and metals.

(i) *Carbon and hydrogen* : The compound is intimately mixed with dry cupric oxide and the mixture then heated in a tube. Carbon is oxidised to carbon dioxide (detected by means of calcium hydroxide solution), and hydrogen is oxidised to water (detected by condensation on the cooler parts of the tube).

(ii) *Nitrogen, halogens and sulphur* : These are all detected by the *Lassaigne method*. The compound is fused with metallic sodium, whereby nitrogen is converted into sodium cyanide, halogen into sodium halide and sulphur into sodium sulphide. The presence of these sodium salts is then detected by inorganic qualitative methods.

(iii) *Phosphorus* : The compound is heated with fusion mixture, whereby the phosphorus is converted into metallic phosphate, which is then detected by the usual inorganic tests.

(iv) *Metals* : When organic compounds containing metals are strongly heated, the organic part usually burns away, leaving behind a residue. This residue is usually the oxide of the metal, but in certain cases it may be the free metal, *e.g.*, silver, or the carbonate, *e.g.*, sodium carbonate.

As a rule, no attempt is made to carry out any test for oxygen: its presence is usually inferred from the chemical properties of the compound.

The non-metallic elements which occur in *natural* organic compounds, in order of decreasing occurrence, are hydrogen, oxygen, nitrogen, sulphur, phosphorus, iodine, bromine and chlorine. Halogen compounds are essentially synthetic compounds, and are not found to any extent naturally. Some important exceptions are chloramphenicol (chlorine), Tyrian Purple (bromine) and thyroxine (iodine). In addition to these non-metallic elements, various metallic elements occur in combination with natural organic compounds, *e.g.*, sodium, potassium, calcium, iron, magnesium, copper.

QUANTITATIVE ANALYSIS

The methods used in the determination of the composition by weight of an organic compound are based on simple principles.

(i) *Carbon and hydrogen* are estimated by burning a known weight of the substance in a current of dry oxygen, and weighing the carbon dioxide and water formed. If elements (non-metallic) other than carbon, hydrogen and oxygen are present, special precautions must be taken to prevent their interfering with the estimation of the carbon and hydrogen.

(ii) *Nitrogen* may be estimated in several ways, but only two are commonly used.

(a) *Dumas' method* : This consists in oxidising the compound with copper oxide, and measuring the volume of nitrogen formed. This method is applicable to all organic compounds containing nitrogen.

(b) *Kjeldahl's method* : This depends on the fact that when organic compounds containing nitrogen are heated with concentrated sulphuric acid, the organic nitrogen is converted into ammonium sulphate. This method however, has certain limitations.

(iii) *Oxygen* is usually estimated by difference. There are, however, some semi-micro methods, *e.g.*, the organic compound is subjected to pyrolysis in a stream of nitrogen, and all the oxygen in the pyrolysis products is converted into carbon monoxide by passage over carbon heated at 1 120°C. The carbon monoxide is then passed over iodine pentoxide, and the iodine liberated is estimated titrimetrically.

(iv) *Chlorine, bromine, iodine, sulphur* and *phosphorus* are usually estimated by the 'oxygen flask' method of combustion. A sample of the compound is wrapped in ashless filter paper and ignited electrically in a flask of oxygen containing the appropriate absorption liquid for the halogen or sulphur oxides produced. Suitable titration (or gravimetric determination) gives the amounts of these elements present. For example, sulphur oxides are absorbed in hydrogen peroxide and the sulphuric acid produced is titrated with standard alkali.

Fluorine is estimated by fusing with sodium or potassium in a nickel bomb, and the alkali fluoride is then determined gravimetrically or by titration.

Quantitative analysis falls into three groups according to the amount of material used for the estimation:

(i) Macro-methods which require about 0.1-0.5 g of material (actual amount depends on the element being estimated).

(ii) Semi-micro methods which require 20-50 mg of material.

(iii) Micro-methods which require 3-5 mg of material.

The micro-methods are almost always used now. Furthermore, with the increasing use of superior methods of separation, there is a trend to use still smaller amounts, the range being from 500 to 30 μg, and there is an increasing tendency to use instrumental techniques.

Empirical formula determination : The empirical formula indicates the *relative numbers* of each kind of atom in the molecule, and is calculated from the percentage composition of the compound. The following example illustrates the procedure.

0.202 g of a substance gave on combustion 0.361g of CO_2 and 0.147g of H_2O. What is the empirical formula of the substance?

$$\text{Wt. of C in sample} = 12/44 \times 0.361 = 0.0985\text{g}$$

$$\text{Wt. of H in sample} = 2/18 \times 0.147 = 0.0163\text{g}$$

$$\text{Wt. of O} = 0.202-(0.0985 + 0.0163) = 0.0872\text{g}$$

(i) *Percentage composition* $\% \text{ C} = \frac{0.0985}{0.202} \times 100 = 48.76$

$$\% \text{ H} = \frac{0.0163}{0.202} \times 100 = 8.07$$

In practice, the percentages of C and H are evaluated and the percentage of O is obtained by subtraction of their sum from 100, *i.e.*,

$$\% = 100-(48.76 + 8.07) = 43.17$$

(ii) *Empirical formula*

(a) If given the weights of sample and combustion products, divide the weights of C, H and O (by difference) by their respective atomic weights. Thus the ratio of atoms is:

$$\text{C : H : O} = \frac{0.0985}{12} : \frac{0.0163}{1} : \frac{0.0872}{16}$$
$$= 0.0082 : 0.0163 : 0.00545$$

To obtain this ratio as whole numbers, divide throughout by the smallest number.

$$\text{C : H : O} = 1.50 : 2.99 : 1$$

(Had these numbers been integers or very near to integers, *e.g.*, 1.02 : 3.04 : 1, then the empirical formula would have been C_2O.)

The procedure is to multiply the ratio by small numbers until a ratio of nearly whole numbers is obtained. Thus, multiplying by 2 gives the ratio 3.00 : 5.98 : 2. The empirical formula is therefore $C_3H_6O_2$.

(b) If given the percentage composition, the procedure is similar to the above, *i.e.*

$$\text{C : H : O} = \frac{48.76}{12} : \frac{8.07}{1} : \frac{43.17}{16}$$
$$= 4.06 : 8.07 : 2.7$$
$$= 1.50 : 2.99 : 1$$

Multiplying by 2 gives 3.00 : 5.98 : 2, and so the empirical formula is $C_3H_6O_2$.

It should be noted that the final ratio is the same as before. This is to be expected since, to obtain the percentage composition, each weight was multiplied by the *same constant*, 100/0.202.

Molecular weight determination : The molecular formula—this gives the *actual* number of atoms of each kind in the molecule—is obtained by multiplying the empirical formula by some whole number which is obtained from consideration of the molecular weight of the compound. In many cases this whole number is *one*.

The methods used for the determination of molecular weights are mainly physical. The standard physical methods are the determination of:

(i) vapour density:

(ii) elevation of boiling point;

(iii) depression of freezing point. These methods are described fully in text-books of physical chemistry. In addition to these standard methods, which are used mainly for relatively simple molecules, there are other physical methods used for compounds having high molecular weights, *e.g.*, rate of diffusion, rate of sedimentation, viscosity of the solution, osmotic pressure. X-ray analysis and mass spectrometry are also used for molecular weight determination.

Determination of structure, *i.e.*, the manner in which the atoms are arranged in the molecule. The usual procedure for elucidating the structure of an unknown compound is to make a detailed study of its chemical reactions. This procedure is known as the *analytical method*, and includes breaking down (*degrading*) the compound into smaller molecules of *known* structure.

When sufficient evidence has been accumulated, a tentative structure which best fits the facts is accepted. Sometimes two (or even more) structures fit the facts almost equally well, and it has been shown in certain cases that the compound exists in both forms which are in equilibrium.

This phenomenon is known as *tautomerism*. Where tautomerism has not been shown to be present, one must accept (with reserve) the structure that has been chosen.

Until recently, structural determinations were based almost completely on purely chemical evidence. Nowadays, physical methods are considered as necessary tools for elucidating structures. They are used on the compound itself, and may also be used in the examination of the fragments obtained by degradative work. Of all the physical methods used, infra-red spectroscopy is probably the most widely applicable.

Spectroscopic methods : For light absorption spectra, the following equation is the starling-point for pure substances:

$$\log(I_o/I) = A = \varepsilon cl$$

I_o is the intensity of the incident radiation and/the intensity of the transmitted radiation; A is the absorbance (*optical density*) and e the *molar absorptivity* (*molar extinction coefficient*), c is the concentration of the solute (moles/litre) and *l* the length (in centimetres) of the medium traversed. Thus the units for ε are litre per mole per cm (1 mol^{-1} cm^{-1}), and if ε (sometimes log ε; or the per cent absorption, *i.e.*, 100 (1 – I/I_o) is plotted as ordinate against wavelength (or frequency) as abscissa, the *absorption curve* or *absorption spectrum* of the compound is obtained, and this is characteristic of a pure compound.

On the other hand, the per cent transmission (100 I/I_o; the ratio I/I_o is called the *transmittance*, T), or the absorbance, A, may be plotted against wavelength. Of particular importance are the values of the absorption maxima and their intensities. These are reported for ultra-violet and visible absorption spectra as, *e.g.*, λ_{max} (EtOH) 280 nm (ε4000), which means that the ethanolic solution of the substance has a maximum absorption of 4 000 at a wavelength of 280 nm.

Infra-red absorption spectra are reported as, *e.g.*, v_{max} (CS_2) 1030 cm^{-1}(s), which means that the substance, in carbon disulphide solution, has a strong absorption maximum at 1030 cm^{-1}. In recording infra-red spectra, it is customary to use the following letters to indicate intensity: m (medium), s (strong), w (weak), v (variable), vs (very strong).

Infra-red spectroscopy (4000-650 cm^{-1}). The study of infra-red spectra leads to a great deal of information, *e.g.*, the presence of various functional groups, hydrogen bonding (intramolecular and intermolecular), the identification of *cis* and *trans* isomers, conformational orientations, orientation in aromatic compounds, etc.

The essential requirement for a substance to absorb in the infra-red region is that vibrations in the molecule must give rise to an unsymmetrical charge distribution. Thus, it is not necessary for the molecule to possess a *permanent* dipole moment. Just as electronic transitions are quantised, so are rotational and vibrational energy levels also quantised. Absorption in the near infra-red is due to changes in vibrational energy levels.

A non-linear molecule can undergo a number of vibrational motions, the two main types being *stretching* (vibration along the bonds) and *deformation* (bending; displacements perpendicular to bonds). Fig. 1 illustrates possible modes for a non-linear molecule (asym. = asymmetrical; def. = deformation; str. = stretching; sym. = symmetrical; and the plus and minus signs represent relative movement perpendicular.

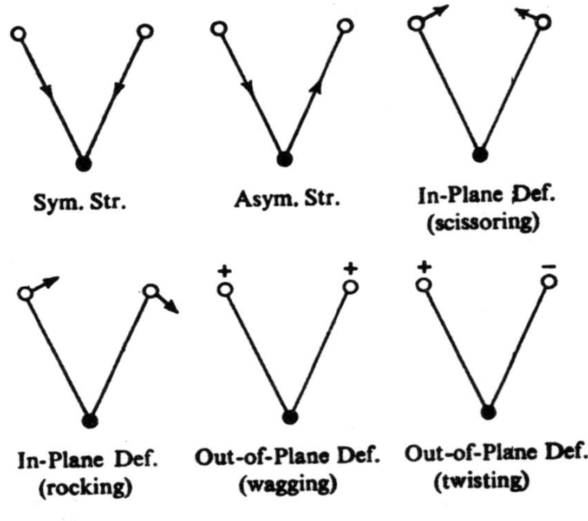

Fig. 1

The stretching regions have higher frequencies (shorter wavelengths) than the deformation regions, and the intensities of the former are much greater than those of the latter. Although, the masses of the bonded atoms predominantly influence the frequency of the absorption, other effects, *e.g.*, environment (*i.e.*, the nature of neighbouring atoms), steric effects, etc., also play a part. Thus, in general, a particular group will not have a fixed maximum absorption wavelength, but will have a *region* of absorption, the actual maximum in this region depending on the rest of the molecule. The spectrum also depends on the physical state of the compound: gas, liquid (as a thin film), solid (as a thin film or as a mull) or solution (preferably dilute; CCl_4, $CHCl_3$, CS_2).

The absorption regions of functional groups have been obtained empirically; many of these regions are described in the text (see also the Index under Infra-red Spectra). Most of the values have been taken from Cross *et al.*

In the initial examination of the spectrum, the usual practice is to look for the presence of the various functional groups. In this way, it may be possible to assign the compound to some particular structural class (or classes). Knowledge of the molecular formula will often help to reject some of the alternatives, and chemical reactions of the compound will help further in this direction. Identification of a compound is carried out by comparison with published spectra (or with the spectrum of an authentic specimen). The region 1 400-650 cm^{-1} is known as the 'finger-print region'; this is the region usually checked for identification, since it is the region associated with vibrational (and rotational) energy changes of the *molecular skeleton*, and so is characteristic of the compound.

If a band has been found which corresponds to a particular group, the presence of this group should be confirmed by ascertaining the presence of another band which is also characteristic of the group, *e.g.*, saturated aliphatic esters show a strong band in the region 1750-1735 cm^{-1} (C=O str.) and another strong band in the region 1250-1170 cm^{-1} (C–O str.).

Furthermore, the absence of a band which is characteristic of a particular group is not conclusive evidence that this group is not present in the molecule. One cause for this is that groups in a molecule may interact, and the result is that both regions are now different from the 'expected' individual regions. It is therefore always desirable to have chemical information about the compound and also spectroscopic data obtained from other methods (uv and NMR).

Various spectra (with wave numbers) have been given in the text. The reader will find it worth his while to make a list of infra-red absorption regions (described for different functional groups, etc.), and to examine the spectra with the legends covered. In this way, he will become familiar with the positions of some of the more important bands.

Raman spectroscopy : The information obtained is similar to that from infra-red spectroscopy. *Nuclear magnetic resonance (NMR)*• *spectroscopy*. In order to explain certain observations in rotational spectra, it was suggested that atomic nuclei spin about their axes. Now, since a rotating charged sphere has associated with it a magnetic moment, then all the charged particles in a nucleus will cause that nucleus to behave (to a first approximation) like a small bar magnet, with its magnetic moment along the axis of rotation.

Nuclei are composed of protons and neutrons, the former carrying a unit positive charge, and the latter being electrically neutral. Magnetic properties occur with those nuclei which have (a) odd atomic and odd mass numbers, *e.g.*, $_1^1H$, $_7^{15}N$, $_9^{19}F$, $_{15}^{31}P$; (b) odd atomic number and even mass number, *e.g.*, $_1^2H$ (D), $_7^{14}N$; (c) even atomic number and odd mass number, *e.g.*, $_6^{13}C$. Nuclei which have no magnetic moment are those with even atomic and even mass numbers, *e.g.*, $_6^{12}C$, $_8^{16}O$, $_8^{18}O$, $_{16}^{32}S$.. It has been assumed that the particles in such nuclei are paired, *i.e.*, spinning in opposite directions, with the result that there is no resultant spin and consequently no magnetic moment. In those nuclei where the magnitude of the spin is not zero, the nuclear spin quantum number, I, may assume any of the values 1/2, 1, 3/2, 2, etc.

Nuclei possessing a resultant spin will thus behave as spinning magnets, and so will tend to orient themselves in an applied magnetic field, and the number of possible energy levels, *i.e.*, orientations with respect to the applied field, is given by 2I + 1. The simplest example of a spinning nucleus is that of the proton. Here, I = 1/2 (cf. the electron), and in this case there are only two orientations possible, lined up with or against the direction of the applied field.

Since work must be done to turn a magnet against a magnetic field, each orientation corresponds to a different energy state of the nucleus. These levels are quantised (cf. ultra-violet and infra-red spectroscopy), and so it should be possible to find electromagnetic radiation of a definite frequency which will be absorbed, thereby changing the orientation of the proton from alignment with to against the field, the change being from a lower to a higher energy level.

This electromagnetic radiation is supplied by an oscillator (with *its magnetic field* at right angles to the applied field), and since the position of the absorption peak, *i.e.*, where resonance occurs, depends on the frequency of the oscillator or the strength of the applied field it is possible to change from the lower to the higher energy level by using a variable frequency with a fixed applied magnetic field, or vice versa. In practice, it has been found easier to vary the field rather than the frequency.

The result is that the NMR spectrum is usually a graph of signal intensity (ordinate) against magnetic field (abscissa); expressed in milligauss at a fixed frequency. For a given field, the strength of the signal depends on the magnetic moment of the nucleus, and since the proton has one of the largest moments, proton magnetic resonance (PMR) is of special importance. Other nuclei used in NMR spectroscopic studies in organic chemistry are ^{13}C, ^{19}F, ^{29}Si, and ^{31}P, but to study these nuclei it is necessary to modify the spectrometer.

The difference between the two energy levels, ΔE, for a proton is given by expression (a), where h is Planck's constant, H the strength of

(a) $\Delta E = h\gamma H/2\pi$

(b) $v = \gamma H/2\pi$

the field experienced by the proton, and y is the *magnetogyric* (*gyromagnetic*) ratio for the proton. The frequency of the radiation (v) which induces the transition is given by expression (b), since $\Delta E = hv$. Thus, the position of the energy absorption is a function of both the frequency of the oscillator and the strength of the applied field. When radiation is absorbed, the proton changes from the lower to the higher energy state.

For an applied field of 9400 gauss, the resonance frequency of a proton is about 40MHz (megahertz = megacycles per second). The energy associated with this frequency is about 0.0167 J mol^{-1} (0.004 cal). Thus, ΔE is very small, and it is because of this that radio-frequency radiation can effect these transitions. Since ΔE is very small, the population in the lower state is only *slightly* greater than that in the higher state. This is the situation when the compound is placed in a magnetic field, and so there is net absorption when the radio-frequency is applied, and it is this absorption which is measured.

In order that a PMR signal be observed, the proton must be in a single state for 10^{-2} to 10^{-1} second. It has also been shown that the spectral line width is inversely proportional to the average time the proton occupies the *higher* energy state. Hence, the longer the time spent in this state, the sharper is the line; and conversely, the shorter the time, the broader is the line.

From what has been said above, it might be expected that the resonance frequency for a given field depends only on the nature of the atomic nucleus concerned. This, however, is not the case. The applied field causes electrons round a nucleus to circulate in a plane perpendicular to the field, and these currents produce a field in opposition to the applied field. Thus, the effective magnetic field (H) experienced by the nucleus is smaller than the applied field (H_0), the relationship between the two being given by the expression

$$H = H_0(1 - \sigma)$$

a (which is non-dimensional) is called the shielding or screening constant, and has a positive value, but in certain circumstances it may be negative, *i.e.*, the effective field is larger than the applied field. In this case, the proton is said to be deshielded. Since the numerical value of a depends on the chemical *environment* of a given nucleus, the shielding or deshielding of a nucleus varies with its environment. Shielding causes a shift of the resonance frequency to higher values of the applied field, *i.e.*, the shift is *upfield.*

On the other hand, deshielding causes a shift of the resonance frequency to lower values of the applied field, *i.e.*, the shift is *downfield.* The magnitude of this shift is known as the *chemical shift*. Since the value of the field experienced by the sample cannot be determined accurately, chemical shifts are measured *relative* to some standard which contains the nucleus under consideration. Various reference compounds (which are usually added to the sample) for PMR have been used, but tetramethylsilane (TMS), $(CH_3)_4Si$, is particularly useful since it contains twelve equivalent protons. The PMR spectrum of this compound shows a single sharp line which occurs at a higher field than any protons in most of the common organic compounds, *i.e.*, most PMR signals occur downfield with respect to TMS.

The chemical shift may be reported in various ways. Since the resonance frequency is dependent on the strength of the applied field, the shift may be reported as field units (milligauss). However, because the field can be expressed in terms of frequency (see expression (b) above), the shift may also be expressed in terms of Hz (c.p.s.).

This separation in Hz is also proportional to the frequency of the oscillator, *e.g.*, if the separation between a proton signal and TMS is 60 Hz at 40 MHz, the separation at 60 MHz becomes 90 Hz [60 × 60/40 = 90]. Hence it is desirable to be able to report chemical shifts in units which are independent of the operating conditions of the spectrometer. This has been done by defining the chemical shift, δ, by the expression

$$\delta = \frac{\text{separration in Hz}}{\text{oscillator frequency}} \times 10^6$$

The factor 10^6 is introduced in order to record the chemical shift as a convenient value. This is usually in the range 1^{-10}, and is quoted in parts per million (p.p.m.). The independence of the chemical shift of the oscillator frequency is shown by the following example. A separation of 60 Hz at an oscillator frequency of 40 MHz becomes 90 Hz at an oscillator frequency of 60 MHz (see above). However, δ remains unchanged: $\delta = (10^6 \times 60)/\ (40 \times 10^6)$ = 1-5 p.p.m.; $(10^6 \times 90)/(60 \times 10^6)$ = 1.5 p.p.m.

It is now becoming common practice to express chemical shifts in τ (tau)-values, defined by the expression

$$\tau\ 10 - \delta$$

where 10 p.p.m. is the value assigned to the line of TMS. Most protons have positive τ-values (*i.e.*, $\delta < 10$); strongly acidic protons, however, have negative τ-values (*i.e.*, $\delta > 10$). The greater the shielding of the nucleus, the larger is its τ-value (the smaller is δ). Since the degree of shielding depends on the electron density round the proton, any structural feature that decreases this density will cause a decrease in shielding, with consequent lowering of the τ-value (the chemical shift moves downfield). Halogens are electron-attracting (electronegative groups), and when joined to a methyl group, the electron density round each proton is decreased.

Consequently, the presence of halogens weakens the induced opposing field, *i.e.*, deshields these protons and so the τ-value will be expected to be lowered (the chemical shift is moved downfield). This prediction is-observed in practice. The order of electronegativity of the halogens is I < Br < Cl < F, and the τ-values of the methyl protons are: CH_3, 7.83;' CH_3Br, 7.35; CH_3Cl, 6.98; CH_3F, 5.70 p.p.m. Similarly, since the order of electronegativity of carbon, nitrogen, and oxygen is C < N < O, the τ-values of the methyl protons in CH_3–C, CH_3–N, CH_3–O are, respectively, 9.12, 7.85, 6.70 p.p.m. Silicon is less electronegative than carbon, and so methyl protons in TMS are more shielded than those in a methyl group attached to carbon, *i.e.*, the protons in TMS absorb upfield with respect to protons in most of the common organic compounds, Table 3.2.

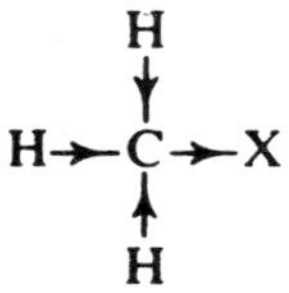

```
        H
        ↓
   H → C → X
        ↑
        H
```

Fig. 2 summarises the terminology described in the foregoing account of NMR spectroscopy.

Measurements of NMR spectra are normally carried out on liquids or solutions (5–20 per cent concentration). The best solvents are those which do not contain protons, *e.g.*, deuterio-chloroform and other deuterated compounds, carbon tetrachloride, etc. However, because of solubility problems, solvents containing protons may also be used, *e.g.*, chloroform.

A difficulty with respect to solutions is that i-values may change with the nature of the solvent, particularly aromatic solvents. Chemical shifts may also change with concentration in a given solvent. Protons attached to carbon are very little affected, but when attached to atoms such as O, N, S, the chemical shift is very much affected.

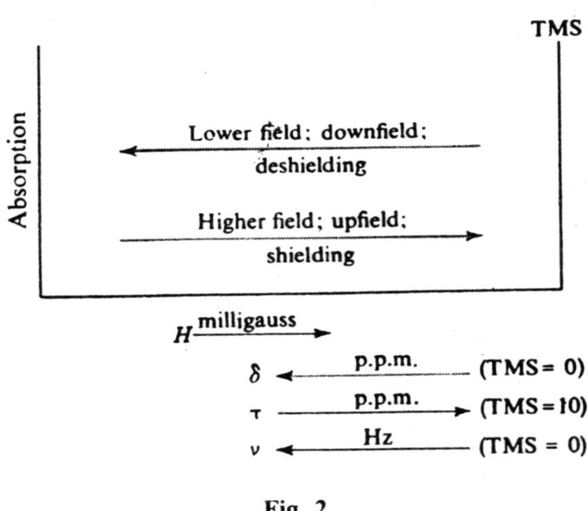

Fig. 2

In the latter case, the changes are due to changes in the degree of hydrogen-bonding, which causes a downfield shift relative to the unbonded state. The study of NMR spectra of liquids and solutions is known as *high resolution NMR*. The study of solids, since they give spectra which consist of broad resonance lines, is referred to as broad line resonance.

Fig. 3 is the NMR spectrum of ethanol, (liquid; 60 MHz), carried out at low resolution. The position of each peak is characteristic of the environment of a particular proton, and the areas under the curves of the peaks have been shown to be in the ratio 1:2:3.

This ratio corresponds to the number of protons in OH, CH_2 and CH_3, respectively, and so it is therefore possible to 'count' the protons in various environments.

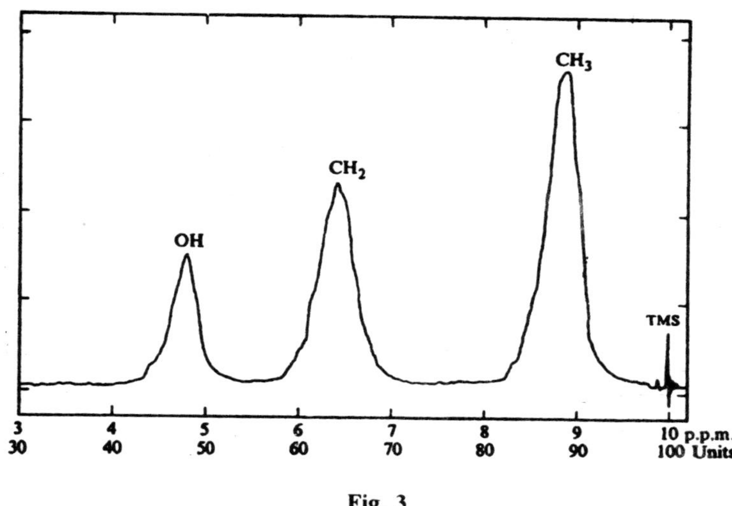

Fig. 3

These areas are now evaluated by means of electronic integrators. Integration produces a trace which rises in steps as each proton signal is passed, and the height (between steps) is proportional to the number of protons in that signal (Fig. 4).

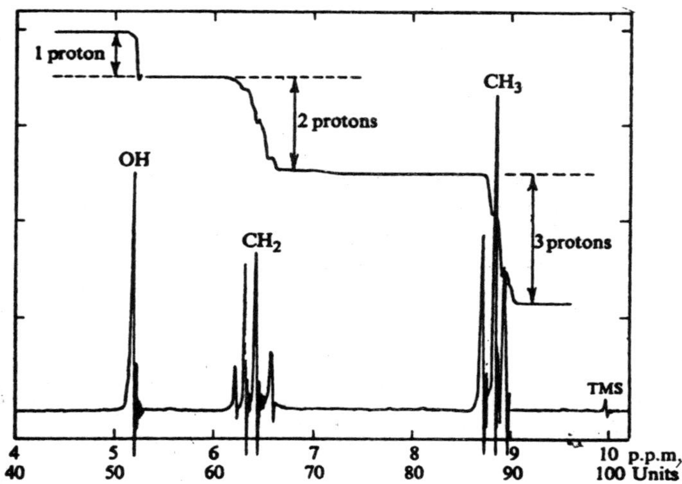

Fig. 4

In this way, the ratios of the numbers of protons in each signal are obtained, and if one number is known, all the others can be estimated. For example, if we know that the compound under examination is a monohydric alcohol then, knowing the position of a hydroxylic atom, we now know that the area of this signal is equivalent to one proton. On the other hand, if we know the molecular formula of the compound, we can also calculate the actual numbers of different protons.

In practice, because of the experimental difficulties, it is unusual to obtain integers from the integration trace. However, the numerical results are usually sufficiently accurate to permit 'counting' of the different types of protons (Fig. 4 the values are 54:37:19).

Spin-spin coupling. Fig. 4 is the high-resolution NMR spectrum of ethanol (20 per cent in CCl_4; 60 MHz), containing a trace of hydrochloric acid. Instead of the broad bands shown in Fig. 1.3, two have been split into multiplets. The total areas under the multiplets are still in the same ratio as before, *i.e.*, 2:3. This fine structure has been explained as being due to shielding of protons by protons on adjacent carbon atoms.

First let us consider the influence of the methylene group, CH_2, on the methyl group, CH_3. Each proton in the methylene group may have its magnetic moment lined up with or against the applied field. If we represent the former alignment by an arrow pointing upwards and the latter alignment by an arrow pointing downwards, then there are the following four possible combinations:

Thus, the shielding effect will depend on the type of combination operating. However, combinations (ii) and (iii) will have the *same* shielding effect on the adjacent methyl group. Hence there are *three different* shielding combinations, and statistically it can be expected that at any given moment, 25 per cent of the methylene protons will be in combination (i), 25 per cent in (iv) and 50 per cent in (ii) and (iii). The net result is that the methyl proton signal is split into a triplet, the ratio of the areas being 1:2:1.

If we apply the same argument to the effect of the protons of the methyl group on the methylene protons, then there are eight possible combinations:

↑↑↑ ↓↑↑ ↑↓↑ ↑↑↓ ↓↓↑ ↓↑↓ ↑↓↓ ↓↓↓

(v) (vi) (vii) (viii)

As before, each combination in (vi) gives rise to the *same* shielding effect; this is also the case for (vii). The net result is that there are four different shielding combinations, resulting in the splitting of the methylene signal into a quartet, the ratio of the areas being 1:3:3:1.

This fine structure within a particular signal is called *spin-spin splitting*, and the magnitude of the separations between peaks in a multiplet (arising from spin-spin couplings) is called the *spin-spin coupling* constant, and is denoted by the symbol J; its values are given in Hz. The magnitude of J is independent of the oscillator frequency, but the *spacings between* signals is not. Hence, a change in frequency changes the signal separations, but not the spacings in a multiplet.

Thus, by measuring the NMR spectra of a given compound at two different oscillator frequencies, if the spacings of a group of lines remain unchanged, the lines are components of a multiplet. If the spacings change, the group of lines arises from non-equivalent protons. It has been stated above that the origin of multiplets is due to spin-spin coupling between groups of protons.

This has been demonstrated by, *e.g.*, the examination of the following deuterated ethanols: (a) CD_3CH_2OH, (b) CH_3CD2OH, (c) CH_3CH_2OD. Since the coupling constant of deuterium with protons is small (~ 1/7 of J for hydrogen coupling), the result is that single peaks (slightly broadened) are given by protons when adjacent to deuterium.

Deuterium signals are far removed from the normal proton signals, and are not observed under the operating conditions (for proton resonance). Thus, all three deuterated ethanols give two signals only: (a) gives a doublet and a triplet, (b) two singlets, and (c) a triplet and a quartet. Because the introduction of deuterium into a molecule leads to a simplified spectrum, this is used as a general method for studying NMR spectra, *i.e.*, for making assignments to various spectral lines.

Since the signal of a methyl group, *e.g.*, in CH_3CD_2OH or TMS, consists of a single line, this means that there is apparently no magnetic coupling between the hydrogen atoms attached to the same carbon atom (geminal hydrogens). The three hydrogen atoms are equivalent, all three having the same chemical shift. This leads to the general rule that protons with the same chemical shift do not give rise to observable splitting. Only when protons are non-equivalent is splitting possible.

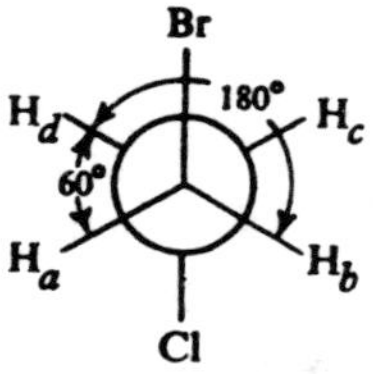

(I)

Fig. 5

There are two types of equivalent protons, *chemically equivalent* and magnetically equivalent. *Chemically equivalent protons* are those which occupy chemically equivalent positions, *i.e.*, are in identical chemical environments. Such protons have the *same* chemical shift. A simple test for chemical equivalence of two (or more) protons is to replace each proton one at a time by substituent Z, and if by doing so, the same compound (or its mirror image) is obtained, then these protons are chemically equivalent.

Let us consider ethane, CH_3CH_3, as an example. Replacement of each proton, one at a time, on one carbon atom gives CH_2ZCH_3, which is also obtained when the other carbon atom is treated in the same way ($C2CH_2Z$). Thus, the three hydrogens on each carbon atom are chemically equivalent, and *all six* hydrogens are also chemically equivalent. Hence, all have the same chemical shift, and consequently there is no splitting *within* a methyl group and none due to coupling *between* the groups. The observed spectrum of ethane consists of one signal which is a singlet.

A group of two (or more) protons are said to be *magnetically equivalent* when not only do they have the same chemical shift, *i.e.*, are chemically equivalent, but also the set of coupling constants to all other protons is identical for each member of the group. This may be illustrated with l-bromo-2-chloroethane, which is represented by the Newmann formula shown, (I).

If we assume, for the moment, that this conformation is fixed, it can be seen that H_a and H_b are chemically equivalent, as also are H_c and H_d (for each pair, the chemical environments are identical). The coupling constants, J_{ad} and J_{bd} however, are not equal, and hence H_a and H_b are not magnetically equivalent.

To understand non-magnetic equivalence, we must now consider the problem of the magnitude of the coupling constant. It has been found that the value of J decreases very rapidly with increase in the number of bonds connecting the interacting protons. Only two cases are important, the methylene type (two bonds), H–C–H (geminal hydrogens) and the vicinal type (three bonds), H–C–C–H (adjacent carbon atoms). Geminal hydrogens, *e.g.*, H_aH_b or H_cHd, couple with splitting only if their environments are different. If coupling occurs through four or more bonds, the coupling is referred to as *long range coupling*. This is usually negligible for saturated compounds, but may be large in *unsaturated* systems.

In addition to depending on the number of intervening bonds, J is also *angular dependent*, *i.e.*, its magnitude varies with the angle between the methylene protons (when these are not chemically equivalent), and the *dihedral angle* for vicinal hydrogens. In practice, rotation about a C–C single bond takes place quite readily. If we take the eclipsed conformation as the starting point, *i.e.*, in (I), the CH_2Br group is kept stationary and the CH_2Cl group is rotated until Cl and Br are in line, and consequently H_aH_c and H_bH_d are also in line (as pairs). In this *eclipsed* conformation, the dihedral angle is 0°, and then keeping the CH_2Br group stationary, the other group can rotate through 360° before the molecule reaches its starting position.

It has been found that the value of J, for a given pair of vicinal hydrogens, depends on the dihedral angle (angle of rotation), being largest when the angle is 0° or 180°, and very small (or zero) when the angle is 90°. Thus, since the dihedral angle for H_aH_d in (I) is 60° and that for H_bH_f is 180°, J for the latter will be larger than that for the former. Hence, these coupling constants for H_a and H_b with H_d are different, and consequently H_a and H_b are not magnetically equivalent.

One theory for the transmission of coupling is as follows. A given proton, because of its magnetic moment (due to spin), affects the spins of the electrons forming the covalent C–H bond. This change affects the spins of the C–C electron pair, and this in turn affects the spins of the electron pair in the adjacent C–H bond. Thus, the spin effects of one proton are transmitted through the covalent bonds to the other spinning proton. Since these effects depend only on the structure and geometry of the molecule and are not due to the presence of the applied field, they are independent of the strength of the applied field.

Rules for determining multiplicity : Some simple rules for determining the multiplicity of a signal have been developed, but these usually apply only when $(\Delta v/J) \geq 6$ (where Av in Hz is the separation of the signals of the interacting groups). Spectra of this type are said to be *first-order spectra.*

(i) Equivalent protons, when coupled, do not cause splitting.

(ii) The spacings (*i.e.*, J values) in a multiplet are equal and are also equal to the spacings of a multiplet arising from mutual coupling, *e.g.*, in the NMR spectrum of ethanol, the spacings in the methyl triplet and the methylene quartet are all equal.

(iii) The multiplicity of a group of equivalent protons is equal to (n + 1), where n is the number of equivalent protons in the group which are coupled to the first group, *e.g.*, in $-O-CH_2-CH_3$. the two equivalent methylene protons are coupled to *three* equivalent protons in the methyl group. Hence, n = 3, and so there will be *four* lines shown by the two methylene protons. Similarly, the three methyl protons give rise to three lines due to coupling with the two (n = 2) methylene protons. Furthermore, the relative intensities of the individual lines of a multiplet correspond (ideally) to the numerical coefficients of the terms in the binomial expansion $(1 + x)^n$. Thus, the lines of the quartet (n = 3) have relative intensities 1:3:3:1, and those of the triplet (n = 2), 1:2:1. Also, the relative intensities of a multiplet are symmetrical (ideally) about the mid-point of the multiplet.

(iv) When a group of equivalent protons is coupled to groups of equivalent protons, n_a in one group, n_b in another, etc., the multiplicity of that group is equal to $(n_b + 1)$ $(n_b + 1)$ .... To illustrate this rule, let us consider the NMR spectrum of *very pure* ethanol. The methyl group is coupled to the methylene group (n = 2), and so the methyl signal will be a triplet. The hydroxyl proton is also coupled to the methylene group, and its signal is therefore a triplet. The methylene group is coupled to both the methyl group (n_a = 3) and to the hydroxyl proton (n_a = 1). Hence, the multiplicity of the methylene signal is given by (3 + 1)(1 + 1) = 8, *i.e.*, an octet. This is the spectrum observed in practice (for very pure ethanol;).

When dealing with first-order spectra, it is possible to construct them diagrammatically. Let us consider ethanol (plus a trace of acid) as our example. This spectrum is shown in Fig. 4; the diagrammatic spectrum is shown in Fig. 1.5. The signal of the methylene group is a quartet of relative intensities 1:3:3:1, and so the total intensity of this signal = 1 + 3 + 3 + 1 = 8 units. Since these 8 units represent two protons, one unit is equivalent to 0.25 proton, and 3 units to 0-75 proton.

Hence these four components of the quartet may be drawn as vertical lines, the lengths of which are proportional to the number of protons they represent, *e.g.*, using 5 mm to be equivalent to 0.25 proton (the weakest line of the multiplets in all signals),-the quartet is then represented by four lines, 5 mm, 15 mm, 15 mm, 5 mm, equally spaced and centred about the τ-value of this signal. In the same way, the intensities of other multiplets in the spectrum may be estimated and represented as vertical lines whose lengths are also based on the arbitrarily chosen unit of 5 mm = 0.25 proton. For ethanol, the t-values of the signals are known, but if the values are not known, it is usually possible to estimate them from tables of τ-values.

The spectrum given in Fig. 6 is idealised. In practice, distortion of intensities may occur, inner lines increasing at the expense of outer lines (cf. Fig. 4). This becomes more pronounced as the ratio Δv/J decreases. Since this distortion occurs between coupled signals, it may be used to show which groups of protons are coupled.

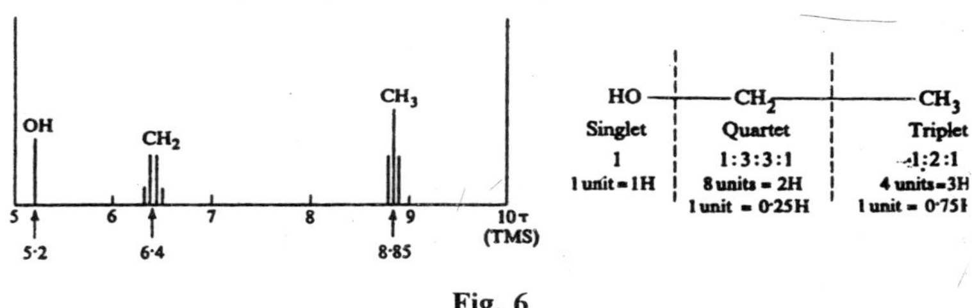

Fig. 6

As we have seen, the simple rules of splitting are applicable only when (Δv/J) ≥ 6. When Δv ≅ J, the simple rules no longer hold good. The spectra are now complicated, and it is usually difficult to recognise the pattern of the lines. Mathematical techniques, however, have been developed for analysing complicated spectra. If the chemical shifts and coupling constants of all the magnetic nuclei present in the molecule are known, it is possible to predict the NMR spectrum of the compound. Alternatively, it may be possible to evaluate the chemical shifts and coupling constants from the observed spectrum.

One cause of the appearance of complicated spectra is that the components of a multiplet may overlap. For the purpose of analysing NMR spectra, a notation has been introduced for discussing spin-spin coupling. Protons which have similar chemical shifts are designated by the letters A, B, C, ..., and protons which have similar chemical shifts that are quite different from those of set A, B, C,..., are designated by the letters ... X, Y, Z. When a proton has a chemical shift that lies between the above two sets and is well separated from both, it is designated by a middle letter of the alphabet, *e.g.*, M. The number of protons of the *same* type is then indicated by a subscript (an integer).

The earlier letters of the alphabet are usually chosen to represent those protons which absorb at lower fields, and the choice of letters depends on which protons are different, on how different they are, and whether they are coupled or not. One practice is to designate protons as A and X if $\Delta v/7 \geq 6$, and if less than this, as A and B. The scheme described so far designates the same letter to protons with the same chemical shift, *i.e.*, to chemically and magnetically equivalent protons. One method of indicating that the protons in one group are not magnetically equivalent is to repeat the same letter with one of them primed, *e.g.*,

CH_3CH_2OH	CH_3NHCHO	$ClCH_2CH_2Br$
A_3M_2X	A_3XY	AA'BB'

The term *spin system* is used to describe groups of nuclei that are spin-spin coupled among each other, but not with any nuclei outside the spin system. It is not necessary, however, for all the nuclei within the spin system to be coupled to all the other nuclei. In many cases, the spin system embraces the complete molecule, *e.g.*, $CH_3CH{=}CH_2$ is a six-spin system.

On the other hand, a molecule may consist of two (or more) parts 'insulated' from each other, and thereby gives rise to two (or more) independent spin systems, *e.g.*, $CH_3CH_2{-}O{-}CH_2CH_2CH_3$. This contains two spin systems, the five-spin ethyl group and the seven-spin n-propyl group. The latter is an example of the case where not all nuclei in the spin system are coupled (coupling between OCH_2 and CH_3 is negligible).

Now let us analyse the spectrum of l-bromo-3-chloropropane. The τ-values (at 40 MHz in $CDCl_4$) are: CH_2Br, 6.30; $-CH_2-$, 7.72; CH_2Cl, 6.45; J value for the first pair is ~6-2 Hz, ~6-2 Hz for the second pair, and zero for the first and third groups. Since $\delta(\text{p.p.m.}) = (\text{Hz/r.f.}) \times 10^6$,

Hz = δ × r.f. × 10^{-6}. Therefore, for the first pair, Δv = 1.42 × (40 × 10^6) × 10^{-6} = 57. Hence, Δv/7 = 57/6.2 = ~9.2. This value is greater than 6; and this also is the case for the second pair (51/6.2 = ~8.2). This spectrum is therefore first-order.

Analysis of this spectrum may now be carried out in a similar manner to that used for ethanol, but here we shall deal with an alternative method, the graphical method. Each component of a multiplet may be represented by a line whose length is proportional to its intensity (cf. Fig. 6), but here we shall indicate relative intensities by numbers (Fig. 7). The single lines in (i) represent each group of equivalent protons, and their spacings are proportional to the differences in chemical shifts.

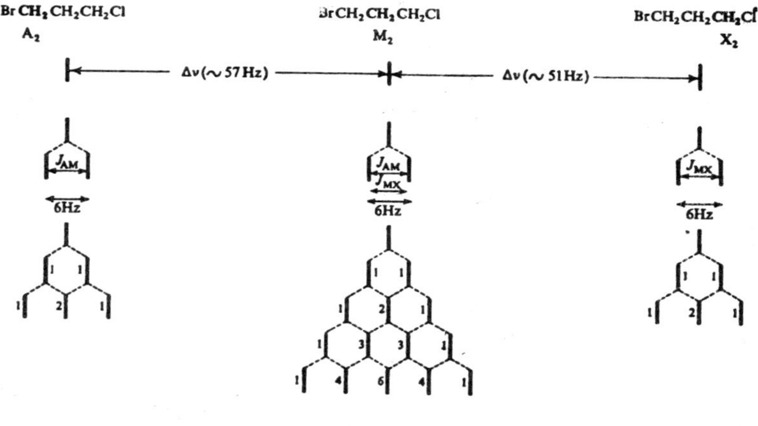

Fig. 7

If there were no coupling, then each methylene group would have given a single line signal, all three signals being of equal intensity. Since, however. A_2 and M_2 are coupled, as are also M_2 and X_2, splitting does occur. Let us first consider the signal given by the A_2 group of protons, always bearing in mind the two possible orientations of a coupled proton. In the absence of coupling, the signal is the single line shown in (i). If we now consider the coupling with one proton in the M_2 group, this produces two lines of equal intensity, and they are separated by J_{AM} (Fig. 7). However, since A_2 is also coupled with the *other* proton in M_2, each component of the doublet is split into a second doublet, each pair being separated by J_{AM}. Hence, the 'inner' line of one doublet overlaps with that of the other doublet, thereby producing a triplet with equally spaced lines and with relative intensities 1:2:1 (Fig. 7iii). In the same way, it can be shown that couplings of the protons X_2 with M_2 also produce a triplet. In both cases, using the simple rules of multiplicity, with n = 2, the number of lines in the multiplet is 3 (= n + 1), and their relative intensities are 1 : 2 : 1 (the coefficients of the expansion $(1 + x)^2$).

Now let us consider the signal of M;. According to the simple rules, their signal would be expected to be a multiplet of *nine* lines [(2 + 1)(2 + 1) = 9]. However, by using the scheme outlined for the coupling between A; and M,, the diagram for the signal of M_2 is a multiplet of five lines (Fig. 2.7iii). It should be emphasised that the pattern of this multiplet is the same whether the M protons are coupled first to the A group, followed by coupling with the X group, or vice versa.

Because all J values are (almost) equal, overlapping of 'inner' components occurs, and consequently the observed number of lines is smaller than that calculated from the simple rules. This situation is generally the case when the J values between different groups of coupled protons are all the same or very nearly the same. Also, in these cases, the total number of lines of the multiplet is given by N = $(b_A + n_x + 1)$ + In the example discussed, $n_A = n_x = 2$; hence N = 5. Furthermore, the relative intensities of the lines are given by the coefficients of the expansion $(1 + x)^{(n_A + n_X)}$ and the lines are equally spaced.

It can be seen from Fig. 4 that the signal from the proton of the hydroxyl group is not split, *i.e.*, there is no evidence of spin-spin coupling between the OH proton and the CH_2 protons, since had there been coupling, the OH proton signal would be expected to be a triplet. This absence of coupling has been explained on the basis that, in the presence of a trace of acid (in this case, hydrochloric acid), there is a rapid exchange of protons between ethanol molecules. This may be represented as shown:

$$C_2H_5{-}OH_A + H_B^+ \rightleftharpoons C_2H_5{-}\overset{+}{O}(H_A)(H_B)$$

$$C_2H_5{-}OH_C + C_2H_5{-}\overset{+}{O}(H_A)(H_B) \rightleftharpoons C_2H_5{-}\overset{+}{O}(H_A)(H_C) + C_2H_5{-}OH_B$$

Since a proton has two spin orientations which have almost the same energy levels, and consequently the populations in these levels are practically equal the exchanged protons (say H_A and H_B) have about the same chance of having the same or opposite spin orientations.

The result is that the CH_2 protons experience both spin couplings to the same extent, the overall effect being a *time-average coupling effect* that is zero. Thus, the CH_2 signal is not split by the OH proton. Because this time-average effect is reciprocal, the OH proton signal is also not split by the CH_2 protons, and consequently gives a single sharp line signal. Thus, the CH_2 and OH protons are effectively, *decoupled.*

The exchange reaction described is an example of *chemical exchange*, and this is common for protons attached to oxygen, nitrogen, or sulphur. However, for this decoupling to take place, the rate of exchange must be rapid. Calculations have shown that if the rate of exchange per second is much greater than the separation (in Hz) of the two separate signals (had there been no exchange), then decoupling occurs. For pure ethanol the rate of exchange is very much slower than Δv (the separation of the two signals), and so coupling occurs, the CH_2 appearing as an octet and the OH as a doublet. Since the rate of chemical exchange is accelerated by a rise in temperature, spin decoupling may sometimes be observed by raising the temperature at which the NMR spectrum is measured.

Similarly, spin coupling may be observed by lowering the temperature. It also might be noted here that complications arising from a proton attached to oxygen (or nitrogen) may be removed by deuterium exchange. This is readily carried out by shaking a solution of the compound in deuteriochloroform with a small amount of D_2O. Signals arising from OH (or NH) protons will no longer be observed.

From what has been said, it can be seen that in exchange reactions, the shape of a band depends on the rate of exchange. When the rate is very fast, the signal of the exchanged proton is a single sharp line, and as the rate becomes slower, a point is reached when this begins to broaden, and finally splits into the number of components required for coupling with the vicinal protons. Thus, analysis of the shape of the band affords a means of evaluating the rate of exchange.

As we have seen above, the simple rules for determining multiplicity usually apply only when $(\Delta v/J) \geq 6$. We have also seen that Δv (the spacing between the signals) is dependent on the oscillator frequency, whereas J is independent of it. Hence, by increasing the oscillator frequency, $\Delta\nu$ can be increased, and if increased sufficiently the conditions for obtaining a first-order spectrum $(\Delta\nu/J) \geq 6$, may be fulfilled, thereby resulting in a simplified spectrum. Oscillator frequencies that have been used are 100 MHz, and better still, 220 MHz.

Double nuclear resonance : It has been pointed out above that spin-spin coupling may be annihilated by rapid exchange. This annihilation may also be effected as follows. Suppose we have two protons, A and B, in different environments and are coupled. Now suppose the resonance frequency of A is being measured. Then, at the same time as this is being done, a strong radio-frequency field whose frequency is the resonance frequency of B is also applied. This latter field produces both absorption and emission by B many times a second, and consequently coupling of B with A is now prevented, *i.e.*, B is decoupled with respect to A, and the time average of coupling between A and B is zero.

Splitting of A by B has been prevented, but it is not possible under the conditions of the experiment to record the resonance of B. By applying double nuclear resonance to each type of proton, each signal can be made to collapse into a singlet line. In this way, the double resonance method produces a much simpler spectrum, and hence makes interpretation easier (cf. spectra of deuterated ethanol, above).

Uses of NMR spectroscopy : NMR absorption spectroscopy has found great use in the study of chemical structure, configurations and conformations of molecules, and the study of rates of chemical exchange. It has also been used in quantitative work, *e.g.*, the determination of enol content in keto-enol mixtures. Barcza (1963) has shown that NMR spectroscopy may be used to determine molecular weights with greater accuracy than the usual methods such as cryoscopic, ebullioscopic, etc. Mass spectrometry. When a compound, in a high vacuum, is bombarded with electrons in a mass spectrometer, it is converted into positive ions by loss of an electron (positive ions may also be produced by other methods).

$$M + e \rightarrow M^{+} + 2e$$

The positive ion, M^{+} (or P^{+}), is known as the *molecular ion* (or the *parent ion*), and is formed when the energy of the electrons is equal to that of the ionisation potential (usually 10-15 electronvolts). In practice, the energy of the electrons is 50-70 eV, and under these conditions the molecular ion is formed with an excess of energy, large enough for it to break down into a mixture of neutral and positively charged fragments, and the latter, if they have an excess of energy, also undergo further fragmentation, etc.

Most ions carry a unit positive charge, but some may carry a double (or greater) positive charge (and some ions may be negatively charged by electron *capture*). All positive ions are accelerated in an electric field and separated by their passage through an electric field and then a magnetic field. In this way, ions which have the same mass/charge (m/e) ratio are collected into beams, and fall on a collector plate. Thus, the ions are sorted out according to their mass charge ratios, and so the masses of the ions can be determined. Since most ions have a unit positive charge, m/e is equivalent to m. To be of value, however, the instrument must be capable of at least separating, *i.e.*, *resolving*, adjacent beams of m/e and (m + 1)/e. Mass spectrometers are now available which are capable of very high resolution, being able to differentiate between ions whose masses differ in the third decimal place.

If a positive ion is doubly-charged, it will behave, as far as collection in a beam is concerned, in the same way as a singly-charged ion of half the mass. There is also the problem of isotopes, *e.g.*, at low resolution, $^{12}C_2H_2$ and $^{12}C^{13}CH$ both have m/e of 26; but if these differ in the third decimal place, it is possible to differentiate between them by high resolution. Since *individual* ions are collected in the mass spectrometer, then the molecular ion will also give rise to a number of peaks, *e.g.*, bromine exists as ^{79}Br, and ^{81}Br.

Table 1

Element	*Mass*	%	*Element*	*Mass*	%
H	1	99.985	F	19	100
D	2	0.015	S	32	95.0
C	12	98.89	S	33	0.74
C	13	1.11	S	34	4.24
N	14	99.63	Cl	35	75.4
N	15	0.37	Cl	37	24.6
O	16	99.76	Br	79	50.6
O	17	0.04	Br	81	49.4
O	18	0.20	I	127	100

Thus, the molecular ion of methyl bromide will appear (low resolution) at peaks of m/e 94 ($^{92}CH_3Br$), 95 ($^{13}CH_3{}^{79}Br$), 96 ($^{12}CH_3{}^{81}Br$), and 97 ($^{13}CH_3{}^{81}Br$). The relative intensities of these M^+, $(M + 1)^+$, ... peaks depend on the abundance ratios of the isotopes in the molecule (Table 1). It can therefore be seen that mass spectrometry affords a means of determining accurate isotopic abundance ratios, molecular weights and consequently molecular formulae (these are obtained from tables).

The mass spectrum is a plot of ion beam intensities (ordinate) against mass/charge (m/e), and for any pure compound is characteristic of that com pound, *i.e.*, a pure compound is characterised by its cracking pattern (and its molecular ion, if it has one). It should be noted that by cracking pattern is meant not only the fragmentation pattern, but also that the relative abundance of the peaks are fixed ratios. Since mass spectra are dependent on the conditions of the experiment, these spectra are reproducible only if the conditions are identical. Hence it is best to use the same instrument and the same conditions for the purpose of comparisons.

The largest peak in a mass spectrum corresponds to the most abundant ion, and is known as the base peak. The base peak is used to report mass spectra in a standardised form; its intensity is arbitrarily given the value of 100, and all other peaks are reported as percentages of the base peak. This series of calculated values is the cracking pattern, and it may be described in the form of a line diagram (bar graph; Fig. 8), or may be reported in tabular form in which mass numbers and relative abundances are listed.

The interpretation of a mass spectrum is difficult, complications arising from various sources. The main source of difficulty is that the molecular ion may undergo rearrangement to give fragmentation patterns not anticipated from the structure of the compound, *e.g.*,

$$ABC + e \rightarrow [ABC]^+ + 2e$$

$$[ABC]^+ \rightarrow [AB]^+ + C.$$

or $AB. + C^+$

or $[BC]^+$ A.

or $[AC]^+$ + B., etc.

Metastable ions : If an ion (molecular or fragment), m_1^+ is accelerated before it breaks down, then, when it decomposes into m_2^+ and m_3, part of the kinetic energy of m_1^+ is lost to the neutral fragment m_3, and m_2^+ continues to be accelerated and is then collected.

$$m_1^+ \rightarrow m_2^+ + m_3$$

Ion m_2^+, *produced in this way*, is not recorded as mass m*, but as mass m*, where $m^* = m_2^2/m_1$. This ion, known as a *metastable ion*, is usually recorded as a weak broad peak, and is not (usually) an integral value. It is evaluated from the recorded masses m_1 and m_2, which arise from m_1^+ ions undergoing decomposition and acceleration in the normal way. The presence of metastable peaks is very useful for deducing fragmentation mechanisms, since they indicate the conversion of m_1^+ into m_2^+ in one step; *e.g.*, three ions, m/e 32, 31, and 30 were recorded. This suggests loss of a hydrogen atom one at a time. A broad peak (metastable peak), however, was also recorded at 28.1, and since 28.1 = 302/32, this means that ion m/e 32 was converted into ion m/e 30 in one step. It should be emphasised, however, that the absence of a metastable peak does not indicate that ion m/e 32 $\rightarrow$ ion m/e 30 does not occur, and it is also possible that ion m/e 31 $\rightarrow$ ion m/e 30 occurs.

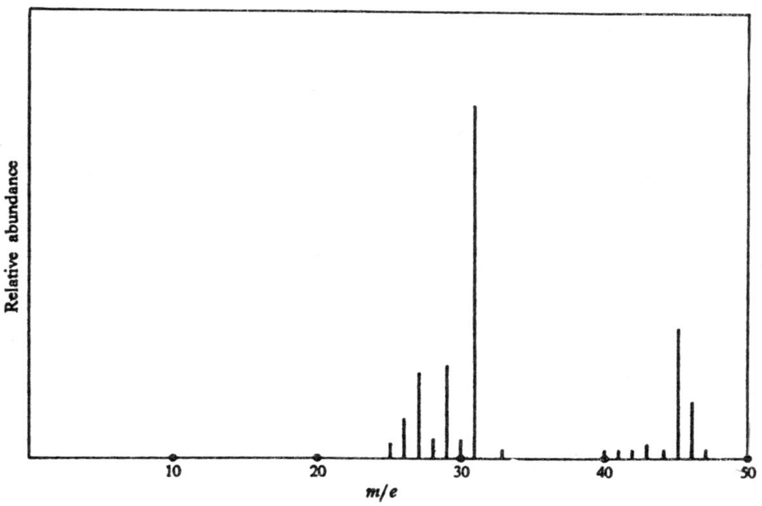

Fig. 8

Mass spectrometry may be used with gases, liquids, and solids, and only very small amounts of material are necessary (a few μg). It has proved extremely valuable for the determination of accurate molecular weights, obtaining molecular formulae, elucidation of structure, quantitative analysis of mixtures, ionisation potentials, and bond strengths.

Cracking patterns are largely dependent of the relative liabilities of bonds, relative stabilities of possible fragment ions and neutral molecules, etc. Many examples are discussed in the text (see Index: Mass spectrometry), but some general principles are mentioned here. When alternative fragmentations are possible, splitting usually occurs in all the alternative ways, but the direction leading to the most stable carbonium ion and/or free radical is the one that predominates.

Most molecules show a peak for the molecular ion, the stability of which is usually in the order: aromatics > conjugated acyclic polyenes > alicyclics > n-hydrocarbons > ketones > ethers > branched-chain hydrocarbons > alcohols. The ease of formation of a molecular ion depends on the type of electron removed; the order is usually n (lone-pair electron) > π (pi-electron) > σ (sigma-electron). However, since the bombardment electron energy is 50-70 eV, all types of electrons may be removed, the one most easily removed being that from the atom with the lowest ionisation potential. Rearrangements occur most readily when a hydrogen atom is involved in a six-membered cyclic transition state or when 1,2-shifts are involved.

As an example of some of the principles discussed, we shall describe the mass spectrum of ethanol. The line diagram (bar graph), and shows the most intense lines. Since there are many decomposition paths, only some of these will be considered. Single headed arrows are used to denote the transfer of one electron (*homolytic fission*), and when the position of the positive charge is known, a plus sign is placed above that atom. When, however, the position of the positive charge is uncertain, the ion is enclosed in square brackets with the symbol + as a superscript. Since the oxygen atom (with two lone pairs) in ethanol is the one with the lowest ionisation potential, one form of the molecular ion (M^+ m/e 46) will have the positive charge on the oxygen atom.

However, in order to propose various paths for the decompositions, it will be necessary to consider other forms of the molecular ion in which the position of the positive charge is uncertain (as we saw above, all types of electrons, n, π, and σ may be removed). It should be also noted that in some cases a path may be postulated which involves the transfer of two electrons (*heterolytic fission*), but this appears to be uncommon

$CH_3—CH_2—\ddot{O}H + e \rightarrow CH_3—CH_2—\overset{+}{O}H + 2e$ m/e 46 (M^+)

$CH_3—CH_2—\overset{+}{O}H \rightarrow CH_3. + CH_2 = \overset{+}{O}H$ m/e 31 (M-15)

$CH_3—\underset{\displaystyle H}{\underset{|}{CH}}—\overset{+}{O}H \rightarrow H. + CH_3—CH = \overset{+}{O}H$ m/e 45 (M-1)

$CH_3—CH_2—OH + e \rightarrow [CH_3—CH_2—OH]^+ + 2e$ (M^+)

$[CH_3—CH_2—OH]^+ \rightarrow HO. + CH_3—CH_2^+$ m/e 29 (M-17).

Although, it may often be possible to .predict the mass spectrum of a known compound, it is usually much more difficult, if not impossible in many cases, to elucidate the structure of a compound from its observed mass spectrum.

Hence, in general, other information is required, and this is usually spectroscopic data: ir, uv, and NMR. High resolution mass spectrometry permits elucidation of molecular formulae, and this knowledge then enables the double bond equivalents (D.B.E.), *i.e.*, the number of double bonds and/or rings, in the compound to be ascertained. If the general formula of the compound is $C_aH_bN_cO_d$, then

$$D.B.E. = a + 1 - (b - c)/2$$

e.g., (i) Benzene is C_6H_6.

D.B.E. = 6 + 1 –6/2 = 4 (3 double bonds; 1 ring).

(ii) Allylamine is C_3H_7N.

D.B.E. = 3 + 1 – (7–1)/2 = 1 (1 double bond).

Univalent elements, such as halogens, may be replaced by one hydrogen atom, and bivalent elements may be ignored (cf. oxygen, above).

The major peaks and peaks near the molecular ion are examined, and m/e values are listed in terms of M-m/e. By checking m/e and M-m/e values against a list of common fragments (Table 2), and knowing the fragmentation patterns of compounds containing various functional groups, it is possible to make a great deal of progress towards solving the structure of an unknown compound. It should be noted that it is not necessary to identify every peak.

Table 2 lists some of the more common fragments which appear as ions and/or lost as radicals or molecules (only the lowest isotopic values are given).

Table 2

m/e	*Fragment*	*m/e*	*Fragment*	*m/e*	*Fragment*
1	H	41	C_3H_5	71	C_5H_{11}
2	H_2	42	C_3H_6	72	$C_4H_8NH_2$
14	CH_2	43	C_3H_7,CH_3CO	73	$CO_2C_2H_5$
16	O,NH_2	44	CO_2,CH_2=CHOH, $C_2H_4NH_2$	74	CH_2=C(OH)OCH_3
17	OH			77	C_6H_5
18	H_2O,NH_4	45	C_2H_5O,CH_3CHOH, CO_2H	78	C_6H_6
19	F			79	C_6H_7,Br
20	HF	46	NO_2	80	HBr
26	C_2H_2,CN	51	C_4H_3	83	C_6H_{11}
27	C_2H_3, HCN	53	C_4H_5	85	C_6H_{13}
28	C_2H_4,CO,N_2	55	C_4H_7	88	CH_2=CH(OH)OC_2H_5
29	C_2H_5CHO	56	C_4H_8	91	C_7H_7
30	CH_2NH_2,NO	57	C_4H_9,C_2H_5CO	93	C_6H_5O
31	CH_2OH,CH_3O	58	$C_3H_6NH_2$, CH_2=C(OH)CH_3	94	C_6H_6O
32	CH_3OH	59	CO_2CH_3,C_2H_5CHOH, CH_2=C(OH)NH_2	97	C_7H_{13}
33	SH			99	C_7H_{15}
34	H_2S	60	CH_2 CO_2H_2	105	C_6H_5CO
35	Cl	65	C_5H_5	127	I
36	HCl	66	C_5H_6	128	HI
39	C_3H_3	69	C_5H_9		
40	CH_2CN	70	C_5H_{10}		

Electron spin resonance (ESR) is particularly useful for the detection and estimation of free radicals.

DIFFRACTION METHODS

(a) *X-ray diffraction* offers a means of determining the complete structure (molecular and spatial) of any crystalline compound. With the advent of computers, calculations from X-ray data can now be quickly performed, and so the use of this method for structural determination will become common. This method will therefore, in principle, render chemical methods superfluous. A particularly interesting example is the alkaloid thelepogine, $C_{20}H_{31}NO$; its structure has been determined entirely by X-ray analysis, no chemical work being carried out at all (Fridrichsons *et al.*, 1960). X-ray analysis may also be used for the determination of molecular weights.

(b) *Electron diffraction* offers a means of determining the complete structure of fairly simple molecules, its main use being the examination of gases or compounds in the vapour state.

Optical properties : Optical rotation (rotation at a single wavelength) and rotary dispersion (rotation at different wavelengths) offer a means of determining relative and absolute configurations, and are also very useful in the study of conformations.

Dipole moments : This method gives information on configuration and conformation of fairly simple molecules, and information on electronic displacements.

Synthesis of the compound : The term synthesis means the building up of a compound, step by step, from a simpler substance of known structure. The term complete synthesis means the building up of a compound, step by step, starting from its elements (and any others that may be necessary). In either case (synthesis or complete synthesis), *the structure of each intermediate compound is taken as proved by its synthesis from the compound that preceded it.*

The synthesis of a compound is necessary to establish its structure beyond doubt. There is always the possibility of one or more steps not proceeding 'according to plan'. Hence, the larger the number of syntheses of a compound by *different* routes, the more reliable will be the structure assigned to that compound. However, as has been pointed out above, physical methods—especially X-ray analysis—make synthesis *as a means a structure determination* less important than previously. Nevertheless, synthesis will still be an extremely important problem in the production of organic compounds, both natural and synthetic.

SOME ORGANIC COMPOUNDS

Lemery published his famous *Cows de Chymie*, in which he divided compounds from natural sources into three classes: *mineral, vegetable* and *animal*. Although organic substances such as sugar, starch, alcohol, resins, oils, indigo, etc., had been known from earliest times, very little progress in their chemistry was made until about the beginning of the eighteenth century. This classification was accepted very quickly, but it was Lavoisier who first showed, in 1784, that all compounds obtained from vegetable and animal sources always contained at least carbon and hydrogen, and frequently, nitrogen and phosphorus. Lavoisier, in spite of showing this close relationship between vegetable and animal products, still retained Lemery's classification. Lavoisier's analytical work, however, stimulated further research in this direction, and resulted in much-improved technique, due to which Lemery's classification had to be modified. Leuery had based his classification on the *origin* of the compound, but it was now found (undoubtedly due to the improved analytical methods) that in a number of cases the same compound could be obtained from both vegetable and animal sources.

Thus, no difference existed between these two classes of compounds, and hence it was no longer justifiable to consider them under separate headings. This led to the reclassification of substances into two groups; all those which could be obtained from vegetables or animals, *i.e.*, substances that were produced by the *living organism*, were classified as *organic;* and all those substances which were not prepared from the living organism were classified as *inorganic*.

At this stage of the investigation of organic compounds it appeared that there were definite differences between organic and inorganic compounds, *e.g.*, complexity of composition and the combustibility of the former. Berzelius (1815) thought that organic compounds were produced from their elements by laws different from those governing the formation of inorganic compounds. This then led him to believe that organic compounds were produced under the influence of a *vital force*, and that they could not be prepared artificially.

In 1828, Wöhler converted ammonium cyanate into urea, a substance hitherto obtained only from animal sources. This synthesis weakened the distinction between organic and inorganic compounds, and this distinction was completely ended with the synthesis of acetic acid from its elements by Kolbe in 1845, and the synthesis of methane by Berthelot in 1856. A common belief appears to be that Wohler's synthesis had little effect on the vital-force theory because it did not start with the elements. Wohler had prepared his ammonium cyanate from ammonia and cyanic acid, both of which were of animal origin. Partington (1960), however, has pointed out that Priestley (1781) had obtained ammonia by reduction of nitric acid, which was later synthesised from its elements by Cavendish (1785). Also, potassium cyanide was obtained by Scheele (1783) by passing nitrogen over a strongly heated mixture of potassium carbonate and carbon, and since one form of carbon used was graphite, this reaction was therefore carried out with inorganic materials. Since potassium cyanide is readily converted into potassium cyanate, Wöhler's synthesis is one which starts from the elements.

Since the supposed differences between the two classes of compounds have been disproved, the terms organic and inorganic would appear to be no longer necessary. Nevertheless, they have been retained, but it should be appreciated that they have lost their original meaning. The retention of the terms organic and inorganic may be ascribed to several reasons:

(i) All so-called organic compounds contain carbon;

(ii) The compounds of carbon are far more numerous (over 3000000) than the known compounds of *all* the other elements put together;

(iii) carbon has the power to combine with other carbon atoms to form long chains. This property, known as *catenation*, is not shown to such an extent by any other element.

Hence, organic chemistry is the chemistry of the carbon compounds.

This definition includes compounds such as carbon monoxide, carbon dioxide, carbonates, carbon disulphide, etc. Since these occur chiefly in the inorganic kingdom (*original meaning*), they are usually described in text-books of inorganic chemistry.

CATALYTIC HYDROGENATION

Many functional groups are reduced catalytically the most common catalysts being nickel, platinum, palladium, rhodium, and ruthenium; other catalysts used are copper chromite and copper-barium-chromium oxide. The catalytic activity of a given metal is dependent on its method of preparation, *e.g.*, nickel. It also depends on other factors;

(*a*) the presence of other compounds which may either increase (*promoters*) or decrease or inhibit (*poisons*) catalytic activity;

(*b*) the nature of the solvent: neutral, basic, or acidic.

In addition to having different reactivities, catalysts have different selectivities, *e.g.*, carboxylic acids can be hydrogenated to alcohols with a Ru---C catalyst but not with a Pt catalyst. In general, catalysts are deposited on the surface of a support such as charcoal, alumina, calcium carbonate, etc., and the activity of the catalyst may be changed by changing the support.

5. Primary, secondary and tertiary alcohols may be prepared by means of a Grignard reagent and the appropriate carbonyl compound.

6. A number of alcohols are obtained by fermentation processes.

7. In recent years synthetic methods have become very important for preparing various alcohols commercially:

(i) By the hydration of alkenes .

(ii) By heating a mixture of carbon monoxide and hydrogen under pressure in the presence of a catalyst, *e.g.*, zinc chromite plus small amounts of alkali metal or iron salts. A mixture of alcohols containing methyl, ethyl, n-propyl, isobutyl and higher-branched alcohols is obtained, the individuals being separated by factional distillation (see below).

(iii) The **Oxo process** (*Oxo synthesis, carbonylation* or *hydroformylation reaction*). A mixture of alkene, carbon monoxide and hydrogen, under pressure and elevated temperature, in the presence of a catalyst, forms aldehydes. A common catalyst is cobalt carbonyl hydride, $[CoH(CO)_4]$, and the product is a mixture of isomeric straight-chain and branched-chain aldehydes (the former predominating). These are reduced catalytically to the corresponding alcohols, *e.g.*, propene gives a mixture of n- and isobutanols:

$$2MeCH = CH_2 + 2CO + 2H_2 \longrightarrow Me(CH_2)_2CHO + Me_2CHCHO$$

$$\xrightarrow[Cu-Zn]{H_2} Me(CH_2)_2\ CH_2OH + Me_2CHCH_2OH$$

The aldehydes are first separated by fractional distillation.

(*iv*) Methanol, ethanol, propanols and butanols are prepared industrially by the oxidation of natural gas. Most of the methods given above can be used for the preparation of any particular class of alcohol: it is only a question of starting with the appropriate compound, *e.g.*, all three types may be prepared via Grignard reagents; *primary* alcohols from formaldehyde ($R^1 = R^2 = H$), *secondary* from aldehydes ($R^1 = R$; $R^2 = H$), and *tertiary* from ketones (R^1 and R^2 = alkyl groups).

$$R^1C(=O)R^2 + R^3MgX \longrightarrow R^1C(OMgX)(R^3)R^2 \xrightarrow{H^+} R^1R^2R^3COH$$

General Properties of the Alcohols

The alcohols are neutral substances: the lower members are liquids, and have a distinctive smell and a burning taste; the higher members are solids and are almost odourless.

In a group of isomeric alcohols, the primary alcohol has the highest boiling point and the tertiary the lowest, with the secondary having an intermediate value. The lower members are far less volatile than is to be expected from their molecular weight, and this is due to association through hydrogen bonding *extending over a chain of molecules,* thus, giving rise to a large molecule' the volatility of which would be expected to below:

$$\underset{\substack{|\\R}}{O}-H-\underset{\substack{|\\R}}{O}-H-\underset{\substack{|\\R}}{O}-H-\underset{\substack{|\\R}}{O}-H$$

The lower alcohols are very soluble in water, and the solubility diminishes as the molecular weight increases. Their solubility in water is to be expected, since the oxygen atom of the hydroxyl group in alcohols can form hydrogen bonds with the water molecules.

In the lower alcohols the hydroxyl group constitutes a large part of the molecule, whereas as the molecular weight of the alcohol increases the hydrocarbon character of the molecule increases, and hence the solubility in water decreases.

This, however; is not the complete story; the structure of the carbon chain also plays a part, *e.g.*, n-butanol is fairly soluble in water, but t-butanol is miscible with water in all proportions.

General Reactions of the Alcohols

1. Alcohols react with organic and inorganic acids to form *esters*:

$$R^1CO_2H + R_2OH \longrightarrow R^1CO_2R^2 + H_2O$$

Esters of the halogen acids are, as we have seen, the alkyl halides.

The order of reactivity of an alcohol with a given acid is primary alcohol > secondary > tertiary, provided the mechanism is bimolecular for all cases. With a given alcohol, the order of reactivity of the halogen acids in the bimolecular mechanism is HI > HBr > HCl.

Let us consider the following example:

$$EtOH + H^{+} + X \overset{fast}{\rightleftharpoons} X^{-} + EtOH_2^{+} \rightleftharpoons \overset{\delta-}{X}\cdots Et\cdots\overset{\delta+}{O}H_2 \xrightarrow{fast} X{-}Et + H_2O$$

The order of reactivity of the halogen acids is attributed to the variation in polarisability of the anion; the greater the size of the anion, the greater is its polarisability. What this amounts to is that the larger the anion, the more readily it can donate a lone pair to form a covalent bond. Also, the nature of the alkyl group will influence the mechanism of halide formation (in aqueous solution) in the same way as it does their hydrolysis.

Protonated t-alcohols readily eliminate a molecule of water to form a carbonium ion and so react by the S_N1 mechanism. Straight-chain primary alcohols will favour the S_N2 mechanism, but if branched-chain, then the S_N1 mechanism may operate. Secondary alcohols would be expected to react by both methanisms.

2. Alcohols react with phosphorus halides to form alkyl halides.

3. Alcohols combine with phenyl isocyanate to form phenyl-substituted urethans:

$$C_6H_5NCO + ROH \longrightarrow C_6H_5NHCO_2R$$

4. Strongly electropositive metals (K, Na, Mg, Al, Zn) liberate hydrogen from alcohols to form *alkoxides*, *e.g.*, sodium reacts with ethanol to form sodium ethoxide:

$$2C_2H_5OH + 2Na \longrightarrow 2C_2H_5O^{-}Na^{+} + H_2$$

Alkoxides are white deliquescent solids, readily soluble in water with hydrolysis:

$$RO^{-}Na^{+} + H_2O \rightleftharpoons ROH + NaOH$$

Alkoxides react with carbon disulphide to form xanthates:

$$RONa + S{=}C{=}S \longrightarrow RO{-}C({=}S){-}\bar{S}Na^{+}$$

The fact that alcohols liberate hydrogen by the action of metals shows that alcohols can behave as acids, but since they do not affect the pH of water, they are weaker acids than water. This accounts for alkoxides being hydrolysed by water and for the fact that the ethoxide ion (the cB of EtOH) is a stronger base than the hydroxide ion (the cB of H_2O).

Since alkyl groups have a +I effect, there will be an increased electron displacement towards the oxygen atom in going from primary to secondary to tertiary alcohol. This may be represented (qualitatively) as follows:

$$Me \rightarrow CH_2 \rightarrow \overset{\delta}{O} \rightarrow H \qquad \begin{matrix} Me \searrow \\ \\ Me \nearrow \end{matrix} CH_2 \twoheadrightarrow \overset{2\delta^-}{O} \twoheadrightarrow H \qquad Me \rightarrow \begin{matrix} Me \\ \searrow \\ \nearrow \\ Me \end{matrix} CH_2 \ggg \overset{3\delta^-}{O} \ggg H$$

The greater the negative charge on the oxygen atom, the closer is the covalent pair in the O—H bond driven to the hydrogen atom and consequently separation of a proton becomes increasingly difficult. Thus the acid strengths of alcohols will be in the order: prim > s > t. This is the order of reactivity of alcohols towards metals, and with t-alcohols the reaction with sodium is so slow that it is better to use the more electropositive potassium for these alcohols, *e.g.*, potassium t-butoxide is the usual salt prepared from t-butanol.

From what has been said above, it can be seen that the tendency for the C—O bond to break will be the reverse of that for the O—H bond, *i.e.*, reactions involving the breaking of the C—O bond will follow the order of reactivity: t > s > prim. This order would be expected from the consideration of the stabilities of the carbonium ions produced.

If we now consider β-substituted alcohols containing strong –I groups, the acidity of the alcohol is increased, *e.g.*, pK_a of ethanol of ethanol is ~ 18 whereas that of trifluoroethanol is 12.4.

$$Me \rightarrow CH_2 \rightarrow O - H \qquad CF_3 \lll CH_2 \lll O - H$$

The Me group is weakly + I, whereas the CF_3 group is strongly –I.

Thus the covalent pair is pulled away from the oxygen atom in the C—O bond in the latter compound, and consequently the covalent pair i the O—H bond is drawn towards the O atom, thereby facilitating release of the hydrogen as a proton.

A number of alkoxides are important as synthetic reagents; *e.g.*, sodium ethoxide, C_2H_5ONa; aluminium ethoxide, $(C_2H_5O)\,Al$; aluminium t-butoxide $[(CH_3)_3\,CO]_3\,Al$. The aluminium alkoxides may be conveniently prepared by the action of aluminium amalgam or aluminium shavings o the alcohol.

5. Primary and secondary alcohols may be acetylated with acetyl chloride, *e.g.*, ethanol gives ethyl acetate:

$$CH_3COCl + C_2H_5OH \longrightarrow CH_3CO_2C_2H_5 + HCl$$

With tertiary alcohols the reaction is often accompanied by dehydration of the alcohol to alkene or by the formation of a tertiary alkyl chloride; *e.g.*, t-butanol gives a good yield of t-butyl chloride:

$$(CH_3)_3\ COH + CH_3COCl \longrightarrow (CH_3)_3\ CCl + CH_3CO_2H$$

However, in the presence of a base such as dimethylaniline, the ester is produced in good yield.

$$CH_3COCl + (CH_3)_3COH \xrightarrow{PhNMe_2} CH_3CO_2C(CH_3)_3 + PhNMe_2\ HCl \quad (63\text{–}68\%)$$

A possible explanation for this is that t-butyl esters are very readily decomposed by hydrochloric acid as follows:

$$MeCOCl + Me_3COH \longrightarrow MeCOOCMe_3 + HCl \longrightarrow MeCO_2H + Me_3CCl$$

In the presence of a base, the HCl is removed as base· HCl, and consequently the second step cannot occur.

Acetylation can also be carried out with acetic anhydride in the presence of a catalyst, *e.g.*, pyridine. This method is successful for prim and s-alcohols, but fails with t-alcohols, *e.g.*, t-butanol is not acetylated. However, t-alcohols can be acetylated if *p*-toluenesulphonic acid is used as acetalyst.

6. Alcohols may be oxidised, and the products of oxidation depend on the class of the alcohol and on the nature of the oxidising agent.

Halogens, in *aqueous* solution, oxidise alcohols to carbonyl compounds. According to Swain *et al.* (1961), the mechanism involves hydride ion removal:

$$Br—Br\ \ H—CHMe—\ddot{O}—H \longrightarrow Br^- + HBr + MeCH{=}\overset{+}{O}H \xrightarrow{-H^+} MeCHO$$

In *alkaline* solution, the mechanism proposed is:

$$MeCH_2O—H\ \ \bar{O}H \rightleftharpoons H_2O + MeCH_2—O^-$$

$$MeCH_2—\bar{O} + Br_2 \longrightarrow MeCH{=}O + HBr + Br^-$$

In both cases, the carbonyl compound is then halogenated under the catalytic influence of acid or alkali.

On the other hand, alcohols are oxidised by hydrogen peroxide and ferrous sulphate to diols which, according to Coffman *et al.* (1960), are produced by dimerisation of free-radical intermediates, *e.g.*, t-butanol forms, 2,5-dimethylhexane-2,5-diol.

$$Fe^{2+} + H2O2 \longrightarrow Fe^{3+} + HO^{-} + HO\cdot$$

$$(CH_3)_2C(OH)CH_3 \xrightarrow[-H_2O]{\cdot OH} (CH_3)_2C(OH)CH_2\cdot \xrightarrow{2} (CH_3)_2C(OH)CH_2CH_2C(OH)(CH_3)_2$$

t-Butanol is particularly useful since all the alkyl hydrogens are equivalent and so only one free radical is produced.

7. Alcohols may be dehyrated to alkenes, the ease of dehydration being t > s > prim. Dehydration may be effected by heat alone (400–800°C), but in the presence of a catalyst, lower temperatures may be used, *e.g.*, when passed over heated alumina, t-alcohols are dehydrated at about 150°C, s-alcohols at 250°C and prim.-alcohols at 350°C.

Also, primary alcohols are dehydrated by concentrated sulphuric acid at about 170°C, and secondary and tertiary alcohols by boiling dilute sulphuric acid (this is used to avoid polymerisation of the alkene).

Isomerisation usually occurs with dehydration by acid or alumina, but in the latter case, isomerisation is suppressed by the addition of a small amount of pyridine.

The mechanism of dehydration with acid is described on p. 108 and, on the basis that a carbonium ion is formed as an intermediate, it can be seen why the ease of dehydration is t > s > prim.; this is the order of stabilities of the carbonium ions.

With secondary ad tertiary alcohols, dehydration may occur in two ways, *e.g.*,

$$CH_3CH_2CH(OH)CH_3 \xrightarrow[(1:1H_2SO_4)]{-H_2O} \begin{cases} CH_3CH_2CH{=}CH_2 \\ CH_3CH{=}CHCH_3 \end{cases}$$

Experiment shows that hydrogen attached to the adjacent carbon atom joined to the least number of hydrogen atoms is eliminated most easily. Thus, in the above reaction, the main product is but-2-ene (65--80 per cent). This elimination therefore occurs i accordance with Saytzeff's rule for the dehydrohalogenation of alkyl halides, and the reason is the same.

However, when thorium oxide or lanthanide metal oxides are used as the dehydrating agents, the yield of 1-alkene from 2-alcohols is generally above 98 per cent (Lundeen *et al.*, 1963). As mentioned above, rearrangement often occurs with acid-catalysed dehydration. All three types of alcohol may behave in this way via a carbonium ion that may undergo methyl and/or hydride ion 1,2-shift. In the following examples, only methyl shifts are shown (all the possible hydride ion shifts lead finally to the same products resulting from some of the competing methyl shifts). The major product is in accordance with Saytzeff's rule. Another point to note is that the rearrangements occur extremely rapidly. In fact, in the examples given, rearrangement is faster than proton elimination. Furthermore, Phelan *et al.* (1967) have shown that ease of anion migration is Ph >> Me > H.

When a double bond is produced in the product and is accompanied by a 1,2-shift, the reaction is said to be a *retropinacol rearrangement*. This type of rearrangement, when occurring in *open chain* compounds, is also sometimes called the *Wagner rearrangement*. Alcohols may also be converted into alkenes via their esters or xanthates, but in these cases no rearrangements occur, *e.g.*, $Me_3CCH(OAc)Me \longrightarrow Me_3CCH{=}CH_2$ (*cf.* the action of acid on the alcohol, above).

Dehydration may be used as the first step in the conversion of a primary alcohol of suitable structure into a secondary or tertiary alcohol, or a s-alcohol of suitable structure into a t-alcohol, *e.g.* (*N. B.* HI additions are in accordance with Markownikoff's rule):

$$(i)\ MeCH_2CH_2OH \xrightarrow[350°C]{Al_2O_3} MeCH{=}CH_2 \xrightarrow[\text{(ii) AgOH}]{\text{(i) HI}} MeCHOHMe$$

$$(ii)\ Me_2CHCH_2OH \xrightarrow[350°C]{Al_2O_3} Me_2C{=}CH_2 \xrightarrow[\text{(ii) AgOH}]{\text{(i) HI}} Me_3COH$$

$$(iii)\ Me_2CHCHOHMe \xrightarrow[350°C]{Al_2O_3} Me_2C{=}CHMe \xrightarrow[\text{(ii) AgOH}]{\text{(i) HI}} Me_3C(OH)CH_2Me$$

The 'reverse' procedure, *i.e.*, the conversion t → s → prim. alcohol can be carried out via hydroboration, *e.g.*,

$$Me_2C(OH)CH_2Me \xrightarrow{-H_2O} Me_2C{=}CHMe \xrightarrow{B_2H_6} (Me_2CHCHMe{-})_2BH \xrightarrow{heat} (Me_2CHCH_2CH_2{-})_2BH$$

$$(Me_2CHCHMe{-})_2BH \xrightarrow{H_2O_2;\ OH^-} Me_2CHCHOHMe$$

$$(Me_2CHCH_2CH_2{-})_2BH \xrightarrow{H_2O_2;\ OH^-} Me_2CHCH_2CH_2OH$$

It is also possible to step down the alcohol series by first dehydrating to the alkene, which is then subjected to ozonolysis, *e.g.*,

$$(a)\ RCH_2CH_2OH \xrightarrow[350°C]{Al_2O_3} RCH{=}CH_2 \xrightarrow[\text{(ii) } Zn/H^+]{\text{(i) } O_3} RCHO \xrightarrow[\text{cat.}]{H_2} RCH_2OH$$

$$(b)\ RCHOHCH_3 \xrightarrow[300°C]{ThO_2} RCH{=}CH_2 \dashrightarrow RCH_2OH$$

8. Alcohols combine with acetylene in the presence of mercury compounds as catalyst to form acetals:

$$2ROH + CH{\equiv}CH \xrightarrow{Hg^{2+}} CH_3CH(OR)_2$$

If, however, the reaction is carried out in the presence of potassium alkoxides at high temperature and under pressure, vinyl ethers are obtained.

4

Nature of Organic Reactions

INTRODUCTION

A chemical equation indicates the initial and final products of a reaction; rarely does it indicate how the reaction proceeds. Many reactions take place via intermediates which may or may not have been isolated. When the products of a reaction are formed by a single collision o the reactant molecules, *i.e.*, the reaction proceeds without any intermediates, the reaction is said to be a one-step (or *elementary*) reaction. Most reactions, however, are complex, *i.e.*, they occur via a number of reaction steps. Mechanisms are, in general, theories that have been devised to explain the facts which have been obtained experimentally. Mechanistic studies also include the nature of the *transition state* leading to intermediates (real or postulated) and to the products (see below). The interpretation of experimental data may not be clear-cut; several mechanisms may appear to fit equally well. In any case, the acceptability is strengthened when a particular mechanism can be used to make predictions which are borne out in practice. Many methods are used to elucidate mechanisms, some of the commoner ones being:

1. Kinetics. Kinetic studies are concerned with rates of reactions and provide the most general method for determining reaction mechanisms.
2. The identification of all the products of a reaction.
3. The detection, or better still (if it is possible), the isolation of intermediates. A special method of detection of intermediates is the trapping experiment.
4. The effect on reaction rates of changing the structure of the reactants.

5. The effect on reaction rates of changing the solvent.
6. Stereochemical evidence.
7. The use of isotopes. This method is particularly useful for tracing the part played by a particular atom in a reaction.
8. The use of crossed-experiments.

Applications of these methods are discussed in the text, but before ending this discussion, there are two other points of interest. One is the *principle of microscopic reversibility*. According to this principle, the mechanism of any reaction, under a given set of conditions, is identical in microscopic detail to that of the reverse reaction under the same conditions, except that it proceeds in the opposite way. With this principle it has been possible to deduce mechanisms where the forward or backward reactions do not lend themselves to kinetic studies.

The other point is that when the various possible products of a reaction are not interconvertible under the conditions of the reaction, then the product formed most rapidly will be the one that predominates in the products. Such reactions are called *kinetically* controlled reactions, and the most rapidly formed product is known as the kinetically controlled product. If, however, the possible products of the reaction are interconvertible under the reaction conditions, then the most stable product will predominate in the final products provided that enough time is given for equilibrium to be established. Reactions such as these are called *equilibrium controlled reactions*, and the most stable product is known as the thermodynamically *controlled* product. Kinetically controlled reactions are the mare common ones.

Now let us examine in more detail what happens when molecules containing *covalent* bonds undergo chemical reaction. Consider the reaction

$$Y + R - X \rightarrow Y - R + X,$$

where RX and RY are both covalent molecules. It can be seen that in this reaction, bond R—X has been broken and the new bond Y—R has been formed. The mechanism of the reaction depends on the way in which these bonds are broken. There are three possible ways in which this may occur, and the result of much work has shown that the actual way in which the break occurs depends on the nature of R, X and Y, and the experimental conditions.

(i) Each atom (forming the X—R bond) retains one electron of the shared pair, *i.e.*,

R–X → R.+X. or Y.+R–X → Y–R+X.

To make the representation of this type of bond breaking and making similar to those described below, one could use a half a TOW head:

R–X → R.+X. or Y R–X → Y–R+X.

This gives rise to *free radicals*, and the breaking of the bond in this manner is known as *homolytic* fission (*homolysis*). Free radicals are odd electron molecules, *e.g.*, methyl radical CH_3., triphenylmethyl radical $(C_6H_5)_3C$., etc. The majority are electrically neutral (some free radical ions are known). All possess addition properties, and are extremely reactive; when a free radical is stable, its stability is believed to be due to resonance. Free radicals are *paramagnetic*, *i.e.*, possess a small permanent magnetic moment, due to the presence of the odd (unpaired) electron. This property is used to detect the presence of free radicals. Diradicals are also known; these have an even number of electrons, but *two are unpaired* (see, *e.g.*, methylene, anthracene). In general, free-radical reactions are catalysed or initiated by compounds which generate free radicals on decomposition, or by heat or light. Furthermore, a reaction which proceeds by a free-radical mechanism can be inhibited by the presence of compounds that are known to combine with free radicals. Another important characteristic is that a free-radical mechanism leads to abnormal orientation in aromatic substitution.

(ii) Atom (or group) R retains the shared pair. This may be represented as:

$R–X \rightarrow R^- + X^+$ or $Y\ R–X \rightarrow Y–R + X^+$

This is known as *heterolytic* fission (*heterolysis*), and Y is said to be an *electrophilic* (electron-seeking) or *cationoid* reagent, since it gains a share in the two electrons retained by R. Also an electrophilic reagent is most likely to attack a molecule at the point of highest electron density. When an electrophilic reagent is involved in a *substitution* or a replacement reaction, that reaction is represented by S_E (S referring to substitution, and E to the electrophilic reagent). When $\overline{R}$: is a negative group in which the carbon atom carries the negative charge, *i.e.*, has an unshared pair of electrons, the group is known as a *carbanion*. There is a great deal of evidence to show that the configuration of the carbanion is tetrahedral; it is sp^3 hybridised, with the unshared pair of electrons occupying one orbital.

(iii) Atom (or group) R loses the shared pair, *i.e.*, the shared pair remains with X. This may be represented as:

$$R\text{–}X \rightarrow R^{+}+X^{-} \quad \text{or} \quad Y\ R\text{–}X \rightarrow Y\text{–}R+X^{-}$$

This also is heterolytic fission (heterolysis), and Y is said to be a *nucleophilic* (nucleus-seeking) or *anionoid* reagent, since it *supplies* the electron pair. A nucleophilic reagent is most likely to attack a molecule at the point of lowest electron density. When a nucleophilic reagent is involved in a substitution or a replacement reaction, that reaction is represented by S_N. When R^+ is a positive group in which the carbon atom carries the positive charge, *i.e.*, lacks a pair of electrons in its valency shell, the group is known as a carbonium ion, and is in a trigonal state of hybridisation (but see later). Such a carbonium ion is said to have a 'classical' structure. There are, however, various cases where the ion is better represented as a *bridged carbonium ion*, and in these cases the ions are said to have a 'non-classical' structure.

Because of their charge, carbonium ions and carbanions are very reactive. In many cases they are also very unstable, but they may be stabilised by delocalisation (*i.e.*, by spreading) of the charge by means of solvation. Alternatively,, the ion may be stabilised by delocalisation of the charge within the molecule by inductive and/or resonance effects. It should be noted that if there is spreading of charge, the carbon atom may, in fact, carry very little charge in the carbonium ion or carbanion.

Transition state theory of reactions : According to the *collision theory* of reactions, before molecules can enter into chemical reaction, they must collide and they must be *activated, i.e., they must attain a certain amount of energy (E)* above the average value. However, the rate of a reaction depends not only on the frequency of collisions in which the energy of activation is reached but also on whether the colliding molecules are suitably oriented with respect to each other for effective reaction to occur. This limitation is known as the probability or steric factor, and depends, for a given type of reaction, on the geometry of the reacting molecules. A simple example of the steric factor is that in the reaction

$$2HI \rightarrow H_2+L_2$$

If hydrogen iodide decomposes on collision, then activated molecules can collide in one of two ways, the 'right' way leading to decomposition, and the 'wrong' way leading to merely a 'change in partners'.

'Right way'

$$\begin{array}{ccc} H \cdots\!\!\rightarrow H & & H—H \\ | \qquad\quad | & \longrightarrow & + \\ I \cdots\!\!\rightarrow I & & I - I \end{array}$$

'Wrong way'

$$\begin{array}{ccc} H \cdots\!\!\rightarrow I & & H—I \\ | \qquad\quad | & \longrightarrow & + \\ I \cdots\!\!\rightarrow H & & I—H \end{array}$$

One can expect that when various paths are possible for a given reaction under given conditions, then the path actually followed will be the one requiring the lowest energy of activation. The problem therefore is to try to work out the path that requires the minimum energy of activation.

The *transition state* theory of reactions does not use the simple idea of collision, but considers how the potential energy of a system of atoms and/or molecules varies as the molecules are brought together. Consider the reaction

$$A + BC \rightarrow AB + C$$

This is known as a *three-centre* reaction. London (1929), by making certain approximations, showed that the minimum energy required in a three-centre reaction is when the reaction proceeds by an *end-on approach*, *i.e.*, in the above reaction, the approach of A to BC requiring the minimum activation energy is for A to approach BC along the bonding line of BC and on the Properties of molecules side remote from C:

$$A \;....\!\rightarrow ....\; B—C \rightarrow A—B + C$$

In this three-centre reaction, the value of the activation energy depends on four factors: (i) The strength of the B—C bond. The stronger this is, the greater will be E. (ii) The repulsion between A and BC. The greater this repulsion, the greater will be E. (iii) The repulsion between AB and C. The greater this repulsion, the greater will be E. (iv) The strength of the A—B bond. The greater the strength of this bond, the lower will be E.

Since most reactions are carried out in solution, another factor affecting the value of E is solvation of molecules and ions.

When we consider the mechanism of activation of this three-centre reaction, we can imagine that there are two extreme cases possible: (i) A is forced up against the repulsion of BC until it is close enough to compete with B on equal terms with C, which is finally expelled, (ii)

BC acquires so much energy that the bond B—C is broken, and then A and B combine without any opposition.

Polanyi *et al.* (1931-38) amplified London's ideas into the *transition state theory*. These authors showed by mathematical treatment that the lowest value of E is obtained when the reaction proceeds through a compromise between the two extremes (i) and (ii) mentioned above. A approaches BC along the bonding line of BC remote from C, and is forced against the repulsion of BC, and at the same time bond BC stretches until A and C can compete on equal terms for B. Thus a point is reached when the distances A—B and B—C are such that the forces between each pair are the same. This condition is the transition state (activated complex); in this state neither molecule AB nor BC exists independently. The system can now proceed in either direction to form A and BC or AB and C. This sequence of events may be represented by the following equation (T.S. = transition state):

$$\underset{\text{T.S.}}{A + BC \rightarrow A...B...C} \rightarrow AB + C$$

The above sequence of events may also "be represented graphically by means of an energy profile diagram (Figs. 1 and 2). This diagram is obtained by plotting the potential energy (P.E.) of the system (calculated by a semi-empirical method) against the reaction co-ordinate (the various distances between the nuclei of A, B and C).

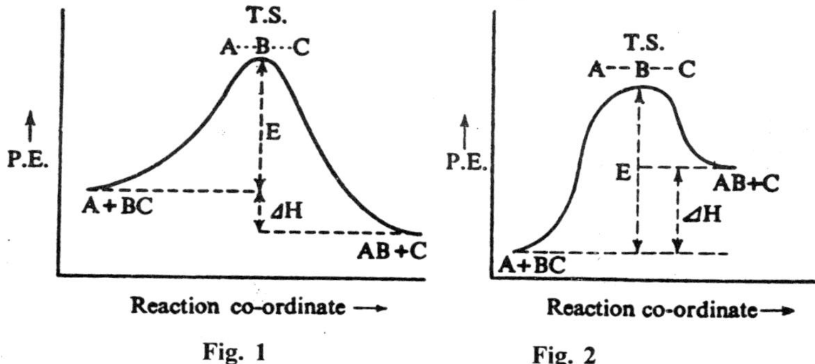

Fig. 1 Fig. 2

E is the activation energy, and AH is the heat of reaction at constant pressure. It is assumed that the reaction rate is given by the rate at which the reactant molecules pass through the transition state. For a given shaped 'hump', the lower it lies (*i.e.*, the lower the energy barrier is), the easier it is for the reactant molecules to enter the transition state. Also, the wider the hump for a given height, the easier it is for the reactant molecules to enter the transition state, since there is now a wider latitude in nuclei positions for the activated complex. Fig. 1 is the energy profile of an exothermic reaction, whereas Fig. 2 is that of an endothermic reaction.

The activated complex is not a true molecule; it contains partial bonds, and the energy content of the system is a maximum. Its life is extremely short, and hence it cannot be isolated; it is always decomposing into reactants or products. The reaction, however, if complex, will proceed through true intermediates which possess some measure of stability, and if this is great enough, the intermediates may be isolated. If the reaction proceeds through a true intermediate (I) (Fig. 3), there will be a minimum in the energy profile diagram. The greater the dip, the more stable will be the intermediate, and conversely, the shallower the dip, the less stable will be the intermediate.

In the extreme case, the dip may be so shallow that the intermediate is indistinguishable from the transition state. It should be noted here that each intermediate has its own transition state.

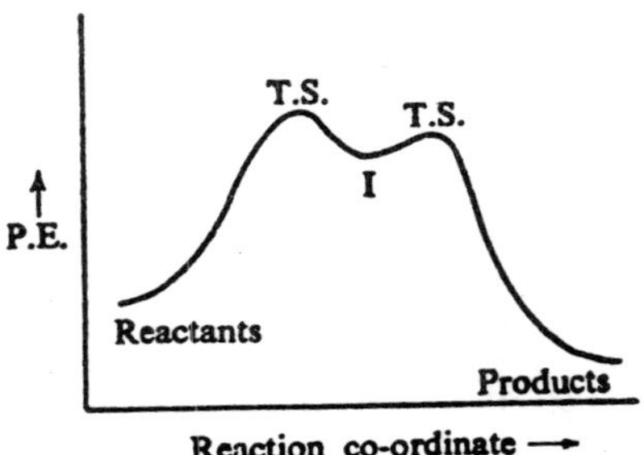

Fig. 3

Since E is the difference in energy content between the T.S. and the reactants, any factor which stabilises the T.S. (relative to the reactants), *i.e.*, lowers the energy content of the T.S., will therefore tend to lower E. Conversely, if any factor stabilises the reactants relative to the T.S., then E will tend to be raised. Resonance and steric effects may operate in these ways.

Structure of the transition state : From Fig. 1 and 2, it can be seen that the structure of the T.S. must be somewhere between that of the reactants and that of the products, and it may be closer to the former or to the latter. If the reaction is strongly exothermic, then generally the T.S. structure resembles that of the reactants, *i.e.*, existing bonds are almost intact and new bonds are very little formed. The reverse is generally true, *i.e.*, the T.S. structure resembles that of the products when the reaction is strongly endothermic. These generalisations also hold good when the reaction proceeds through inter-mediates (Fig. 3), but in this case the intermediate is the 'product' for the preceding T.S. and is the 'reactant' for the T.S. that follows. These generalisations are known as the Hammond principle (1955).

As pointed out above, as the reactant molecules approach, the P.E. of the system increases because of the increasing repulsion between them. This repulsion is overcome by the kinetic energy of the particles, and the higher the K.E. required the higher is E. However, for a given value of E, the larger the number of molecules possessing the necessary K.E. for reaction (*i.e.*, to reach E), the faster will be the reaction. If we consider a gaseous system, then it can be shown that at a given temperature (the system then has a fixed energy content), the molecules have different velocities, and the total K.E. (for 1 mole) is given by K.E. = N_A (1/2 mu^2), where N_A is the Avogadro constant, m the mass of the molecule and u is the root-mean-square (RMS) velocity. Maxwell (1860), using the theory of probability. calculated the distribution of molecular velocities, *i.e.*, the fraction of molecules (ΔN_A) having a particular velocity. Since K.E. is a function of velocity, the equation for the distribution of molecular velocities can be written in the form

$$\Delta N_A = Be^{-(K.E.)/RT}$$

where B is a proportionality constant. Because the relationship between ΔN_A and T is exponential, a small change in T has a large effect on ΔN_A. The results discussed may be represented graphically where $T_2 > T_1$. It can be seen that as T is raised, the maximum is flattened and the curve is shifted to the right. Since the maximum represents the fraction of molecules with average K.E., as T is raised, the fraction of molecules with a K.E. higher than average increases (*e.g.*, AC > AB).

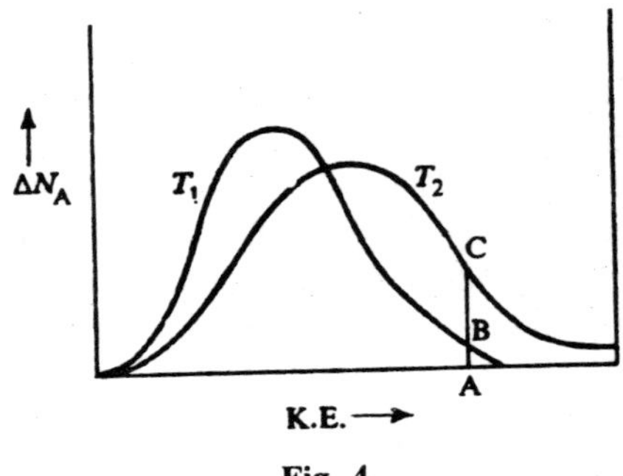

Fig. 4

Arrhenius (1889) investigated the effect of temperature on rates of reaction and derived the empirical relationship

$$k = Ae^{-E/RT} \text{ or } k = A \exp(-E/RT)$$

where A is the frequency factor. It can be seen that this equation has the same form as the K.E. distribution described above, and so k and fare related in the same way as are ΔN_A and K.E., *i.e.*, a small change in f will have a large effect on k, and as T is raised, k will increase rapidly. This follows from the fact that the repulsion between approaching reactant molecules is overcome by the K.E. of the molecules, and since raising T increases the number of molecules possessing the necessary K.E., the rate of the reaction increases. These arguments are based on the assumption that E is independent of T. Although this is not strictly true, it is satisfactory for limited variations of T (up to about 300-400°C).

The Arrhenius equation was later modified to

$$k = PZ \exp(-E/RT)$$

where Z is the total number of collisions per second of the reactant molecules (in unit volume) and P is the probability or steric factor.

Summary : Activation energy is the excess of energy over the average energy that the reactant molecules must acquire for reaction to occur. Values of E lie in the range 41.84-188.3 kJ mol^{-1} (10.45 kcal mol^{-1}) for most reactions. The existence of an energy barrier arises mainly from the fact that there is repulsion between reacting molecules and because bond breaking requires an input of energy. Repulsion is overcome by the kinetic energy of the approaching molecules, and the greater their velocity the more easily the molecules approach each other.

Also, there is an output of energy when new bonds are formed, and this is utilised in assisting the breaking of bonds, since both making and breaking of bonds occur simultaneously in the formation of the T.S. Because of this, E is usually less than the bond energy (of the bond being broken). In addition to these factors, there is a steric factor which may have to be satisfied for reaction to occur. In solution, the nature of the solvent affects rates (and mechanisms) of reaction.

Reaction kinetics : Since arguments based on reaction kinetics are used throughout the book, it is worth while to consider this problem briefly at this point. According to the law of mass action, the rate of a chemical reaction is proportional to the product of the active masses of the reacting substances, the molar concentration generally being taken as a measure of the active mass of a substance. The rate of a reaction is defined as the amount of reactant that is consumed in unit time or, alternatively, the amount of product formed in unit time. The equation relating reaction rate and molar concentrations (represented by square brackets) is called a rate law. The constant k is the rate constant or specific rate (this has a definite value at a given temperature), the exponent of a concentration factor is the reaction order for that substance, and the sum of the exponents of the concentrations is the order of the reaction. A rate law must be established experimentally.

Consider the reaction

$$A + B \rightarrow \text{products}$$

If it were found experimentally that the rate was proportional to the concentration of A and to the square of the concentration of B, the rate law would be written as

$$\text{rate} = k[A][B]^2$$

This reaction is first order with respect to A, second order with respect to B, the reaction itself being therefore a third-order reaction.

The units of k depend on the *order* of the reaction, *e.g.*, for first- and second-order reactions we have k (s^{-1}) and k (1 mol^{-1} s^{-1}). It can be seen from the above rate laws that k is numerically equal to the rate of the reaction when all the concentrations of the reactants are unity. It is thus not equal to the rate (except under the above conditions); it is a measure of the rate.

Cases also arise where the rate law does not contain the concentration of one of the reactants, *e.g.*, in the above reaction, it might be found that

$$\text{rate} = k[B]^2$$

In this case, the reaction is *zero order* with respect to A. When a reaction occurs via a number of steps, the rate of the overall reaction is determined by the slowest step, provided the others are relatively rapid. Thus the kinetics and order of such a reaction are basically those of the slowest step; this is called the *rate-determining* step. Hence, when the reaction [s zero order with respect to one reactant, it means that the mechanism consists of two (or more) steps, the step in which this reactant is involved being relatively fast (see also below).

When two (or more) steps of a complex reaction are fairly slow, the rate law will then likely be complicated.

The order of a reaction has already been defined above, but there is another term, the molecularity of a reaction, that is also used to define the number of reactant molecules participating in the rate-determining step. Order and molecularity may or may not be the same.

A rate law may also be derived mathematically on the basis of a *postulated mechanism*. By comparing the experimentally observed rate law with various derived rate laws, it may be possible to formulate a reasonably acceptable mechanism. Additional evidence to support this mechanism may then be sought by the application of other methods.

Now let us consider the *reversible* reaction

$$A+B \underset{k_{-1}}{\overset{k_1}{\rightleftharpoons}} C+D$$

and let us suppose that forward and reverse reactions are all first order with respect to each reactant. Then, at equilibrium, *i.e.*, when the forward rate is equal to the reverse rate,

$$k_1[A][B] = k_{-1}[C][D]$$

$$k_1/k_{-1} = [C][D]/[A][B] = K$$

The *equilibrium constant*, K, can thus be determined experimentally from this equation.

When a reaction is complex, the step with the highest activation energy will be rate-determining (r/d), *i.e.*, the step leading to the highest point in the energy profile. The concentrations of reactants up to this point are involved in the rate law, but the rate law is independent of concentrations after this point. Let us consider the two-step reaction with reactants A, B, and C to form product D and involving an inter- mediate, I. The most common reaction of this type is:

(i) $$A+B \underset{k_{-1}}{\overset{k_1}{\rightleftharpoons}} I$$

(ii) $$I+C \xrightarrow{k_2} D$$

Equilibrium involving the formation of I is reached slowly : Since the first step is r/d, the overall rate of reaction is given by the rate at which A and B are consumed, *i.e.*,

$$\text{rate} = k_1[A][B] - k_{-1}[I]$$

To solve this equation, it is necessary to know [I]. This cannot be measured, but since the rate of reaction of I to form the products is very much faster than the reversible formation of A and B (*i.e.*, $k_2 >> k_1$ and k_{-1}), then I is used up too quickly for the reversible reaction to occur to any appreciable extent (*i.e.*, [I] = 0). In these circumstances, the rate law will be

$$\text{rate} = k_1[A][B]$$

Thus, the rate law is independent of the concentration of C when the first step is rate-determining.

Equilibrium involving the formation of I is reached rapidly : Since it is now the second step which is r/d, the overall rate of reaction is given by

$$\text{rate} = k_2[I][C]$$

Also, because the equilibrium is reached rapidly (*i.e.*, k_1 and $k_{-1} >> k_{-2}$), [I] is given by

$$K = [I]/([A][B]);\ i.e.,\ [I] = K[A][B]$$

Hence, the rate of the reaction is

$$\text{rate} = k_2K[A][B][C] = k[A][B][C]$$

Thus, the rate law involves the concentrations of all three reactants when the second step is r/d. It should also be noted that the rate constant k is not k_2 (it is the product of k_2 and K).

Summary, (i) If a reactant does not appear in the rate law, it is involved in reaction after the r/d step.

(ii) If all reactants appear in the rate law, then (a) all reactants are involved in the r/d step (a termolecular reaction), or (b) some are involved in a rapid equilibrium step which precedes the r/d step. Unless other information is available, it is not possible to distinguish between (a) and (b). (iii) Any proposed mechanism must give the observed rate law.

The energy profiles of the two reactions discussed above will be of the types shown in Figs. 4.5(a) and (b), respectively.

In both profiles, the activation energies for the separate steps are E_1 and E_2, but the activation for the *overall* reaction is E.

This is the extra energy over the average energy of the reactants which must be acquired before the reactants can surmount the highest energy barrier that separates them from the products.

It should also be noted that in Fig. 5(a), $E_3 > E_2$, whereas in Fig. 4.5(b), $E_3 < E_2$ (where E_3 is the activation energy of the reverse reaction: I $\rightarrow$ A + B).

Since k = PZ exp (–E/RT), it follows that the rate of the reaction: A + B $\rightarrow$ C is given by

$$\text{rate} = k\,[A]\,[B] = PZ\,\exp(-E/RT)\,[A]\,[B]$$

Thus, the rate of a reaction is dependent on the concentrations of the reactants, P, Z, T, and E. As pointed out above, for a given value of E, small changes in T cause large changes in the number of molecules reaching E.

On the other hand, P and Z are almost independent of T. Consequently, for a given reaction which can proceed by only one mechanistic path, the rate is predominantly affected by the exp(–E/RT) factor.

When a reaction can proceed by different mechanisms, the path followed (exclusively or predominantly) will be the one with the lowest activation energy. However, since rate is dependent on concentration, changes in concentration of a reactant may speed up the slow mechanistic path to such an extent that this is now the path through which the reaction proceeds.

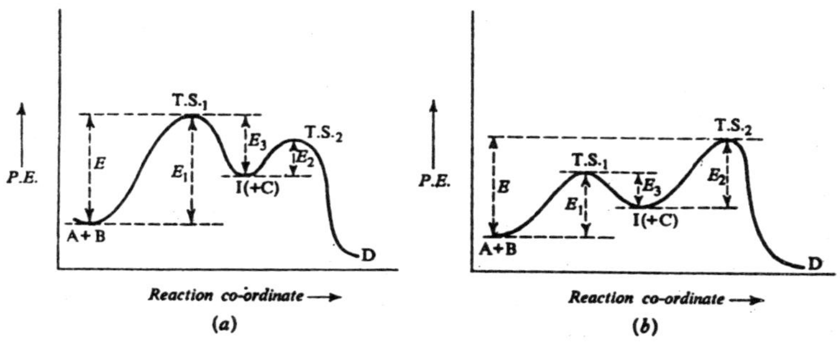

Fig. 5

In addition to depending largely on T, E and concentration of reactants, the rate of a chemical reaction also depends on solvent and on pressure. Pressure effects on reactions in solution, however, are appreciable only at very high pressures (several thousand atmospheres). The effect of pressure on rate arises chiefly from volume changes that may occur when reactants form the transition state, and hence this relationship can often provide a great deal of information on the nature of the T.S.

Thermodynamic properties : Thermodynamics is the study of energy changes and equilibrium positions involved in physical and chemical transformations. The method is independent of any theory of the structure of matter and is not concerned with the time taken for equilibrium to be reached. Thus, chemical thermodynamics is mainly concerned with the position of equilibrium in chemical reactions and is independent of mechanism, whereas chemical kinetics is concerned with rates and mechanisms of reactions.

The study of thermodynamics involves the use of various functions (properties), *e.g.*, P, V, T, U, H, G, S. These are variables whose values depend only on the state of the system and are independent of the path by which this is reached. Functions such as these are known as state functions.

Thermodynamic properties are of two types: (i) extensive properties; these depend on quantity, *e.g.*, mass, V, U, H, G, S. (ii) Intensive properties; these are independent of quantity, *e.g.*, P, T.

Let us now consider some thermodynamic applications. If U is the internal energy of a system at pressure P (atmospheres) and volume V (litres), then H, the enthalpy (heat content) of the system, is defined by the equation

$$H = U + PV$$

When a reaction is carried out at constant temperature, the enthalpy change, ΔH, is the difference between the enthalpy of the final and initial states. If heat is evolved, AH is negative. In this case the final state of the system has a smaller enthalpy than the initial state, and such a reaction is said to be exothermic. If, however, heat is absorbed, AH is positive, and the reaction is said to be endothermic and the final state has a larger enthalpy than the initial state. ΔH and ΔU are practically identical for reactions involving solids and liquids (ΔV is very small).

In order to evaluate enthalpies of compounds, it is necessary to know the absolute enthalpies of the elements present in these compounds. Since, in practice, it is not possible to measure absolute enthalpies, the difficulty is overcome by dealing with the changes in energy which accompany chemical reactions. This treatment requires an arbitrary choice of zero enthalpy, and is carried out as follows. The standard state of an element or compound is defined as its most stable physical form at a pressure of one atmosphere and at a specified temperature. Then, by convention, each element in its standard state is given an enthalpy of zero.

The enthalpy which accompanies the formation of one mole of a compound in its standard state from its elements in their standard states is called the *standard enthalpy of formation* at temperature T K. T K is usually 298 K (*i.e.*, 25°C), and if the temperature is not specified, then it is assumed to be 298 K. Properties of elements and compounds in their standard states are indicated by the superscript symbol e, *e.g.*,

$$\begin{array}{llll} C(s) & + \; O_2(g,\ 1\ \text{atm}) & \rightarrow & CO_2(g,\ 1\ \text{atm}) \\ O & \quad O & & \qquad -393\text{-}3\ \text{kJ} \end{array}$$

Thus, the standard enthalpy of formation, $\Delta H^{\ominus}{}_{f}$, is –393-3 kJ at 25°C. (N.B. s, g, *l* stand for solid, gas and liquid, respectively.) It follows from what has been said above that compounds for which $\Delta H^{\ominus}{}_{f}$ is negative are thermally stable (with respect to their elements); and vice versa. Although this is generally true, there are exceptions, *i.e.*, some endothermic compounds are thermally stable with respect to their elements.

In reactions that proceed in steps, the enthalpy of the overall reaction is given by the sum of the enthalpies of each step (Hess's law), *e.g.*,

$$A \xrightarrow{\Delta H_1} B \xrightarrow{\Delta H_2} C \xrightarrow{\Delta H_3} D;$$

$$\Delta H = \Delta H_1 + \Delta H_2 + \Delta H_3$$

By means of Hess's law it is possible to calculate one AH if all except .this one are known.

Another property used in thermodynamics is *entropy*, S (JK^{-1} mol^{-1}; cal K^{-1} mol^{-1}). The physical interpretation of entropy is not easy, but one way is to associate entropy with the disorder (or randomness) of the system. In a crystal, the molecules (or ions) have fixed positions, and the entropy of the system (at 25°C) is small. When the solid is liquefied, the molecules can now move everywhere. The entropy is therefore greater than in the solid state, and is very much greater when the liquid is vaporised. This idea of disorder can also be applied to molecules, *e.g.*, n-hexane, a straight-chain compound, can take up many conformations (shapes), whereas cyclohexane, a ring compound, can take up much fewer conformations.

Thus, the entropy of n-hexane (388-3 JK^{-1} mol^{-1}) is greater than that of cyclohexane (289-3 JK^{-1} mol^{-1}). Hence, in general terms, a system changing from order to disorder is accompanied by an increase in entropy (ΔS is positive); and *vice versa*. A particularly important example of this general statement is that in chemical reactions in which the number of product molecules is greater than the number of reactant molecules, there will be an increase in entropy (the number of degrees of freedom has increased); and vice

The Gibbsfree energy, G, is defined by the equation

$$G = H - TS$$

Free energy is the energy available from the system at constant temperature and pressure for useful work, whereas the product TS gives the energy unavailable for useful work. The *change* in free energy, ΔG, of a system is given by

$$\Delta G = \Delta H - T\Delta S$$

When a system at constant pressure and temperature is in equilibrium, its free energy is a *minimum* (at equilibrium, ΔG, *i.e.*, the change in free energy, is zero). Thus, a system not in equilibrium will tend to change irreversibly, thereby lowering its free energy so as to reach equilibrium.

Entropies have been assigned absolute values, and free energies standard values:

$$AS^{\ominus} = \Sigma S^{\ominus} \text{ (products)} - \Sigma S^{\ominus} \text{ (reactants)}$$

$$AG^{\ominus} = \Sigma G^{\ominus} \text{ (products)} - \Sigma G^{\ominus} \text{ (reactants)}.$$

Here, the entropy of a perfect crystal of any element or compound is zero at 0 K, and the free energy of any element in its standard state is zero. The standard state for free energy is a *solution at molar concentration or a gas at one atmosphere pressure*, at a specified temperature.

H, S, and, G are not temperature-independent, but their variation is very small over a limited range of temperature (cf. activation energy). Hence, at different temperatures, the main change in $\Delta G^{\ominus}$ arises from the term $T\Delta S^{\ominus}$, *i.e.*, to a first approximation:

$$AG^{\ominus}_{T} \cong \Delta H^{\ominus}_{298} - T\Delta S^{\ominus}_{298}$$

If $\Delta G^{\ominus}$ is the standard free energy change of the reaction (at temperature T) whose equilibrium constant is K (at temperature T), it can be shown that (In = loge; $R = 8.314\ JK^{-1}\ mol^{-1} = 1.987 cal\ K^{-1}\ mol^{-1}$).

$$\Delta G^{\ominus} = -RT \text{ In } K = -2\text{-}3 \text{ RT log } K$$

Thus, if $\Delta G^{\ominus}$ is negative, K is greater than unity, and the more negative $\Delta G^{\ominus}$ is, the greater K is. Here the reaction is thermodynamically favourable. If $\Delta G^{\ominus}$ is positive, K is less than unity, and the reaction is thermodynamically unfavourable.

Although, a chemical reaction can proceed spontaneously (*i.e.*, without the assistance of any external agency) only in the direction of loss of free energy (*i.e.*, $\Delta G^{\ominus}$ is negative), it does not necessarily mean that it will proceed at a measurable rate. The presence of catalyst does not increase the yield, but increases the rate to the equilibrium position of a reaction which is thermodynamically favourable. As we have seen above, rates of reaction depend on the activation energy, E, and the frequency factor, A. Since catalysts provide alternative routes in which they generally participate to form intermediates, this means that either E is less and/or A is greater for the catalysed than for the uncatalysed reaction. It appears that, in general, both E and A vary. However, for closely related reactions, A will very likely be little changed, the changes in rates then being due to changes in E.

Now let us consider the equation

$$\Delta G^{\ominus} = \Delta H^{\ominus} - T\Delta S^{\ominus}$$

When $\Delta H^{\ominus}$ is negative and $\Delta S^{\ominus}$ is positive $\Delta G^{\ominus}$ is negative and the reaction will be thermodynamically favourable. If, however, both $\Delta H^{\ominus}$ and $\Delta S^{\ominus}$ are negative, the term $T\Delta S^{\ominus}$ may be versa, larger or smaller than $\Delta H^{\ominus}$ and so $\Delta G^{\ominus}$ will be positive or negative, respectively. In the former case, the reaction will not be thermodynamically favourable. In the same way, if $\Delta H^{\ominus}$ is positive, then the sign of $\Delta G^{\ominus}$ will depend on the sign and magnitude of $\Delta S^{\ominus}$. For very closely similar reactions, if it be *assumed* that $\Delta S^{\ominus}$ remains sensibly constant, then

$$\Delta G^{\ominus} = \Delta H^{\ominus} + \text{constant}$$

In these circumstances, $\Delta G^{\ominus}$ is a function of enthalpy. We may therefore write

$$\Delta H^{\ominus} = \Delta G^{\ominus} = -RT \text{ In } K$$

Thus $\Delta H^{\ominus}$ may be used (but with caution) as a criterion for deciding whether a reaction is thermodynamically favourable. Usually, reasonable estimates of $\Delta H^{\ominus}$ may be obtained from a knowledge of bond energies in the reactants and products:

$$\Delta H^{\ominus} = \underset{\text{(negative)}}{\Sigma H^{\ominus} \text{ (bonds formed)}} - \underset{\text{(positive)}}{\Sigma H^{\ominus} \text{ (bonds broken)}}$$

This arbitrary approach of equating $\Delta G^{\ominus}$ with $\Delta H^{\ominus}$ is made necessary by the fact that the relevant values of $\Delta S^{\ominus}$ are very often unavailable, It is also difficult to arrive at a reasonable estimate for these values. One way out of the difficulty has been to make assumptions about entropy changes, *e.g.*, as was done above for closely similar reactions. This may be a very bad estimate of the situation under consideration, and if so, it usually manifests itself by giving an answer in disagreement with experiment. On the other hand, where we get a good answer, we may be too hasty in accepting that our assumptions were sound.

What has been said above applies to estimating (qualitatively) $\Delta G^{\ominus}$ in order to arrive at some idea about the position of equilibrium of various reactions. For rates of reaction, we need to know the relevant energies of activation. E may, in many cases, be determined experimentally from the equation:

$$2.3 \log k = 2.3 \log A - E/RT$$

If k is measured at several temperatures, the graph of log k against l/T is usually a 'good' straight line whose slope is –E/2.3R and intercept is 2.3 log A. Furthermore, since the slope is either zero or negative, the value of E is *zero or positive*.

Now let us consider the following reaction in terms of transition-state theory, where AB represents the transition state:

$$A + B \rightleftharpoons AB^{\ddagger} \rightleftharpoons \text{products}$$

Treating the T.S. as a molecular species, we have:

$$K^{\ddagger} = [AB^{\ddagger}]/[A][B]$$

and $\Delta G^{\ominus\ddagger} = \Delta H^{\ominus\ddagger} - \Delta S^{\ominus\ddagger} = RT \ In \ K^{\ddagger}$

The quantities , $\Delta G^{\ominus\ddagger}$, $\Delta H^{\ominus\ddagger}$ and $\Delta S^{\ominus\ddagger}$ are called the *standard free energy of activation, heat* (or enthalpy) and entropy of activation. Now, it can be shown from transition-state theory that

$$k_{T.S.} = \frac{kT}{h} K^{\ddagger} = \frac{kT}{h} \exp (\Delta G^{\ominus\ddagger}/RT)$$

$$= \frac{kT}{h} \exp (\Delta S^{\ominus\ddagger}/R) \ . \ \exp (-\Delta G^{\ominus\ddagger}/RT)$$

where $k_{T.S.}$ is the rate constant for the formation of the T.S., k is Boltzmann's constant (= R/N_A), and h is Planck's constant. Also, since

$$k = A \exp(-E/RT)$$

we may therefore equate $\Delta S^{\ominus\ddagger}$ with A and $-\Delta H^{\ominus\ddagger}$ with –E. It has been found that $\Delta S^{\ominus\ddagger}$ is very close to the value of E for liquid and solid systems, and $\Delta S^{\ominus\ddagger}$ can be calculated from the experimental rate constant and activation energy.

If the T.S. is formed very early, its structure will closely resemble that of the reactants and so $\Delta S^{\ominus\ddagger} \cong 0$. On the other hand, if the T.S. is formed fairly late, since it is more organised than the reactants, $\Delta S^{\ominus\ddagger}$ would be expected to be negative, and the more organised the T.S., the more negative will be $\Delta S^{\ominus\ddagger}$, Hence, an evaluation of $\Delta S^{\ominus\ddagger}$ will give information on the structure of the T.S.

Because the rate constant involves entropy of activation in T.S. theory, the rate constant can be interpreted in terms of entropy. Furthermore, since the entropy factor may outweigh the enthalpy factor, some authors prefer to use the free-energy profile rather than the P.E. profile. In this case, $\Delta G^{\ominus\ddagger}$ replaces the activation energy and $\Delta G^{\ominus}$ replaces $\Delta H^{\ominus}$.

One other problem will be discussed here, viz., the effect of resonance on chemical equilibria. It has been assumed that for very closely similar reactions, ΔH is a measure of ΔG, and also that, in *solution*, ΔH and ΔU are practically identical. We may therefore conclude that, under these conditions, ΔU is a measure of ΔG, *i.e.*,

$$\Delta U = AG^{\ominus} = -RT \text{ In } K$$

Since $$AU = \Sigma U\ [(\text{products}) - \Sigma Y\ (\text{reactants})$$

the lower the value of ΣU (products), the more negative will be ΔU. Thus, $\Delta G^{\ominus}$ will be more negative, and consequently the greater will be K. As we have seen, a resonance hybrid has lower internal energy than any of its contributing resonating structures. Thus, if the products of a reaction have greater resonance stabilisation than the reactants, the position of equilibrium will be displaced towards the formation of the products; and vice versa. In other words, resonance can be a driving force or an inhibiting force in chemical reactions. This argument has considered resonance only. However, reactions in solution involve solvated molecules and ions, and solvation energies play an important part in equilibria positions.

Use of isotopes in organic chemistry : In recent years the use of isotopes has been extremely helpful in the study of reaction mechanisms and rearrangements, in the elucidation of structures, and also in quantitative analysis. The application of isotopes in biochemistry has also been particularly fruitful, since they offer a means of identifying intermediates and the 'brickwork' of the final products. The common isotopes that have been used in organic chemistry are: deuterium (^{2}H, D; stable), tritium (^{3}H T; radioactive), ^{13}C (stable), ^{14}C (radioactive), ^{15}N (stable), ^{18}O (stable), ^{32}P (radioactive), ^{35}S (radioactive), ^{37}Cl (stable), ^{82}Br (radioactive), ^{131}(radioactive).

Various methods of analysis are used. Radioactive isotopes are usually analysed with the Geiger-Muller counter, and the stable isotopes by means of the mass spectrograph. Deuterium is often determined by means of infra-red spectroscopy; and there are also the older methods for deuterium and ^{18}O of density, or refractive index measurements (of the water produced after combustion of the compound). Nuclear magnetic resonance is now also used, being applicable only to those isotopes having nuclear magnetic moments.

Isotopes are usually used as tracers, *i.e.*, the starting material is labelled at some particular position, and after reaction the labelled atom is then located in the product. This does not mean that labelled compounds contain 100 per cent of the isotope, but that they usually contain an abnormal amount of the isotope. Many examples of the use of isotopic indicators will be found in the text (see Index, Isotopic indicators).

The use of isotopes, stable or radioactive, is based on the fact that the chemical behaviour of any particular isotope is the same as that of the other atoms isotopic with it (the chemical properties of an element depend on the nuclear positive charge and the number of electrons surrounding the nucleus, and not on the number of neutrons in the nucleus). This identity in chemical behaviour is essentially true for the heavy atoms, but in the case of the lightest elements, reactions involving heavier isotopes are slower, but so long as identical paths are followed, the final result is unaffected.

This difference in rates of reaction of isotopes is known as the kinetic isotope effect, and the magnitude of such effects depends on the weight ratio of the isotopes involved. Thus the kinetic isotope effect is greatest with H and D (and T). The kinetic effect has been widely used to study reaction mechanisms, since this difference in rate is significant when the bond attaching the isotopic atom is stretched in the activated state, *i.e.*, if substitution by the isotope changes the rate, then breaking of that bond is involved in the rate-determining step.

It was generally assumed at first that differences in rate (and equilibrium) constants would arise only in reactions involving making and breaking of bonds with respect to isotopic atoms. It has now been found, however, that the presence of an isotope may affect rate or equilibrium constants even though *'isotopic bonds'* are neither broken nor formed during the reaction. This phenomenon is known as the secondary isotope effect, and is much weaker than the kinetic isotope effect.

Ion accelerators and nuclear reactors produce, as by-products, artificial isotopes, particularly those which are radioactive. These isotopes are then supplied in the form of some compound from which a labelled compound can be synthesised, *e.g.*, ^{14}C is usually supplied as $Ba^{14}CO_3$, ^{15}N as $^{15}NH_4Cl$, etc. A simple example of the synthesis of a labelled compound is that of acetic acid (in the following equations, which involve a Grignaid reagent, $\overset{*}{C}$ is ^{14}C; this is a common method of representing a tracer atom, provided its nature has been specified).

$$Ba\overset{*}{C}O_3 + 2HCl \longrightarrow BaCl_2 + H_2O + \overset{*}{C}O_2$$

(i) $\overset{*}{C}O_2 + CH_3MgI \longrightarrow CH_3\overset{*}{C}O_2H$

(ii) $\overset{*}{C}O_2 + 3H_2 \xrightarrow{\text{Catalyst}} H_2O + \overset{*}{C}H_3OH \xrightarrow{I_2/P} \overset{*}{C}H_3I \xrightarrow{Mg} \overset{*}{C}H_3MgI$

$\overset{*}{C}H_3MgI \xrightarrow{CO_2} \overset{*}{C}H_3CO_2H \qquad \overset{*}{C}H_3MgI \xrightarrow{\overset{*}{C}O_2} \overset{*}{C}H_3\overset{*}{C}O_2H$

Fig. 6

In a number of cases, an exchange reaction is a very simple means of preparing a labelled compound, *e.g.*, dissolving a carboxylic acid in water enriched with deuterium.

$$RCO_2H + D_2O \rightleftharpoons RCO_2D + DHO$$

Because of the possibility of this exchange reaction, it is often necessary to carry out control experiments.

Isotopes are also very useful in the analysis of mixtures, particularly for the determination of the yield of products in a chemical reaction when isolation is difficult. A simple method is that of isotopic dilution. The labelled compound is prepared, and a known amount is then added to the mixture to be analysed. A portion of the substance is now taken and analysed for its isotopic content. From a knowledge of the isotopic content of the labelled compound added and recovered, and the weight of the labelled compound added, it is thus possible to calculate the weight of the labelled compound in the mixture. This method can only be used as long as there is no isotopic exchange during the isolation.

Nomenclature of labelled compounds : Several methods have been used, *e.g.*, the positions of isotopes are indicated by arabic numerals placed within square brackets:

[l—^{2}H] ethanol, C_3CH^2HOH; [l—^{14}C,2—^{13}C]acetaldehyde, $^{13}CH_3\,^{14}CHO$.

Acids and bases. The study of the acidity and basicity of organic compounds offers a very good means of examining the theories relating structure with reactivity.

There are various definitions of acids and bases. According to Arrhenius (1884), acids are compounds which ionise in aqueous solution to produce hydrogen ions, and bases ionise to produce hydroxide ions. According to the Brϕnsted-Lowry definition (1923), an *acid is a proton donor, i.e., is protogenic, and a base is a proton acceptor, i.e., is protophilic.*

Thus:

$$B + H^+ \rightleftharpoons BH^+$$
base proton acid

Proton donation and acceptance are reversible, and the proton transfer, which is extremely fast, is known as *protonation*. The acid and base in the above equation are said to be *conjugate* with respect to each other; B is the conjugate base (cB) of the acid BH^+ and BH^+ is the conjugate acid (cA) of the base B. Thus every acid must have its conjugate base, and every base its conjugate acid. The Brφnsted-Lowry definition is more general than the Arrhenius definition, since the former applies to any type of solvent, whereas the latter is restricted to aqueous solutions.

The free proton is encountered only in a vacuum or in a very dilute gas. Thus, since a free proton cannot exist in solution (in measurable concentrations), all acid-base reactions are of the type:

$$A_1 + B_2 \rightleftharpoons B_1 + A_2$$

where A_1—B_1 and A_1—B_2 are conjugate acid-base pairs, *i.e.*, in order that an acid may exhibit its acidic properties, there must be a proton acceptor (base) present. Some examples are:

Acid (A_1)	+ Base (B_2)	Acid (A_2)	+ Base (B_1)
HC1	+ H_2O	H_3O^+	+ Cl^-
HSC_4^-	+ NH_3	NH_4^+	+ $SO_4^=$
NH_4	+ H_2O	H_3O^+	+ NH_3
Ph_3CH	+ NH_2^-	NH_3	+ Ph_3C^-

conjugate

conjugate

Let us now consider the case of water. This is the most important solvent for acids and bases, and acts as an acid or a base according as the other compound acts as a base or an acid. The 'other' compound, however, can be water itself, and this produces the following equilibrium:

$$H_2O + H_2O \rightleftharpoons H_3O^+ + OH^-$$
$$(A_1) \quad (B_2) \quad (A_2) \quad (B_1)$$

This 'self-ionisation' is known as *autoprotolysis*. Many organic compounds, particularly hydroxy compounds, can undergo autoprotolysis, *e.g.*,

$$C_2H_5OH + C_2H_5OH \rightleftharpoons C_2H_5OH_2^+ + C_2H_5O^-$$

We can summarise the situation so far as follows: (i) acids and bases may be neutral or charged; (ii) solvents such as water, which can undergo autoprotolysis, are called amphiprotic solvents; the cA (of the solvent) is often called the tyonium ion, and the cB the lyate ion; (iii) solvents which are neither proton donors nor acceptors are called aprotic solvents.

A third definition of acids and bases is that due to Lewis (1938). Lewis acids are molecules or ions which are capable of co-ordinating with unshared electron pairs, and Lewis bases are molecules or ions having unshared electron pairs available for sharing with acids. Since a reagent must have an unshared electron pair to accept a proton, some reagents may act as bases both in the Brϕnsted-Lowry and Lewis sense. On the other hand, all Brϕnsted-Lowry acids are Lewis acids, since a proton from any proton donor can co-ordinate with an unshared electron pair. The Lewis definition of an acid, however, also includes many reagents such as BF_3, $AlCl_3$, etc., since these can also co-ordinate with unshared electron pairs (to form co-ordination compounds), *e.g.*,

$$:NH_3 + BF_3 \rightarrow H_3\ N^+ — BF_3^-$$

It might be noted here that Lewis acids and bases are respectively electrophilic and nucleophilic reagents.

Strengths of acids and bases : The 'strength' of an acid is a measure of its tendency to lose a proton, and the 'strength' of a base is a measure of its tendency to take up a proton. If we consider the equilibrium between the conjugate acid-base pairs, $A_1—B_1$ and $A_2—B_2$, *i.e.*,

$$A_1 + B_2 \rightleftharpoons B_1 + A_2$$

then the equilibrium will be the more to the right the stronger A_1 and B_2 are, and the weaker B_1 and A_2 are, *i.e.*, the equilibrium always shifts in the direction of the formation of the weaker acid and weaker base. Thus, it is not possible to speak of *absolute* strengths of acids and bases; the strength of acid A_1 will be *relative* to the strength of base B_2. B_2 is usually the solvent, and-water is used for the purpose of measuring relative strengths. There is much evidence to show that the 'aqueous' proton is firmly 'bound' to four water molecules:

$$HA + 4H_2O \rightleftharpoons A^- + H_3O + .3H_2O \text{ or } H^+(H_2O)_4$$

However, usual practice is to write this equation as involving the formation of the hydroxonium (hydronium) ion, H_3O^+:

$$HA + H_2O \rightleftharpoons A^- + H_3O^+$$

Ionisation of an acid is also often written:

$$HA \rightleftharpoons H^+ + A^-$$

it being understood that there are no free protons but protonated solvent molecules.

From what has been said above, it follows that very strong acids have very weak conjugate bases, and vice versa, *e.g.*, if HY is a stronger acid than HZ, then Z^- is a stronger base than Y^- Similarly, very strong bases have very weak conjugate acids, and vice versa.

Ionisation constants (K_a and K_b). These measure the strengths of acids and bases, and they are derived by the application of the law of mass action.

Acids $\quad HA + H_2O \rightleftharpoons H_3 + A^-$

$$\therefore K' = [H_3O^+]\ [A^-]/[CA][H_2O]$$

$$\therefore K_a = [H_3O^+]\ [A^-]/[HA]$$

$[H_2O]$ is effectively constant. Also

$$pK_a = \log_{10} (1/K_a) = -\log_{10} K_a$$

Thus, the stronger an acid is, the larger is its K_a and consequently the smaller is its pK_a.

Bases $\quad B + H_2O \rightleftharpoons BH^+ + OH^-$

$$\therefore K_b = [BH^+COH^-]/[B][H_2O]$$

$$\therefore k_b = [BH^+][OH^-]/[B]$$

Thus, the stronger a base is, the larger is its K_b and consequently the smaller is its pL_b.

Since in the Brϕnsted-Lowry system, acid-base systems involve proton transfer, there is a tendency to replace base ionisation constants, K_b, by the acid ionisation constants, K_a of the conjugate acid, BH^+.

Thus: $\quad BH^+ + H_2O \rightleftharpoons B + H_3O^+$

$$\therefore K_a = [B][H_3O^+]/[BH^+]$$

It therefore follows that for a base and its conjugate acid,

$$K_b . K_a = \frac{[BH^+][OH^-]}{[B]} . \frac{[B][H_3O^+]}{[BH^+]} = [H_3O^+][OH^-] = K_w$$

where K_w, is the autoprotolysis constant of water (= 10^{-14}, or pK_w, = 14; at 25°C). Finally, since

$$pK_a + pK_b = pK_w$$
$$\therefore pK_a = pK_w - pK_b$$

Thus, the *stronger* a base is (*i.e.*, the lower its pK_b), the larger is the pK_a of its conjugate acid.

An important point to note is that the preceding discussion has dealt with *thermodynamic acidity*, *i.e.*, the position of equilibrium of the ionisation of an acid is a thermodynamically controlled reaction (and is measured by K). On the other hand, there is also kinetic acidity, which deals with the rate at which an acid transfers its proton to a base. This is a kinetically controlled reaction (and measured by k).

In the text will be found the pK_a values of various acids and bases. Most of these values have been taken from Albert and Serjeant.

Levelling effect : Most carboxylic acids behave as weak acids in water, but their pK_a values are different. When dissolved in liquid ammonia, however, all these acids form ammonium salts:

$$RCO_2H + NH_3 \rightleftharpoons RCO_2^- + NH_4^+$$

Since the acids are completely ionised (by proton transfer), all the acids now behave as *strong* acids due to the strong basic property of ammonia (relative to water). The basicity of ammonia is said to exert a *levelling effect* on the 'strengths' of acids dissolved in it. Thus, by using very strong bases, very weak acids may be converted into their conjugate bases. One very important example of this is the conversion, in ethanol solution, of ethyl malonate (a very weak acid) into its conjugate base (the ethyl malonate carbanion) by the action of sodium ethoxide (the ethoxide ion is an extremely strong base; it is the conjugate base of the extremely weak acid, ethanol):

$$CH_2(CO_2C_2H_5)_2 + C_2H_5O^- \rightleftharpoons :\overline{C}H(CO_2C_2H_5)_2 + C_2H_5OH$$

It should be noted that many compounds may therefore behave as acids (or bases) in suitable solvents, whereas in water they behave as 'neutral' substances, *e.g.*, methanol (pK_a 15.5). Such pK_a values have been estimated by indirect methods and then related to water.

Many factors control the strengths of organic acids and bases: inductive, resonance and steric effects, solvation, entropy, hydrogen bonding and conformational effects.

THE ELECTRONIC THEORY OF VALENCY

The electronic theory of valency starts with the assumption that valency involves the electrons in the outer shells: in some cases only those in the highest sublevel in the outermost shell; in other cases those in the highest and penultimate sublevels, even though the penultimate sublevel may be in a lower quantum shell.

Lewis (1916) assumed that the electron configuration in the rare gases was particularly stable (since these gases are chemically inert), and that chemical combination between atoms took place by achieving this configuration. The outermost shell of the rare gases always contains an octet of two s and six p electrons. Since both the 5 and p sublevels are completely filled, the octet will be a stable configuration. In the case of helium, however, an octet is impossible; here the stable arrangement is the *duplet*, the two is electrons of which completely fill the first quantum shell.

The octet rule applies only to atoms with 2s and 2p electrons, *i.e.*, to elements in the second period (Li—F). With the other elements, d orbitals may also be used in bond formation, and hence higher covalencies (*i.e.*, expansions beyond an octet) are possible, *e.g.*, PBr_5 (10 electrons), SF_6 (12 electrons) and IF_7 (14 electrons). Since elements in period 2 have only 2s and 2p orbitals, the maximum covalency they can exhibit is 4.

Lewis also suggested that there was a definite tendency for electrons in a molecule to form pairs. This rule of 2, as we have seen, became established by the developments of quantum mechanics. There are also molecules that contain an odd number of valency electrons: where such *odd electron molecules* do exist, unusual properties are found to be associated with them.

There are three general extreme types of chemical bonds: electrovalent, covalent and metallic bonds. In addition to these extreme types,, there are also bonds of intermediate types.

1. *Electrovalency* is manifested by the *transfer* of electrons, and gives rise to the ionic bond. Consider sodium chloride. Sodium is $(1s)^2(2s)^2(2p)^6(3s)$: chlorine is $(1s)^2(2s)^2(2p)^6(3s)^2(3p)^5$. Sodium has completed K and L shells, and is starting the M shell

with one electron. This electron (the 3s electron) is the valency electron of sodium. Chlorine has completed K and L shells, and has seven electrons in the M shell. These M electrons are the valency electrons of chlorine. If the sodium completely transfers its valency electron to the chlorine atom, then each atom will have eight electrons in its outermost shell, and this, as we have seen, is a stable arrangement. Since both atoms were originally electrically neutral, the sodium atom, in losing one electron, will now have a single positive charge, *i.e.*, the neutral atom has become a positive ion.

Similarly, the neutral chlorine atom, in gaining one electron, has become a negative ion. In the sodium chloride crystal these ions are held together by electrostatic forces. If the symbol of an element is used to represent the nucleus of an atom and all the electrons other than the valency electrons, and dots are used to represent the valency electrons, then the combination of the sodium and chlorine atoms to form sodium chloride may be represented as follows:

$$Na. + :\underset{..}{\ddot{C}l}. \rightarrow Na^{+}\ :\underset{..}{\ddot{C}l}:^{-}$$

2. *Covalency :* This type of bonding involves a *sharing* of electrons in pairs, each atom contributing one electron to form a shared pair, each pair of electrons having their spins antiparallel. This method of completing an octet (or any of the other possible values) gives rise to the covalent bond.

   Hydrogen is usually *unicovalent:* occasionally it is uni-electrovalent, *e.g.*, in sodium hydride, hydrogen exists as the hydride anion, formed by accepting an electron from the sodium:

$$Na. + H. \rightarrow Na^{+}:^{-}$$

   Carbon almost invariably forms covalent compounds. The electron configuration of carbon is $(1s)^2(2s)^2(2p)^2$. Since the two 2s electrons are paired, it would appear that carbon is bivalent, only the two single 2p electrons being involved in compound formation. As pointed out previously, carbon is almost always quadrivalent; thus the 2s and 2p electrons must be involved. Just how these four electrons give quadrivalent carbon will be described later; at this stage we shall assume it done, and write quadrivalent carbon as $\cdot\dot{\underset{\cdot}{C}}\cdot$.

In methane the four hydrogen atoms each contribute one electron and the carbon atom four electrons towards the formation of four shared pairs:

$$4H. + \cdot\dot{\underset{\cdot}{C}}\cdot \rightarrow H:\underset{H}{\overset{H}{\ddot{\underset{\cdot\cdot}{C}}}}:H$$

Each hydrogen atom has its duplet (as in helium), and the carbon atom has an octet.

Each pair of shared electrons is equivalent to the ordinary 'valency-bond', and so electronic formulae are readily transformed into the usual structural formulae, each bond representing a shared pair, *e.g.*,

$H:\ddot{\underset{\cdot\cdot}{O}}:H$ or H–O–H; H:C:::C:H or H–C≡C–H; $H:\dot{\underset{\cdot\cdot}{N}}:H$

From these examples it can be seen that there is a very important difference between an electronic formula and its equivalent structural formula. In the former, all valency electrons are shown whether they are used to form covalent bonds or not; in the latter, only those electrons which are actually used to form covalent bonds are indicated. This is a limitation of the usual structural formula. A widely used scheme is to represent structures with ordinary valency bonds and to indicate lone pairs by pairs of dots.

3. *Co-ordinate valency* is a special type of covalency. Its distinguishing feature is that both of the shared electrons forming the bond are supplied by only one of the two atoms linked together, *e.g.*, when ammonia combines with boron trifluoride, it is the lone pair of the nitrogen atom that is involved in the formation of the new bond. In boron trifluoride, the boron has only six electrons in its valency shell; hence it can accommodate two more to complete its octet. Thus, if the nitrogen atom uses its lone pair, the combination of ammonia with boron trifluoride may be shown as (I) or (II). In the latter, a co-ordinate bond is represented by an arrow pointing away from the atom

supplying the lone pair (Sidgwick, 1927).

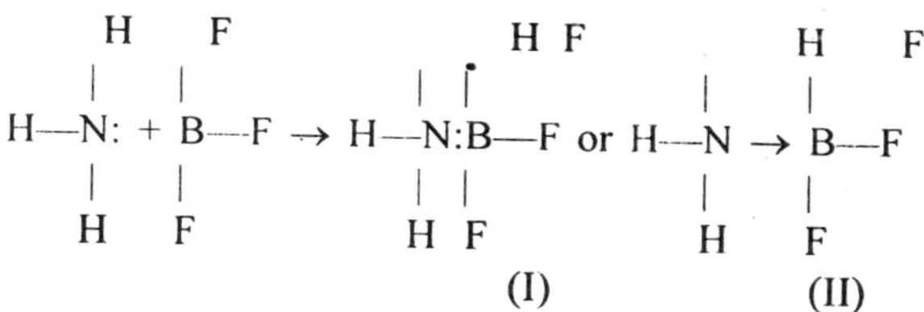

The atom that supplies the lone pair is known as the *donor*, and the atom that receives a share is the *acceptor*. Since it is one atom that donates the lone pair, the co-ordinate bond is also known as the *dative* bond (Sidgwick, 1927).

Before combination, both donor and acceptor are electrically neutral: after combination, the donor has lost a share in the lone pair, and the acceptor has gained a share. Therefore the donor acquires a positive charge and the acceptor a negative charge, and the presence of these charges may be indicated by writing the formula $H_3\overset{+}{N}—\overset{-}{B}F_3$.

Once the co-ordinate bond has been formed, there may be no way of distinguishing it from a covalent bond, but since one atom has supplied the pair of shared electrons, charges are produced in the molecule. When a covalent bond is formed, charges may also be produced in the molecule, giving rise to a dipole (q.v.). Hence, the co-ordinate bond is effectively a covalent bond. The extent of the charge on each atom in a dative (or covalent) bond may be found as follows.

Add the number of unshared electrons to one half of the shared electrons, and compare the result with the number of valency electrons of the neutral atom, *e.g.*, (i) methane, CH_4. Here there are 8 shared electrons; $1/2 \times 8 = 4$ = number of electrons in the neutral carbon atom; therefore-methane is uncharged, (ii) $H_3\overset{+}{N}—\overset{-}{B}F_3$. For the nitrogen atom we have, $1/2 \times 8 = 4$, but since the neutral nitrogen atom has 5 valency electrons, in the compound $H_3\overset{+}{N}—\overset{-}{B}F_3$ the nitrogen has a charge of +1. For boron we have $1/2 \times 8 = 4$; but since the neutral boron atom has 3 electrons, in this molecular compound the boron has a charge of –1.

Electrovalent compounds are good electrical conductors in the fused state or in solution. They are generally non-volatile, and are usually insoluble in hydrocarbons and allied solvents. Covalent compounds are non-electrical conductors, are generally volatile, and are usually soluble in hydrocarbons and allied solvents. Since the covalent bond is directional, stereo*isomerism* (space-isomerism) is possible. Co-ordinated compounds behave very much like covalent compounds, but they are usually less volatile than pure covalent compounds.

Two important atomic quantities are:

(i) *Ionisalion potential* (I.P.); this is the amount of energy absorbed when one electron is *removed* from a neutral (or charged) atom. It is measured in electronvolts (1 eV = 96.44 kJ mol^{-1} = 23.05 kcal mol^{-1}), and the higher its value, the more difficult it is to remove the electron. The I.P.s of carbon are: $C \rightarrow C^+ + e$ (+11.27 eV); $C^+ \rightarrow C^{2+} + e$ (24.38 eV); $C^2 \rightarrow C^{3+} + e$ (47.87 eV). Even the first I.P. is relatively high, and hence carbon shows little tendency to form electrovalent bonds.

(ii) *Electron affinity* (E.A.): this is the amount of energy (in eV) which is *evolved* when a neutral (or charged) atom adds one electron. Most non-metals have relatively high (negative) E.A.s, and consequently easily form negative ions, *e.g.*, $Cl + e \rightarrow Cl^-$ (–13.0); $Br + e \rightarrow Br^-$ (–11.84); $I + e \rightarrow I^-$ (–10.44 eV):

Now let us reconsider the co-ordinate bond. This may be imagined to be formed in two steps

$$H_3N: + BF_3 \rightarrow H_3\overset{+}{N}. + .\bar{B}F_3 \rightarrow H_3\overset{+}{N}—\bar{B}F_3$$

Thus, the lower the I.P. of the donor atom and the higher the E.A. of the acceptor atom, the more readily will the co-ordinate bond be formed. Also, when the bond has been formed, since the donor atom now-has a positive charge, it will be more difficult for it to form a second co-ordinate bond (provided that the atom has more than one unshared pair), *e.g.*, the I.P.s of O are $O \rightarrow O^+$ (13.62) and $O^+ \rightarrow O^{2+}$ (35.15 eV). The latter is too great for its formation under normal conditions, and consequently the hydronium ion, H_4O+, is readily formed (one co-ordinate bond), but not H_4O^{2+} (two co-ordinate bonds).

ELECTRONIC VALENCING THEORY

The electrons are arranged in shells around the nucleus, each shell being able to contain up to a maximum number of electrons, this maximum depending on the number of the shell, n. n is known as the *principal quantum number*, and indicates the main energy level of the electrons in that shell, n has whole number values, 1, 2, 3, 4 . . ., the shells corresponding to which are also denoted by the letters K, L, M, N ...

respectively. In every principal quantum shell there are n energy sublevels, and these are indicated by *l*, the *orbital quantum number* (also known as the *azimuthal* or *serial* quantum number). Just as the principal quantum number n can have values 1, 2, 3 . . ., so can *l* have values 0, 1, 2, 3 . . ., n–1. The energy state corresponding to *l* = 0 is called the s state; *l* = 1, the p state; *l* = 2, the d state; etc.

As we shall see later these s, p and a states are subdivided into a number of *orbitals*. The total number of orbitals that a principal quantum shell can contain is given by n^2. According to modem theory, an atom consists of a *nucleus* which contains *protons* and *neutrons*, and which is surrounded by *electrons*. The mass of a proton is almost the same as that of a neutron, but whereas the proton carries a unit of *positive* charge, the neutron is electrically neutral. The electron has about 1/1850th of the mass of a proton, and carries a unit of negative charge.

Thus, when the principal quantum number is 1 (*i.e.*, the first or K shell), then *l* = 0, *i.e.*, there is a single orbital in this K shell and is of the 5 type and is known as the 1s orbital. When n = 2 (*i.e.*, the second or L shell), then *l* = 0 or 1. This means there are two energy sublevels in the L shell. As pointed out above, the total number of orbitals in a given quantum shell is equal to n^2. Thus, when n = 2, there are four possible orbitals. When *l* = 0, this corresponds to the 2s orbital. When *l* = 1, then there are *three equivalent* orbitals; these are the three 2p orbitals. When n = 3 (*i.e.*, the third or M shell), then the total number of possible orbitals is 9 (3^2). These correspond to one 3s orbital (*l* = 0), three 3p orbitals (*l* = 1) and five 3d orbitals (*l* = 2).

The existence of one s level, three/?, five d and seven/levels of energy was used to explain the existence of spectral lines observed in the spectra of atoms and molecules. It should also be noted that the farther an electron is from the nucleus, the greater is its potential energy. Owing to the phenomenon of penetration of orbitals, *i.e.*, outer electrons can penetrate into the shell of inner electrons, the energy level of an electron is thus not completely determined by its principal quantum number, but also depends on the orbital quantum number (*i.e.*, on the shape of the orbital). Thus there is the following order of increasing energy: Is, 2s, 2p, 3s, 3p, 4s, 3d, 4p, 5s, 4d, 5p.

In addition to the energy levels of an electron described by quantum numbers n and *l*, electrons have spin about their axis, some spinning in one direction and others in the opposite direction. This is indicated by the *spin quantum number* (s), and can have values of +1/2 and –1/2. Finally, an electron also has a *magnetic quantum number* (m), and this gives the allowed orientations of the orbitals in an external magnetic field. Thus an electron is described by *four* quantum numbers, n, *l*, s and m.

By the fundamental *Pauli Exclusion Principle* (1925), no two electrons, in any system, can be assigned the same set of four quantum numbers. Hence there can be only two electrons in any one orbital, and these must be differentiated from each other by their spins, which *must* be *antiparallel*, *i.e.*, in the opposite sense. Such electrons are said to be *paired*, and a pair of electrons with antiparallel spins in the same orbital is represented by the symbol ↓↑. Since a moving charge is accompanied by a magnetic field, a- spinning electron behaves as a small bar-magnet, and consequently two paired electrons will give a zero resultant magnetic field.

The hydrogen atom consists of one proton and one electron. When the hydrogen atom is in the *'ground'* state, *i.e.*, the state of lowest energy, its electron will be in the lowest energy level, *i.e.*, the 1s level, and is represented by (1s). When hydrogen is in an *'excited'* state, its electron will occupy a higher energy level, the actual level depending on the amount of excitation'.

Helium has two electrons; hence its electron configuration in the ground state is represented as $(1s)^2$. Lithium has three electrons. Since the maximum number of electrons in the K shell (n = 1, *l* = 0) is two, the third electron must start the L shell (n = 2, *l* = 0, 1). Electrons occupy lowest energy levels first. Thus this third electron occupies the 2s orbital, and not the 2p, because the 2p is a higher energy level than the 2s. Hence the electron configuration of lithium is $(1s)^2(2s)$. Thus the K shell is filled first. Then the electrons enter the L shell until that is filled. In this shell the s level is filled before the p. In fitting electrons into shells containing orbitals of equivalent energy, Hund's rules are used to assign the electrons to their orbitals.

These rules are: (i) electrons tend to avoid being in the same orbital as far as possible; (ii) two electrons, each singly occupying a given pair of *equivalent* orbitals, tend to have their spins parallel when the atom is in the ground state. Thus carbon, with six electrons, may be represented as $(1s)^2(2s)^2(2p)^2$. The K shell is filled first; the L shell is filled next, the 2s orbital being doubly filled before a higher level is used; then singly two of the 2p orbitals. Nitrogen, with seven electrons is $(1s)^2(2s)^2(2p)^2$: all three 2p orbitals each contain one electron. Oxygen, with eight electrons, is $(1s)^2(2s)^2(2p)^4$: here one of the 2p orbitals is doubly filled.

DIPOLE MOMENTS

When a covalent bond is formed between two identical atoms, *e.g.*, H—H, Cl—Cl, etc., the two electrons forming the covalent bond may be regarded as being symmetrically disposed between the two atoms. The centres of gravity of the electrons and nuclei therefore coincide. With two dissimilar atoms the two electrons are no longer symmetrically disposed, because each atom has a different electronegativity, *i.e.*, attraction for electrons.

Chlorine has a much greater electronegativity than hydrogen that when chlorine and hydrogen combine to form covalent hydrogen chloride, the electrons forming the covalent bond are displaced towards the chlorine atom without any separation of the nuclei:

$$H. + :\underset{..}{\ddot{C}}1. \rightarrow H:\underset{..}{\ddot{C}}1: \quad \text{or} \quad \overset{\delta+}{H}—\overset{\delta-}{C}1$$

The hydrogen atoms will, therefore, be slightly positively charged, and the chlorine atom slightly negatively charged. Thus, owing to the greater electronegativity of the chlorine atom, the covalent bond in hydrogen chloride is characterised by the separation of small charges in the bond. A covalent bond such as this, in which one atom has a larger share of the electron-pair, is said to possess partial ionic character.

In analogy with a magnet, such a molecule is called a dipole, and the product of the electronic charge, e, and the distance d, between the charges (positive and negative centres) is called the dipole moment, μ; *i.e.*, $\mu = e \times d$; e is of the order of 10^{-10} e.s.u.; d, 10^{-8} cm. Therefore μ is of the order 10^{-18} e.s.u., and this unit is known as the Debye (D), in honour of Debye, who did a large amount of work on dipole moments.

The dipole moment is a vector quantity, and its direction is often indicated by an arrow parallel to the line joining the points of charge, and pointing towards the negative end, *e.g.*, H—Cl. The greater the value of the dipole moment, the greater is the polarity of the bond. The terms polar and non-polar are used to describe bonds, molecules and groups, and the reader is advised to make sure he appreciates how the terms are applied in each case under consideration.

Not only do polar bonds contribute to the dipole moment of a molecule, but so do lone pairs, *e.g.*, the lone pair in ammonia makes a large contribution to the dipole moment.

Table 1 gives some electronegativity values, and it will be seen that electronegativity increases from left to right and decreases from top to bottom of the periodic table. The values are not absolute values; they are relative values and this is a satisfactory scheme since their use involves differences in electronegativities for the measurement of relative electronegativities). Furthermore, it should be noted that electronegativity deals with electron attraction within a molecule, whereas electron affinity deals with the attraction of an electron from outside the atom (and has an absolute value).

Table 1

				H 2.1			
Li	Be	B	C		N	O	F
1.0	1.5	2.0	2.5		3.0	3.5	4.0
Na	Mg	Al	Si		P	S	Cl
0.9	1.2	1.5	1.8		2.1	2.5	3.0
K	Ca				As	Se	Br
0.8	1.0				2.0	2.4	2.8
			Sn		Sb		I
			1.7		1.8	2.5	

ELECTRON DISPLACEMENTS IN A MOLECULE

1. *Inductive effect :* Consider a carbon chain in which one terminal carbon atom is joined to a chlorine atom: —C_3—C_2—C_1—Cl. Chlorine has a greater electronegativity than carbon; therefore the electron pair forming the covalent bond between the chlorine atom and C_1 will be displaced towards the chlorine atom. This causes the chlorine atom to acquire a small negative charge, and C_1 a small positive charge. Since C_1 is positively charged, it will attract towards itself the electron pair forming the covalent bond between C_1 and C_1. This will cause C_1 to acquire a small positive charge, but the charge will be smaller than that on C_1 because the effect of the chlorine atom has been transmitted through C_1 to C_2. Similarly, C_3 acquires a positive charge which will be

smaller than that on C_1. This type of electron displacement along a chain is known as the *inductive effect*; it is permanent, and decreases rapidly as the distance from the source increases. From the practical point of view, it may be ignored after the second carbon atom. It is important to note that the electron pairs, although permanently displaced, *remain in the same valency shells*.

This inductive effect is sometimes referred to as a transmission effect, since it takes place by a displacement of the intervening electrons in the molecule. There is also another effect possible, the direct or field effect, which results from the electrostatic interaction across space or through a solvent of two charged centres in the same molecule, *i.e.*, the direct effect takes place independently of the electronic system in the molecule (Ingold, 1934). Apparently it has not been possible to separate these two modes of inductive effect in practice. The inductive effect may be represented in several ways. The following will be adopted in this book: $-C \rightarrow C \rightarrow C \rightarrow Cl$.

Inductive effects may be due to atoms or groups, and the following is the order of decreasing inductive effects:

+I O^-, CO_2^-, $(CH_3)_3C$, $(CH_3)_2CH$, $(CH_3)CH_2$, CH_3

–I NR_3^+, SR_2^+, NH_3^+, NO_2, SO_2R, CN, CO_2H, F, Cl, Br, I, OR, OH

For measurement of relative inductive effects, hydrogen is chosen as reference in the molecule CR_3–H as standard. If, when the H atom in this molecule is replaced by Z (an atom or group), the electron density in the CR_3 part of the molecule is less in this part than in CR_3—H, then Z is said to have a –I effect (electron-attracting or electron-withdrawing). If the electron density in the CR_3 part is *greater than* in CR_3—H, then Z is said to have a +1 effect (electron-repelling or electron-releasing) *e.g.*, Br is –I; C_2H_5 is +I.

This terminology is due to Ingold (1926); Robinson suggests the opposite signs for I, *i.e.*, Br is + 1; C_2H_5, –I. Ingold's terminology will be used in this book. The inductive effect may be measured with respect to hydrogen by the effect of substituents on the change in acid strength of substituted acetic acids.

2. *Electromeric effect* : This is a *temporary* effect involving the *complete transfer* of a shared pair of electrons to one or other atom joined by a multiple bond, *i.e.*, a double or triple bond.

The electromeric effect is brought into play only at the requirements of the attacking reagent, and because of this, *the direction of the electromeric effect is always that which facilitates reaction.*

The electromeric effect is represented as follows:

$$A = B \rightleftharpoons \overset{+}{A} - \overset{-}{B}$$

The curved arrow shows the displacement of the shared electron pair, beginning at the position where the pair was originally, and ending where the pair has migrated. Since A has lost its share in the electron pair and B has gained this share, A acquires a positive charge and B a negative charge.

The electromeric effect is represented by the symbol E, and is said to be +E when the displacement is away from the atom or group, and –E when towards the atom or group (cf. the I effect).

The displacement of the electron pair forming a covalent bond when a unit charge is brought up is a measure of the *polarisability* of that bond. It is not a permanent polarisation since, when the charge is removed, the electron displacement disappears. Thus, the electromeric effect is a polarisability effect and operates in the excited state.

3. *Mesomerism or Resonance :* The theory of mesomerism was developed on chemical grounds. It was found that no structural formula could satisfactorily explain all the properties of certain compounds, *e.g.*, benzene. This led to the idea that such compounds exist in a state which is some combination of two or more electronic structures, all of which seem equally capable of describing most of the properties of the compound, but none of describing all the properties. Ingold (1933) called this phenomenon mesomerism ('between the parts', *i.e.*, an intermediate structure).

   Heisenberg (1926), from quantum mechanics, supplied a theoretical background for mesomerism; he called it resonance, and this is the name which is widely used. The chief conditions for resonance are:

   (i) The positions of the nuclei in each structure must be the same or nearly the same.

(ii) The number of unpaired electrons in each structure must be the same.

(iii) Each structure must have about the same internal energy, *i.e.*, the various structures have approximately the same stability.

Let us consider carbon dioxide as an example. The electronic structure of carbon dioxide may be represented by at least three possible electronic arrangements which satisfy the above conditions:

$$\underset{\text{(I)}}{\ddot{\text{O}}{::}\text{C}{::}\ddot{\text{O}}} \qquad \underset{\text{(II)}}{:\overset{-}{\ddot{\text{O}}}{:}\text{C}{:::}\overset{+}{\text{O}}:} \qquad \underset{\text{(III)}}{:\overset{+}{\text{O}}{:::}\text{C}{:}\overset{-}{\ddot{\text{O}}}:}$$

Structures (II) and (III) are identical as a whole, since both oxygen atoms are the same. Each structure, however, shows a given oxygen atom to be in a different state, *e.g.*, the oxygen atom on the left in (II) is negative, whereas in (III) it is positive. Although two (or more) of the electronic structures may be the same when the molecule is considered as a whole, each one must be treated as a separate individual which makes its own contribution to the resonance state. Structures (I), (II) and (III) are called the resonating, unperturbed or canonical structures of carbon dioxide, and carbon dioxide is said to be a resonance hybrid of these structures, or in the mesomeric state.

It is hoped that the following crude analogy will help the reader to grasp the concept of resonance. Most readers will be familiar with the rotating disc experiment that shows the composite nature of white light. When stationary, the disc is seen to be coloured with the seven colours of the rainbow. When rotating quickly, the disc appears to be white. The resonating structures of a resonance hybrid may be compared to the seven colours, and the actual state of the resonance hybrid to the 'white'; *i.e.*, the resonating structures may be regarded as super- imposed on one another, the final result being one kind of molecule. *In a resonance hybrid all the molecules are the same; a resonance hybrid cannot be expressed by any single structure.*

In a resonance hybrid the molecules have, to some extent, the properties of each resonating structure. The greater the contribution of any one structure, the more closely does the actual state approach to that structure. At the same time, however, a number of properties differ from those of any one structure. The observed enthalpy of formation of carbon dioxide is greater than the calculated value by 132.2 kJ mol^{-1} (31.6 kcal). In other words, carbon dioxide requires 132.2 kJ more energy than expected to break it up into its elements, *i.e.*, carbon dioxide is more stable than anticipated on the structure O=C=O. How can this be explained? Arguments based on quantum mechanics show that a resonance hybrid would be more stable than any single resonating structure, *i.e.*, the internal energy of a resonance hybrid is less than that calculated for any one of the resonating structures.

The difference between the enthalpy of formation of the actual compound, *i.e.*, the observed value, and that of the resonating structure which has the lowest internal energy (obtained by calculation) is called the resonance energy. Thus the value of the resonance energy of any resonance hybrid is not an absolute value; it is a relative value, the resonating structure containing the least internal energy being chosen as the arbitrary standard for the resonance hybrid. The greater the resonance energy, the greater 4s the stabilisation. The resonance energy is a maximum when the resonating structures have equal energy content, and the more resonating structures there are, the greater is the resonance energy, provided that all contributing structures have similar stabilities.

Bond Energies

There are two types of bond energy: (i) Dissociation energy (designated by D). (ii) Bond energy (designated by E).

(i) *Dissociation energy* (D). This is the energy required to break a particular bond in a polyatomic molecule, in the gaseous phase, into neutral fragments (free radicals), also in the gaseous phase, *i.e.*, D is the energy required for .the reaction:

$$Y—Z(g) \rightarrow Y.(g) + Z.(g)$$

(ii) *Bond energy* (E). In polyatomic molecules, *e.g.*, CH_4, the four bonds are equivalent, but the energy required to break the first bond ($CH_4 \rightarrow CH_3. + H.$) is not the same as that for the second bond ($CH_3. \rightarrow :CH_2 + H.$); etc. Each individual value is, as we have seen, the bond dissociation energy of that particular bond. It can also be seen from Table 2.2 that the values of the dissociation energies of the C—H bond vary in the first four compounds. Thus, the energy of a bond depends on the nature of the rest of the molecule. In practice, it is usual to take the average of all the different values, and this average value is called the bond energy. For diatomic molecules, D and E are identical. Various factors are responsible for the differences in energy of a given bond in different compounds, *e.g.*, steric effects, angular strain. Table 2 lists some bond energies.

Table 2

Bond	*Energy*		*Bond*	*Energy*		*Bond*	*Energy*	
	kJ	(kcal)		kJ	(kcal)		kJ	(kcal)
CH_2—H	426.8	(102)	C≡N	866.1	(207)	O=O	497.9	(119)
$MeCH_2$—H	405.8	(97)	C—F	447.7	(107)	O—O	146.4	(35)
Me_2CH—H	393.3	(94)	C—Cl	326.4	(78)	O—H	464.4	(111)
Me_3C—H	374.5	(89.5)	C—Br	284.5	(68)	S—S	225.9	(54)
C—H(av.)	414.2	(99)	C—I	213.4	(51)	S—H	347.3	(83)
C—C	347.3	(83)	C—S	272.0	(65)	S=O	497.9	(119)
C=C	606.7	(145)	H—H	431.0	(103)	F—F	150.6	(36)
C=C	803.3	(192)	H—N	389.1	(93)	H—F	560.7	(134)
C—O	334.7	(80)	N—N	163.2	(39)	Cl—Cl	242.7	(58)
C=O	694.5	(166)	N=N	418.4	(100)	H—Cl	426.8	(102)
O=C=O	803.3	(192)	N=N	945.6	(226)	Br—Br	188.3	(45)
C—N	284.5	(68)	N—O	200.8	(48)	H—Br	364.0	(87)
C=N	615.1	(147)	N=O	606.7	(145)	I—I	150.6	(36)
						H—I	297.1	(71)

The resonance energy of a molecule is a property of the molecule in the ground state. Most measurements of resonance energies have been obtained from heats of combustion, but some measurements have also been obtained from heats of hydrogenation; the latter method is more accurate than the former. The heat of combustion of the most stable classical structure (*i.e.*, the resonating structure with the lowest internal energy) is, as mentioned above, obtained by calculation. This presupposes that accurate values for bond-energies are known. If these values are not accurate, then one cannot expect to obtain accurate resonance energies. In addition to this problem, there are other difficulties involved with the measurement of resonance energies.

The reference molecule, not being identical with the actual molecule (resonance hybrid molecule), will therefore differ in a number of physical properties, *e.g.*, bond lengths.

Although these differences may be small, nevertheless energy changes are involved in bond compression and bond stretching, and hence these energy terms should also be considered when comparing the reference molecule to the actual molecule. There are also energy terms involving hybridisation changes (see later), steric factors, etc. Now, the list of bond energies has been obtained for compounds in which these factors (steric, resonance, etc.) are absent. In general, the total chemical binding energy of a molecule may be regarded as being comprised of three factors:

(i) The sum of the bond energies;

(ii) Stabilising effects which increase the total binding energy, *e.g.*, resonance;

(iii) Destabilising effects which decrease the total binding energy, *e.g.*, steric effects.

It can therefore be seen that the difference between the energy of the reference molecule (calculated from bond energies) and the actual molecule is not necessarily a measure of the resonance energy. The difficulty here is that there appears to be no unambiguous way of evaluating the individual contributions of the above three factors. Because of this, the term stabilisation energy is often used instead of resonance energy to designate the energy difference between the reference molecule and the resonance hybrid molecule. In this sense, stabilisation energy is the sum of the stabilising and destabilising effects. However, when the S.E. is relatively large, the R.E. greatly outweighs the destabilising effects, and so the term resonance energy is used in this book to represent the overall stabilising effect due to (predominantly) resonance for a further discussion).

Another property of the resonance hybrid which differs from that of any of the resonating structures is the bond length, *i.e.*, the equilibrium distance between atoms joined by a covalent bond. The normal length of the carbonyl double bond (C=O) in ketones is about 1.20Å; the value found in carbon dioxide is 1.15Å. For a given pair of atoms, the length of a single bond is greater than that of a double bond, which, in turn, is greater than that of a triple bond. Resonance, therefore, accounts for the carbonyl bond in carbon dioxide not being single, double or triple. Table 3 gives some bond lengths.

In a resonance hybrid, the electronic arrangement and bond lengths will be different from those of the resonating structures. Consequently the observed dipole moment may differ from that calculated for any one structure.

Table 3

Bond	*Length(Å)*	*Bond*	*Length(Å)*	*Bond*	*Length (Å)*
C—C	1.54	C—S	1.82	C—F	1.42
C=C	1.40	C—O	1.43	C—Cl	1.77
C≡C	1.21	C=O	1.20	C—Br	1.91
C—H	1.12	O—H	0.97	C—I	2.13
C—N	1.47	N—H	1.03		

As we have seen above, in a resonance hybrid all the molecules have the same structure. A difficulty that arises with the resonance theory is the representation of a resonance hybrid. The molecules corresponding to the structures chosen as the resonating structures do not necessarily have an actual existence. Thus, if these resonating structures are fictitious, what fictitious structures are we to choose? The normal way of solving this problem is first to ascertain the structure of the molecule by the usual methods, and then describe it by means of the classical valency-bond formula. Lèt us consider again the case of carbon dioxide.

The classical structure is (I). As we have also seen, it has been found that not all the properties of carbon dioxide are described by this classical formula. Thus the classical structure is an approximation, *and it is in this sense that classical structures are fictitious*. By *postulating* other electronic structures (II) and (III), wave functions can then also be obtained for these fictitious structures. By a linear combination of all three functions, a 'structure' is obtained which describes the properties of carbon dioxide. This 'structure' is called the resonance hybrid of the classical (I) and the two postulated electronic structures (II) and (III),

It is very important to note here that wave-mechanics offers a theoretical method of studying the electron distribution in a molecule, but starts with a knowledge of the relative positions of all the nuclei concerned, *i.e.*, with the 'classical structure'. Theoretically, it is possible to start from a molecular formula, and then solve the structure. The number of possibilities and mathematical difficulties, however, are far too great at present, and so it seems that the chemist, who arrives at the classical structures by chemical and physical methods, will still be 'in business' for a long time to come. Since, however, by means of wave-mechanics one can calculate the density of electronic charge at all points in a molecule (of known classical structure), it is possible from this information to deduce charge distributions, bond lengths and bond angles, and consequently the size and shape of a molecule.

The question that now arises is: Starting with the classical structure, what other electronic structures are we justified in postulating? A very important point in this connection is that *resonance can occur only when all the atoms involved lie in the same plane (or nearly in the same plane)*. Thus, any change in structure which prevents planarity will diminish or inhibit resonance. This phenomenon is known as steric inhibition of resonance.

In practice, then, the conditions described above must be considered when choosing canonical structures. At the same time, the following observations will be a useful guide:

(i) Elements of the first two rows never violate the octet rule (hydrogen, of course, can never have more than a duplet).

(ii) The more stable a structure, then the larger will be its contribution to the resonance state. The stability of a molecule can be found from its bond energies. Generally, the structure with the largest number of bonds is the most stable. This is because the sum of the bond energies in a molecule gives a measure of the stability of the compound.

(iii) If the different resonating structures have the same number of bonds, but some structures are charged, then the charged molecules will be less stable than the uncharged. On the other hand, if charged forms have more covalent bonds than uncharged forms, then these charged forms may make a significant contribution to the resonance. The high energy content of a charged molecule is due to the work put into the molecule to separate unlike charges, and the greater the distance of charge separation, the less stable is that structure. In the same way, structures in which adjacent atoms carry like charges are relatively unstable, since work must also be put into the molecule to bring like charges together. However because charged structures are stabilised by solvation, the greater the polarity of the solvent the greater will be the contributions of charged structures. It should also be noted that the most stable-charged structures are those in which the negative charge is on the most electronegative atom and the positive charge on the least electronegative atom. Furthermore, an electron-releasing group at one end of a molecule and an electron-withdrawing group at the other end tend to stabilise (and hence increase the contribution of) charged structures, *e.g.*, the charged contribution in (V) will be much greater than that in (IV):

(IV) $H_2\ddot{N}$—CH=CH—CH=CH_2 $H_2\overset{+}{N}$=CH—CH=CH—$\bar{C}H_2$

(V) $H_2\ddot{N}$—CH=CH—CH=O $H_2\overset{+}{N}$=CH—CH=CH—$\bar{O}$

When the different resonating structures contain the same number of bonds, the phenomenon is called isovalent resonance, and when a different number of bonds, *heterovalent* resonance (Mulliken, 1959).

The final problem is the method of representing a resonance hybrid. Various methods have been used, and the one used in this book is that introduced by Bury (1935). This consists of writing down the resonating structures with a double-headed arrow between each pair:

$$O{=}C{=}O \leftrightarrow \bar{O}{-}C{\equiv}\overset{+}{O} \leftrightarrow \overset{+}{O}{\equiv}C{-}O$$

Inductive and resonance (mesomeric) effects are permanently operating in the 'real' molecule; collectively they are known as the *polarisation effects*. On the other hand, there are also two temporary (time-variable) effects, the electromeric effect and the *inductomeric effect* (which operates by an inductive mechanism). Both of these are brought into play by the attacking reagent, and collectively they are known as the *polarisability effects*. Remick (1943) has suggested the use of subscripts 5 and d to represent the static, (permanent) and dynamic (time-variable) effects.

Thus the inductive effect may be represented by the symbol I_s, and the inductomeric effect by I_d. Since polarisability effects are brought into play only by the approach of the attacking reagent, they will therefore aid and never inhibit a reaction. Strictly speaking, the term *resonance effect* (R) is not the same as the *mesomeric effect* (M). The mesomeric effect is a permanent polarisation, and the mechanism of electron transfer is the same as that in the electromeric effect, *i.e.*, the mesomeric effect is a permanent displacement of electron pairs which occurs in a system of the type

Z—C=C; *e.g.*, Z=R_2N, Cl:

$R_2\ddot{N}$—C=C; $:\ddot{C}l$—C=C—

Thus the essential requirement for mesomerism is the presence of a *multiple* bond in the molecule. On the other hand, the resonance effect embraces ail permanent electron displacements in the molecule in the ground state, *e.g.*, the hydrogen chloride molecule is a resonance hybrid of two resonating structures:

$$H—Cl \leftrightarrow H^+Cl^-$$

Since there is no multiple bond in this molecule, the mesomeric effect is not possible. When the electronic displacement is *away* from the group the mesomeric (resonance) effect is said to be + M (+ R), and when *towards* the group, – M (– R). The mesomeric effect is particularly important in conjugated systems and the *combined* mesomeric and electromeric effects are known as the *conjugative effect*. This term is also used in the same sense as the resonance effect. Also, since this combined effect was first recognised in connection with tautomerism, it has also been called the tautomeric effect (±T).

Table 4.4

Electronic mechanism	*Polarisation effect (permanent)*	*Polarisability effect (temporary)*
Inductive (±1)	Inductive (I or I_s)	Inductomeric (I_d)
Conjugative (Tautomeric, ±T; or Resonance, ±R)	Mesomeric (M) or Resonance(R)	Electromeric (E)
Fields in which operative	Physical properties	–
	Reaction equilibria	–
	Reaction rates	Reaction rates only

It might be noted here that when a + M effect is present, then a strong electromeric effect will operate if a + E effect is the requirement of the attacking reagent. If, however, the attacking reagent requires a – E effect, then with the permanent + M effect present, this –E effect will be extremely weak or absent. Similarly, for a – M effect, a –E effect, if required, is strong, whereas a + E effect, if required, is extremely weak or absent. The possible polar influences of groups are shown in the following Table 4.4:

As has been pointed out above, resonance describes all permanent electron displacements in the molecule in the ground state. It therefore follows that the I-effect can be described as due to resonance. Thus resonance is the combination of I- and M-effects. It is more convenient, however, from the point of view of the organic chemist, to consider a molecule with respect to its I-effect and 'resonance' (mesomeric) effect separately. Hence, from this point of view, resonance is the *additional permanent* electronic displacements to the I-effect, and it is customary to ignore the latter effect when discussing 'resonance'. In other words, resonance, in this context, is concerned only with the part of the molecule containing multiple bonds and is therefore, strictly speaking, *π-electron resonance*. This is the sense in which the term resonance will be used in this book.

Effect of structure on reactivity : The type of reaction of an organic compound is largely dependent on the nature of the functional group present. It has been found that various structural changes, *e.g.*, the introduction of a given group into different positions in a molecule containing a given functional group, usually affect the rate of a given type of reaction and also the equilibrium position, and may even change the type of mechanism of the reaction.

Much work has been done to try to correlate structure and reactivity (*i.e.*, rate of reaction), and as an outcome of this work, it appears that some sort of quantitative correlation can be made on the basis of consideration of independent contributions of inductive, resonance (mesomeric) and steric effects. When each effect has been assessed, then all three may be combined, and in this way there is obtained a relationship between structure and reactivity. In the text are discussed many cases of the effects on reaction rates and mechanism by polar (I and R) and steric effects.

Reactions in organic chemistry may be classified into the following main types: (i) substitution; (ii) replacement or displacement; (iii) addition; (iv) elimination; (v) isomerisation (rearrangement).

HYBRIDISATION OF BOND ORBITALS

The electron configuration of carbon is $(1s)^2(2s)^2(2p_x,2p_y)$. It therefore appears to be bivalent. To be quadrivalent, the $(2s)^2$ and the $(2p_x,2p_y)$ electrons must be involved. One way is to uncouple the paired 2s electrons, and then *promote* one of them to the empty $2p_z$ orbital. Should this be done, four valencies would be obtained, since each of the electrons could now be paired with an electron of another atom. The resulting bonds, however, would not all be equivalent, since we would now have the component A.O.s 2s, $2p_z$, $2p_y$, $2p_z$. All work on *saturated* carbon compounds of the type Ca_4. indicates that the four valencies of carbon are equivalent. In order to get four equivalent valencies, the four 'pure' A.O.s must be 'mixed' or *hybridised.*

It is possible, however, to hybridise these four 'pure' A.O.s in a number of ways to give four valencies which may, or may not, be equivalent. Three methods of hybridisation are important: (i) tetrahedral (sp^3 bond), (ii) trigonal (sp^2 bond), (iii) digonal (sp bond). The following discussion applies to molecules of the type CH_4 (sp^3), $CH_2 = CH_2$ (sp^2), and CH≡CH (sp).

(i) *In tetrahedral hybridisation*, the (2s) and ($2p_x$, $2p_y$, 2_z,) electrons are all hybridised, resulting in four equivalent orbitals arranged tetrahedrally, *i.e.*, pointing towards the four corners of a *regular* tetrahedron (Fig. 7). The orbitals are greatly concentrated along these four directions (Fig. 7 a shows the shape along one of these directions). Then, by linear combination with the is orbitals of four hydrogen atoms, four equivalent M.O.s are obtained for methane. Because of the large amount of overlapping between the hybridised A.O.s of the carbon and the s A.O. of the hydrogen atom, there will be strong binding between the nuclei.

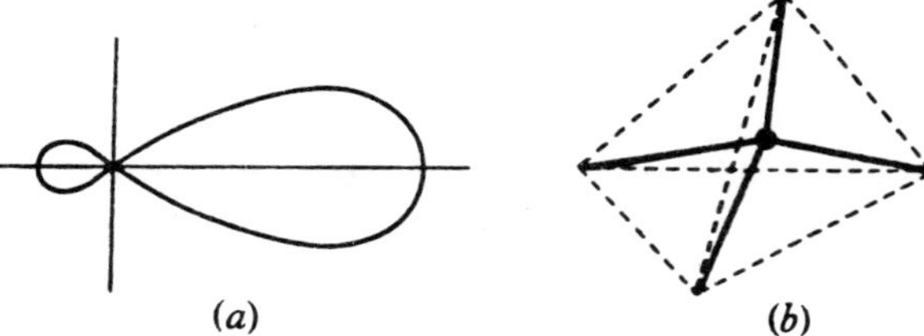

(*a*) (*b*)

Fig. 7

As in the case of the hydrogen molecule, each M.O. is almost completely confined to the region between the two nuclei concerned, *i.e.*, in methane are four localised molecular orbitals. This scheme of localised M.O.s may be satisfactorily applied to all compounds containing single covalent bonds. Bond orbitals which are symmetrical about the line joining the two nuclei concerned are known as *σ-bonds* (sigma-bonds).

(ii) In *trigonal hybridisation*, the 2s, $2p_x$ and $2p_y$ orbitals are hybridised, resulting in three equivalent coplanar orbitals pointing at angles of 120° in the xy plane (Fig. 8a). The remaining orbital is the undisturbed $2p_z$ (Fig. 8b). Thus there will be three equivalent valencies in one plane and a fourth pointing at right angles to this plane (Fig. 8c). The three coplanar valencies form σ-bonds, and the $2p_z$ valency forms the so-called π-bond (pi-bond).

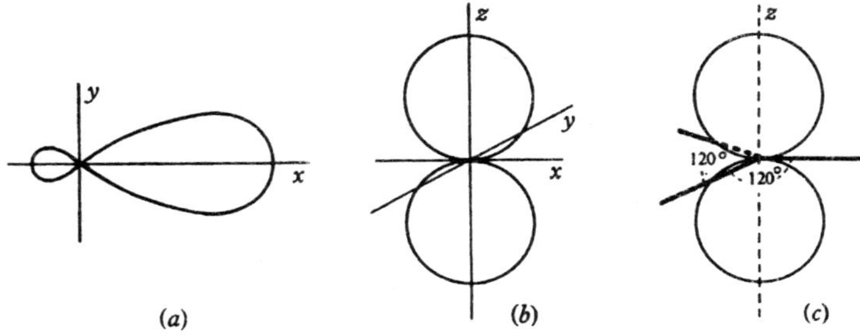

Fig. 8

The $2p_z$ electrons are known as π-electrons, *mobile* electrons or *unsaturation* electrons when they form the π-bond. The trigonal arrangement occurs in compounds containing a double bond, which is regarded as being made up of a strong bond (σ-bond) between two trigonal hybrid A.O.s of carbon, and a weaker bond (re-bond) due to the relatively small overlap of the two pure p_z orbitals in a plane at right angles to the trigonal hybrids. Fig. 9(a) shows the plan, and (b) the elevation of ethylene, $CH_2{=}CH_2$.

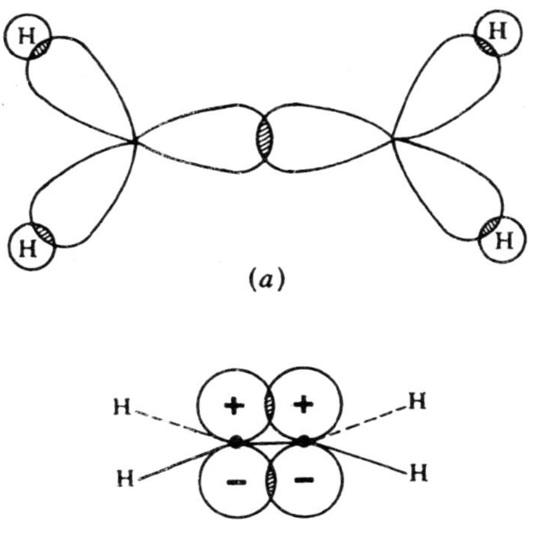

Fig. 9

The H—C—H angle in ethylene has been measured spectroscopically, and it has been found to be 119° 55' (Gallaway *et al.*, 1942). This is in agreement with the value expected for trigonal hybridisation.

It is the n-electrons which are involved in the electromeric and resonance effects.

When a compound contains two or more double bonds, the resulting M.O.s depend on the positions of these bonds with respect to one another.

(iii) In *digonal hybridisation*, only one 2s electron and the $2p_x$ electron are hybridised, resulting in *two equivalent collinear orbitals*; the $2p_y$ and $2p_z$ electrons remain undisturbed. Thus we get two equivalent valencies (forming the σ-type of bond) pointing in opposite directions along a straight line, and two other valencies (each forming a π-type of bond), one concentrated along the y-axis (the $2p_y$ orbital), and the other along the z-axis (the $2p_z$ orbital). The digonal arrangement occurs in compounds containing a *triple* bond, *e.g.*, acetylene.

From the above discussion, it can be seen that the carbon valency angles follow the order sp^3 (109° 28') < sp^2 (120°) < sp (180°). Thus, increasing the p-character of a bond decreases the bond angle, or increasing the p-character of a bond increases the bond angle. At first sight, it might have been anticipated that the bond angle is therefore fixed, *i.e.*, it depends only on the nature of the-hybridisation. In practice, it is found that bond angles are also dependent on the size (which produces a steric effect) and the electronegativity (which changes the amount of s-character in the bonds) of the atoms (or groups) attached to the carbon atom, *e.g.*, in $CHCl_3$, the Cl—C—Cl angles are increased from the normal angle of 109° 28' to about 111° and the Cl—C—H angles decreased to about 108°.

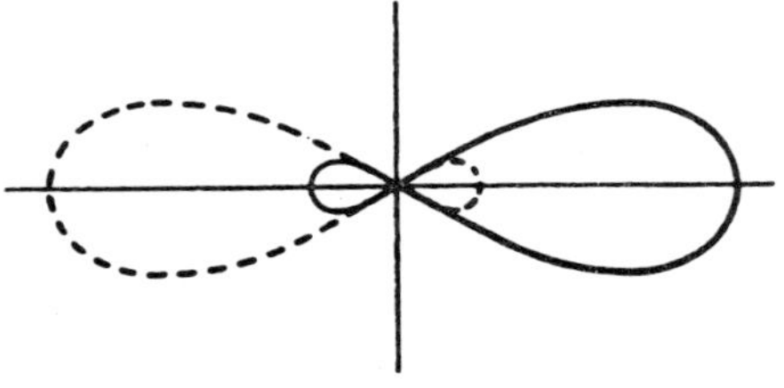

Fig. 10

When the electrons of any atom have been placed in hybridised orbitals, that atom is said 19 be in a 'valence state'. The atom on its own cannot exist in a valence state; energy is required to promote the atom to this condition.

This energy is obtained through the formation of bonds which are stronger with the hybridised orbitals than with the 'pure' orbitals, *i.e.*, more energy is released with the former than with the latter.

Shapes of Molecules

As we have seen, in *saturated* carbon compounds the four sp^3 orbitals point towards the corners of a tetrahedron. In ethylene we have used sp^2 trigonal hybridisation (one σ- and one re-bond) to describe the double bond. It is possible, however, to still use sp^3 hybridisation to describe ethylene. In this case two electrons are in one orbital of 'banana' shape ('tient' bond), and the other two electrons in a second 'banana' orbital, equivalent to the first but the mirror image of it. Thus ethylene may be represented (Fig. 11):

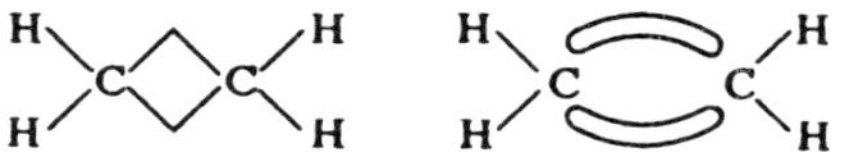

Fig. 11

It is interesting to note that this 'bent' bond method of representing ethylene is equivalent to Baeyer's description of a double bond. In the same way, the triple bond in acetylene (previously described in terms of one σ- and two π-bonds) may also be regarded as made up of three equivalent sp^3 hybrids symmetrically disposed round the C—C axis.

Quantum mechanical arguments show that both methods of representing these multiple bonds are equal to each other, each method having certain advantages. The σ-π bond method is more convenient for describing transitions from one state into another (*e.g.*, in electronic spectra), whereas the 'bent' bond method is more convenient for describing electron distribution in a molecule.

Now let us consider molecules in which the central atom has lone pairs. Consider nitrogen, with electron configuration $(1s)^2(2s)^2(2p)^3$. This has three 2p orbitals, and if each combines with a hydrogen atom, then the molecule of ammonia is $:NH_3$. In this molecule there is a lone pair $(2s)^2$, and the three hydrogen atoms, bound by the overlap with 2p orbitals, will therefore have a valency angle of 90°. The nitrogen atom is tercovalent, and by virtue of its lone pair, can act as a donor to become quadrivalent uni-electrovalent, *e.g.*,

$$:NH_3 + HCl \rightarrow NH_4^+ Cl^-$$

At first sight it might appear that the four hydrogen atoms in the ammonium ion are not equivalent; three bonds are formed from 2p electrons, and one by the lone $2s^2$. However, the fact is that all four hydrogen atoms in ^H^ are equivalent, and also the valency angle is about 109.5° (the tetrahedral value). Moreover, the valency angle in ammonia is about 107° and not 90° (the expected value from 2p bonding).

Now consider oxygen $(1s)^2(2s)^2(2p)^4$. One 2p orbital is doubly filled, and so, in water, one would expect that the two single 2p electrons would combine with hydrogen to form water in which the bond angle HOH is 90°.

Actually the bond angle is about 104.5°. Also, since the water molecule has two lone pairs, it can act as a donor to form the hydroxonium ion, $H_3O:^+$, and it is found that all the three hydrogen atoms are equivalent. The valency angles are nearly those expected of tetrahedral configurations, and the difference from 90° is far too large to be accounted for by repulsion between hydrogen atoms.

A satisfactory explanation is as follows. Sidgwick *et al.* (1940) assumed that lone pairs of electrons and bonding pairs were of equal importance, and that they arranged themselves symmetrically so as to minimise the repulsions between them. Thus pairs of electrons in a valency shell, whether a bonding pair or a lone pair, are always arranged in the same way, and this depends only on the total number of pairs.

Thus two pairs are arranged linearly (*e.g.*, $HgCl_2$), three pairs in the form of an equilateral triangle (*e.g.*, BCl_3), four pairs tetrahedrally (*e.g.*, CH_4), five pairs in the form of trigonal bipyramid and six pairs octahedrally. In all cases it is assumed that all the pairs of electrons occupy hybridised orbitals and only σ-bonds are present.

Now let us consider ammonia and water from this point of view. In each of the valency shells of these central atoms there are four pairs of electrons (nitrogen with one lone pair, and oxygen with two lone pairs). If these four pairs occupy four tetrahedrally hybridised orbitals, then the valency angles should be about 109.5°. As we have seen above, in ammonia the angle is about 107°, and in water about 104.5°. The problem then is to account for these deviations from the anticipated 'regular' shapes. This may be explained by assuming electrostatic repulsions between electron pairs in the valency shell (both bonding and lone pairs) are in the following order:

lone pair—lone pair > lone pair—bond pair > bond pair—bond pair.

This order can be explained on the basis that lone pairs are more concentrated than bonding pairs (the latter are 'stretched' in the formation of covalent bonds). Thus lone-pair electrons exert greater electrostatic repulsions on other lone pairs than on bonding pairs.

Consequently, when lone pairs occupy the valency shell, bonding pairs are 'forced together'. In CH_4 there are four bonding pairs, and so the distribution is symmetrical with a valency angle of 109° 28'. In NH_3, there are three bonding pairs and one lone pair, and since the latter has a bigger repulsive force, the bonding pairs are forced closer together (the bond angle is about 107°). It should be noted that the lone pair is also in a hybridised orbital, and so when the ammonia molecule is converted into the ammonium ion the four bonding pairs are in the same state of hybridisation, and consequently all four hydrogen atoms are equivalent.

In the water molecule (bond angle 104.5°), there are two lone pairs, and consequently the bonding pairs are 'closed up' more than in ammonia.

Finally, it should be noted that when deciding the shape of a molecule, it is the number of σ-electrons and lone pairs around the bonded atom that are counted; π-bonds are fitted in afterwards. The effect of re-bonds is to shorten bond lengths without affecting the basic shape of the molecule.

Bonding and anti-bonding orbitals : When two atomic orbitals overlap, the nature of the M.O. depends on whether the overlap occurs between two regions having the same sign or opposite signs.

When the former occurs, .the result is a bonding orbital, whereas when the latter occurs, the result is an anti-bonding orbital. In some cases, two orbitals of opposite sign cancel out each other; the result is non-bonding. In bonding orbitals, the electron charge is concentrated in the region between the two nuclei, thereby holding the two nuclei together.

On the other hand, in anti-bonding orbitals' electron charge is withdrawn from the region between the two nuclei, thereby resulting in increased repulsion between the two nuclei.

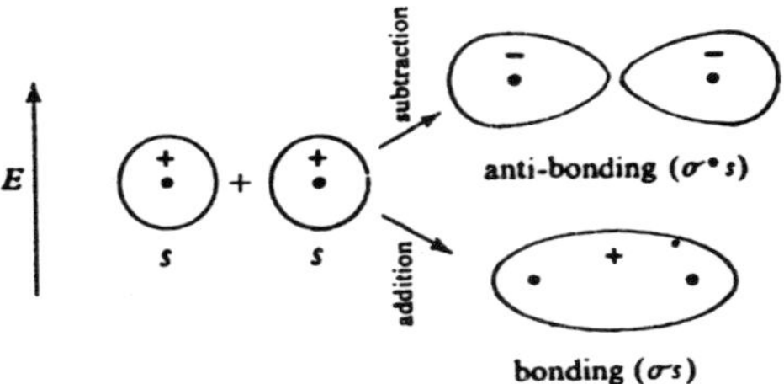

Fig. 12

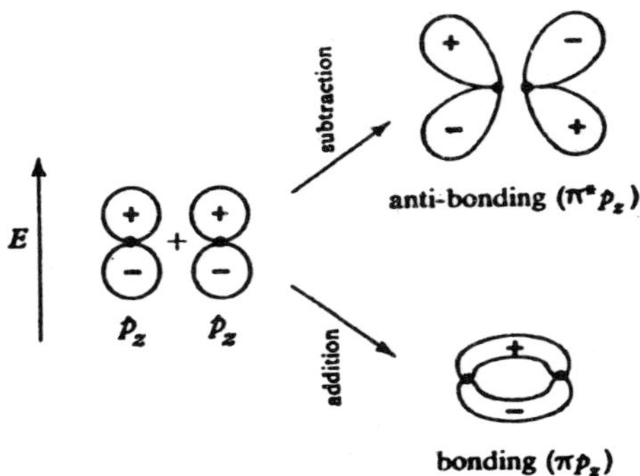

Fig. 13

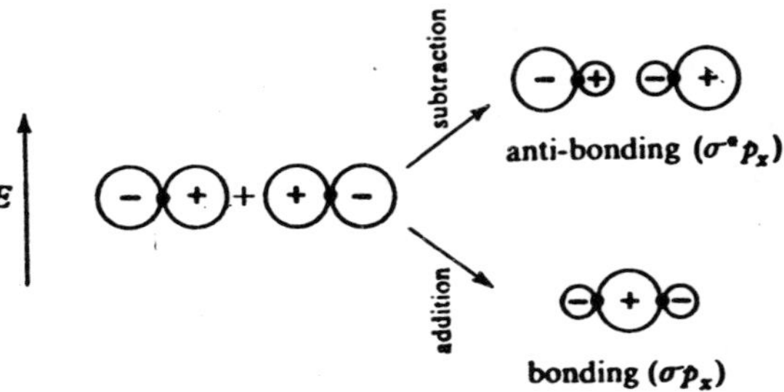

Fig. 14

The normal symbols, σ and π, used to designate bonding orbitals, are starred to designate the corresponding anti-bonding orbitals, *i.e.*, σ*, π*. Combinations of various A.O.s, the anti-bonding orbital involving the reversal of the sign (or signs) in one of the orbitals (arbitrarily the one on the right-hand side). In general, if n A.O.s are combined, then there are n M.O.s. Some of these are bonding, others anti-bonding, and in fewer cases, some are non-bonding.

METHODS OF DISTINGUISHING BETWEEN THE THREE CLASSES OF ALCOHOLS

1. By means of *oxidation*. The nature of the oxidation products of an alcohol depends on whether the alcohol is primary, secondary of tertiary.

(*i*) A *primary* alcohol on oxidation first gives an *aldehyde*, and this, on further oxidation, gives an acid. *Both aldehyde and acid contain the same number of carbon atoms as the original alcohol, e.g.*,

$$CH_3CH_2OH \xrightarrow{[O]} CH_3CHO \xrightarrow{[O]} CH_3CO_2H$$

(*ii*) A *secondary* alcohol, on oxidation, first gives a *ketone with the same number of carbon atoms as the original alcohol.* Ketones are fairly difficult to oxidise, but prolonged atoms than the original alcohol, *e.g.*, methyl-n-propylmethanol gives first pentan-2-one, and then a mixture of acetic and propionic acids:

$$CH_3CHOHCH_2CH_2CH_3 \xrightarrow{[O]} CH_3COCH_2CH_2CH_2CH_3 \xrightarrow{[O]}$$
$$CH_3CO_2H + CH_3CH_2CO_2H$$

(*iii*) *Tertiary* alcohols are resistant to oxidation in neutral or alkaline solution, but are readily oxidised by acid oxidising agents to a mixture of *ketone and acid, each containing fewer carbon atoms than the original alcohol.*

$$(CH_3)_2C(OH)CH_2CH_3 \xrightarrow{[O]} (CH_3)_2CO + CH_3CO_2H$$

The oxidising agents usually used for oxidising alcohols are: acid dichromate, acid or alkaline potassium permanganate, and dilute nitric acid.

The accepted mechanism for the oxidation of primary and secondary alcohols by dichromate and pemanganate involves an ester intermediate, *e.g.*, with acid dichromate:

$$Cr_2O_7^{2-} + H_2O \rightleftharpoons 2HCrO_4^-$$

$$Me_2CHOH + HCrO_4^- + 2H^+ \rightleftharpoons Me_2CH{-}O{-}CrO_3H_2^+ + H_2O$$

$$\underset{\underset{H_2O \quad H}{|}}{Me_2C}{-}O{-}CrO_3H_2^+ \xrightarrow{slow} Me_2CO + H_3O+ + H_2CrO_3$$

The rate-determining step is the removal of hydrogen from the intermediate ester: This has been demonstrated by Westheimer *et. al.* (1949), who showed that Me_2CDOH is oxidised at about 1/7 the rate of either Me_2CHOH or $(CD_3)_2CHOH$. Thus, the r/d step involves the fission of the C—H bond (this accounts for the kinetic isotope effect).

This mechanism cannot operate with t-alcohols, since there is no H—COH group present. According to Sager (1956), the mechanism involves dehydration, *e.g.*,

$$Me_3COH \xrightarrow[\text{(ii) } -H_2O]{\text{(i) } H^+} Me_3C^+ \xrightarrow{-H^+} Me_2C{=}CH_2 \xrightarrow{[O]} Me_2\overset{O}{C{-}}CH_2 \xrightarrow{[O]} Me_2CO + HCO_2H$$

2. The three classes of alcohols differ in their behaviour when the vapour is passed over copper at 300°C:

(*i*) A primary alcohols is dehydrogenated to an aldehyde, *e.g.*,

$$CH_3CH_2OH \xrightarrow[300°C]{Cu} CH_3CHO + H_2$$

(*ii*) A secondary alcohol is dehydrogenated to a ketone, *e.g.*,

$$CH_3CH(OH)CH_3 \xrightarrow[300°C]{Cu} CH_3COCH_3 + H_2$$

(*iii*) A tertiary alcohol is dehydrated to a alkene, *e.g.*,

$$(CH_3)_2C(OH)CH_2CH_3 \xrightarrow[300°C]{Cu} (CH_3)_2C{=}CHCH_3 + H_2O$$

Methyl alcohol, methanol (*carbinol*) is prepared industrially by several methods. The earliest method was the destructive distillation of wood, whereby tar and an aqueous fraction known as *pyroligneous acid* are obtained. Pyroligneous acid contains methanol, acetone and acetic acid, and all three compounds may be obtained by suitable treatment. It was this method which gave rise to the name 'wood spirit' for methanol. The modern methods are synthetic.

(*i*) Synthesis gas is passed at a pressure of 200 atmospheres over a catalyst containing the oxides of copper, zinc and chromium at 300°C:

$$CO + 2H_2 \longrightarrow CH_3OH$$

If the proper precautions are taken, the yield of methanol is almost 100 per cent, an its purity is above 99 per cent. By changing the catalyst and the ratio of carbon monoxide to hydrogen, methanol and a variety of higher alcohols are produced.

(*ii*) By the catalytic oxidation of methane. A mixture of methane and oxygen (ratio by volume of 9 : 1) at a pressure of 100 atmospheres is passed through a copper tube at 200°C:

$$CH_4 + \tfrac{1}{2}O_2 \longrightarrow CH_3OH$$

Methanol is a colourless, inflammable liquid, b.p. 64°C, and is poisonous. It is miscible with water in all proportions, and is also miscible with most organic solvents.

It burns with a faintly luminous flame, and its vapour forms explosive mixtures with air or oxygen when ignited. It combines with calcium chloride to form $CaCl_2{\cdot}4CH_3OH$, and hence cannot be dried this way (*cf.* ethanol).

Methanol is used as a solvent for paints, varnishes, shellac celluloid cements, etc.; in the manufacture of dyes, perfumes, formaldehyde, etc. It is also used for making methylated spirit and automobile antifreeze mixtures.

STRUCTURE OF METHANOL

Analysis and molecular-weight determinations show that the molecular formula of methanol is CH_4O. Assuming that carbon is quadrivalent, oxygen bivalent and hydrogen univalent, only one structure is possible:

$$\begin{array}{c} \text{H} \\ | \\ \text{H—C—O—H} \\ | \\ \text{H} \end{array} \quad \text{or} \quad CH_3OH$$

This is supported by all the chemical reactions of methanol, *e.g.*, (*i*) only one hydrogen atom is methanol is replaceable by sodium; this suggests that one hydrogen atom is in a different state of combination from the other three.

(*ii*) Methanol is formed from methyl chloride by hydrolysis with sodium hydroxide. Methyl chloride can have only the structure CH_3Cl. It is reasonable to suppose that the methyl group in methyl chloride is unchanged by the action of dilute alkali, and that the reaction takes place by the replacement of the chlorine atom by a hydroxyl group.

(*iii*) The presence of the hydroxyl group is confirmed, for example, by the reaction between methanol and phosphorus pentachloride, when methyl chloride, hydrogen chloride and phosphoryl chloride are formed. Thus one oxygen atom (bivalent) and one hydrogen atom (univalent) have been replaced by one chlorine atom (univalent). This implies that the oxygen and hydrogen atoms exist as a univalent group in methanol: the only possibility is as a hydroxyl group, OH. It is the hydrogen of the hydroxyl group which is displaced by sodium.

Infra-red spectroscopic studies of hyroxy-compounds show that they absorb (O—H stretch) in the region of 3650 - 3580 cm^{-1}(v). This is true only if there is no hydrogen bonding. Intermolecular hydrogen bonding produces absorption) broad band in the region 3350 – 3230 cm^{-1}(v), whereas intramolecular hydrogen bonding absorbs in the region 3590 – 3420 cm^{-1} (v), Moreover, the former shows changes in intensity and frequency shifts on dilution, whereas the latter is unaffected by dilution; thus the two may be distinguished.

In addition to O—H stretching vibrations, there are also C—O stretching vibrations which are characteristic of the type of the type of alcohol: primary, close to 1050 (s); secondary, close to 1100 (s); tertiary, close to 1150 cm^{-1}(s). Thus the classes of alcohols may be identified in these regions.

However, a further point that may be mentioned here is the τ-value for methylene protons in the —CH_2—O—group is between 6.5–6.65 p.p.m., and when the alcohol is acetylated, the τ-value changes to ~ 5.9. This change may be used to identify a primary alcohol. Secondary alcohols can also be identified; > CH—O—(τ = ~ 6.1) and the acetylated alcohol (τ = ~ 5.0). These down-field shifts are due to the increased deshiedling effect of the acetoxyl group ($OOCCH_3$) compared to hydroxyl oxygen.

Mass Spectra

The mass spectrum of ethanol but have we shall describe the mass spectra of alcohols from a general point of view. Usually, strong peaks are shown by fragmentation involving β-cleavage (*i.e.*, the C—COH bond), and the fragment containing the hydroxyl group is stabilised by resonance: The alkyl group with the heaviest mass is eliminated preferentially, but ions also appear for alternative eliminations. Further fragmentation may also occur as follows (R^2 = H, R_3= C_2H_5): Hence, both *s*- and *t*-alcohols can also show peaks at *m/e* 31 (the most characteristic ion of alcohols).

Ethyl alcohol, ethanol, is prepared industrially by several methods, *e.g.*, ethylene (from cracked petroleum) is absorbed in concentrated sulphuric acid (98 per cent) at 75-80°C, under pressure (17.5–35kg/cm^2). Ethyl hydrogen sulphate and ethyl sulphate are formed:

$$C_2H_4 + (HO)_2SO_2 \longrightarrow C_2H_5OSO_2OH$$

$$C_2H_5OSO_2OH + C_2H_4 \longrightarrow (C_2H_5O)_2SO_2$$

The reaction mixture is then diluted with about an equal volume of water, and warmed. Hydrolysis takes place and ethanol together with some diethyl ether is formed:

$$C_2H_5OSO_2OH + H_2O \longrightarrow C_2H_5OH + H_2SO_2$$

$$(C_2H_5O)_2SO_2 + 2H_2O \longrightarrow 2C_2H_5OH + H_2SO_4$$

$$(C_2H_5O)_2SO_2 + C_2H_5OH \longrightarrow (C_2H_5)_2O + C_2H_5OSO_2OH$$

The ether is kept to a minimum by separating the ethyl sulphate from the reaction products, and hydrolysing it separately. The hydrolysed liquids are distilled, and the aqueous ethanol distillate is concentrated by fractional distillation.

Ethanol is also manufactured by the *direct* hydration of ethylene with steam under pressure in the presence of a suitable catalyst, *e.g.*, phosphoric acid on a support.

$$C_2H_4 + H_2O \longrightarrow C_2H_5OH$$

Some ether (~ 5 per cent) is formed as a by-product.

The earliest method of preparing ethanol is by fermentation and this is still used for the manufacture of beer, wine, brandy, etc., and also a source of ethanol. The starting material is starch, which is obtained from sources depending on the particular country: common sources of starch are wheat, barely, potato, etc. Molasses is also used as the starting material for ethanol. The grain, *e.g.*, wheat or barley, is mashed with hot water, and then heated with malt (germinated barley) at 50°C for 1 hour. Malt contains the enzyme *diastase* which, by hydrolysis, converts starch into the sugar, maltose (*q.v.)*:

$$(C_6H_{10}O_5)_n + \frac{n}{2}H_2O \xrightarrow{\text{maltase}} \frac{n}{2}C_{12}H_{22}O_{11}$$

If molasses is used, then this step is unnecessary, since it contains carbohydrates already present as sugars which can be fermented.

The liquid is cooled to 30°C and fermented with yeast for 1–3 days. Yeast contains various enzymes, among which are *maltase*, which converts the maltose into glucose, and *zymase*, which converts the glucose into ethanol:

$$C_{12}H_{22}O_{11} + H_2O \xrightarrow{\text{maltase}} 2C_6H_{12}O_6$$

$$C_6H_{12}O_6 \xrightarrow{\text{zymase}} 2C_2H_5OH + 2CO_2$$

The carbon dioxide is recovered and sold as a by-product. The fermented liquor or 'wort', which contains 6-10 per cent ethanol and some other compounds, is fractioned into three fractions:

(*i*) *First runnings*, which consists mainly of acetaldehyde.

(*ii*) *Rectified spirit*, which is 93–95 per cent w/w ethanol.

(*iii*) *Final runnings* or *fusel oil*, which contains n-propyl, n-butyl, isobutyl, n-amyl, isoamyl and 'active] amyl alcohol.

Industrial alcohol is ordinary rectified spirit. *Methylated spirit* is of two kinds: (*a*) *Mineralised methylated spirit* is 90 per cent rectified spirit, 9 per cent methanol and 1 per cent petroleum oil, and a purple dye. (*b*) *Industrial methylated spirit* is 95 per cent rectified spirit and 5 per cent methanol, whose purpose is to '*denature*' the rectified spirit, *i.e.*, make it unfit for drinking purposes.

Absolute alcohol is 99.5 per cent ethanol, and is obtained from rectified spirit. When an aqueous solution of ethanol is fractionated, it forms a constant-boiling mixture containing 96 per cent ethanol from which 100 per cent ethanol may be obtained by adding a small amount of benzene, and then distilling. The first fraction is the ternary azeotrope, *i.e.*, a constant-boiling mixture containing three constituents, b.p. 64.8°C (water, 7.4 per cent; ethanol, 18.5 per cent; benzene, 74.1 per cent). After all the water has been removed, the second fraction that distils over is the binary azeotrope, b.p. 68.2°C (ethanol, 32.4 per cent; benzene, 67.6 per cent). After all the benzene has been removed, pure ethanol, b.p. 78.1°C, distils over.

Ethanol cannot be dried by means of calcium chloride, since a compound (an *alcoholate*) is formed, *e.g.*, $CaCl_2 \cdot 3C_2H_5OH$ (*cf.* methanol). Distillation of rectified spirit over calcium oxide, and then over calcium, gives absolute alcohol. This method is often used in the laboratory, and was formerly used industrially. Ethanol is a colourless, inflammable liquid, b.p. 78·1°C. It is miscible with water in all proportions, and is also miscible with most organic solvents. Ethanol ad methanol resemble each other very closely, but they may be distinguished :

(*i*) by the fact that ethanol gives the haloform reaction (*q.v.*), whereas methanol does not; and

(*ii*) ethanol gives acetic acid on oxidation; methanol gives formic acid. These two acids are readily distinguished from each other.

Ethanol is used for the preparation of esters, ether, chloral, chloroform, etc. It is also used as a solvent for gums, resins, paints, varnishes, etc., and as a fuel.

Structure of Ethanol

Analysis and molecular-weight determinations show that the molecular formula of ethanol is C_2H_6O. Assuming that carbon is quadri-valent, oxygen bivalent, and hydrogen univalent, two structures are possible.

$$\underset{\text{(I)}}{CH_3-CH_2-OH} \qquad \underset{\text{(II)}}{CH_3-O-CH_3}$$

(*i*) Only one hydrogen atom in ethanol is replaceable by sodium or potassium. This indicates that one hydrogen atom is in a different state of combination from the other five. In (I), one hydrogen atom differs from the other five, but in (II) all the hydrogen atoms are equivalent.

(*ii*) When ethanol is treated with hydrochloric acid or phosphorus pentachloride, one oxygen atom (bivalent) and one hydrogen atom (univalent) are replaced by one chlorine atom (univalent) to give ethyl chloride, C_2H_5Cl. This implies the presence of a hydroxyl group (*cf.* methanol).

(*iii*) When ethyl chloride is hydrolysed with dilute alkali, ethanol is obtained. This reaction also indicates the presence of a hydroxyl group in ethanol.

(*iv*) Ethanol may be prepared as follows:

$$C_2H_6 \xrightarrow{Cl_2} C_2H_5Cl \xrightarrow{NaOH} C_2H_5OH$$

The arrangement of the six hydrogen atoms in ethane is known, and it is reasonable to suppose that five retain their original arrangement in ethyl chloride and ethanol, since these five hydrogen atoms do not enter (presumably) into the above reactions. Thus there is an ethyl group C_2H_5—in ethanol. This is so in (I), but not in (II).

(*v*) Structure (II) is definitely eliminated, since it can be shown that it is the structure of dimethyl ether (*q.v.*), a compound that has very little resemblance, physically or chemically, to ethanol. Hence, (I) is accepted as the structure of ethanol, and it accounts for all the known properties of ethanol.

It can be seen from the various examples given on structure determination, *e.g.*, methane, butanes, ethylene, methanol and ethanol, that the method of approach follows certain definite lines. First the molecular formula is obtained. Then, if the compound is simple–in the sense that it contains a small number of unlike atoms, and that the total number of atoms is also small–the valencies of the atoms present are assumed, and various possible structures are written down. If the compound is 'simple' the number of possible structures will not be large (four or five at the most). Then by considering the chemical properties of the compound in question, the structure which best fits the observed facts is accepted as the correct one.

If the compound is not 'simple', the procedure is to detect the presence of as many functional groups as possible; to degrade the compound into simpler substances whose structures are already known or which may be determined by further degradation; to build up structural formulae based on the facts obtained; and then to choose the structure which best fits the facts. Finally a synthesis is attempted, and if successful, will usually give proof of the correctness of the structure suggested.

It should be noted that the arguments used for structure determination by purely chemical methods are based on the assumption that no rearrangement have occurred in the series of reactions performed. Thus, the possibility of rearrangement must always be borne in mind when carrying out chemical reactions.

As was pointed out physical methods are being used nowadays for structural determination. Once the molecular formula is obtained, functional groups may be identified, etc. In this way, the structures of many compounds have now been ascertained with a minimum of chemical evidence.

Propyl Alcohols

Two isomeric propyl alcohols are possible, and both are known.

*n-Propyl alcohol, propan-*1*-ol, n-propanol*, was originally obtained from fusel oil (see above), but it is now produced by the hydogenation of carbon monoxide. A more recent method is by the catalytic reduction of propargyl alcohol (from acetylene and formaldehyde).

$$CH{\equiv}CCH_2OH + 2H_2 \xrightarrow{Ni} 2CH_3CH_2CH_2OH$$

n-Propanol, b.p. 97.4°C, is miscible with water, ethanol and ether. It is used in the preparation of propionic acid, toilet preparations such as lotions, etc.

Isopropyl alcohol, propan-2-ol, isoprpanol, is prepared industrially by passing propene (from cracked petroleum) into concentrated sulphuric acid, then diluting with water, and distilling off the isopropanol. Isopropyl ether is obtained as a by-product (*cf.* ethanol):

$$CH_3CH{=}CH_2 + (HO)_2SO_2 \longrightarrow (CH_3)_2CHOSO_2OH$$

$$(CH_3)_2CHOSO_2OH + CH_3CH{=}CH_2 \longrightarrow [(CH_3)_2CHO]_2SO_2$$

$$(CH_3)_2CHOSO_2OH + H_2O \longrightarrow (CH_3)_2CHOH + H_2SO_4$$

$$[(CH_3)_2CHO]_2SO_2 + 2H_2O \longrightarrow 2(CH_3)_2CHOH + H_2SO_4$$

$$[(CH_3)_2CHO]_2SO_2 + (CH_3)_2CHOH \longrightarrow [(CH_3)_2CH]_2O + (CH_3)_2CHOSO_2SO_4$$

Isopropanol is also prepared by *direct* hydration by passing a mixture of propene and steam at 220-250°C, and under pressure (220 atm), over a catalyst of tungsten oxide plus zinc oxide, on a silica carrier. Isopropanol, b.p. 82.4°C, is soluble in water, ethanol and ether. It is used for preparing esters, acetone, keten, as a solvent, and for high-octane fuel.

Butyl Alcohols

Four isomers are possible, and all are known.

*n-Butyl alcohol, butan-*1*-ol, n-butanol*, b.p. 117.4°C, is prepared industrially from propene by the Oxo process. It is widely used as a solvent.

Isobutyl alcohol, 2*-methylpropan-*1*-ol, n-butanol*, b.p. 108°C, is also obtained by the Oxo process. It behaves as a primary alcohol, but it readily rearranges due to the presence of a branched chain near the COH group, *e.g.*, when it is treated with hydrochloric acid, isobutyl chloride and t-butyl chloride are obtained:

$$(CH_3)_2CHCH_2OH \xrightarrow{HCl} (CH_3)_2CHCH_2Cl + (CH_3)_3CCl$$

*s-Butyl alcohol, butan--*2*-ol, s-butano*, b.p. 100°C, is prepared industrially by the hydration of 1- or 2-butene (from cracked petroleum) by means of concentrated sulphuric acid (*cf.* ethanol and isopropanol). It is used for the preparation of butanone, esters, and as a lacquer solvent.

t-Butyl alcohol, t-butanol (*trimethylmethanol*), m.p. 25.5°C, is prepared synthetically by the hydration of isobutene (from cracked petroleum). It is mainly used as an alkylating agent.

Amyl Alcohols

Eight isomers are possible, and all are known:

1. $CH_3CH_2CH_2CH_2CH_2OH$, n-amyl alcohol, pentan 1-ol, n-pentanol, p.b. 138°C.
2. $(CH_3)_2CHCH_2CH_2OH$, isoamyl alcohol, isopentanol, p.b. 130°C.
3. $CH_3CH_2CH(CH_3)CH_2OH$, 'active' amyl alcohol, 2-methylbutan-1-ol, b.p. 128°C.
4. $(CH_3)_3CCH_2OH$, neopentyl alcohol, t-butylcarbinol, b.p. 113°C.
5. $CH_3CH_2CH_2CH(OH)CH_3$, pentan-2-ol, methyl-n-propylcarbinol, p.b. 120°C.
6. $CH_3CH_2CH(OH)CH_2CH_3$, pentan-3-ol, diethylcarbinol, b.p. 117°C.
7. $(CH_3)_2CHCH(OH)CH_3$, 3-methylbutan-2-ol, isopropylmethyl-carbinol, b.p. 114°C.
8. $(CH_3)_2C(OH)CH_2CH_3$, 2-methylbutan-2-oil, t-amyl alcohol, t-pentanol, ethyldimethylcarbinol, b.p. 102°C.

There amyl alcohols, *viz.*, n-pentanol, isopentanol and 'active' amyl alcohol, have been isolated from fusel oil (*q.v.*). The last two are the chief constituents of fusel oil, and all three are produced by the fermentation of protein matter associated with the carbohydrates in starch. This mixture of amyl alcohols (from fusel oil) is used for the preparation of esters (for artificial essences), scents, and as a laboratory reducing agent with sodium it is better than ethanol owing to its higher boiling point (this mixture of amyl alcohols will be referred to in future as isopentanol).

A mixture of amyl alcohols, known as *pentasol*, is prepared industrially by chlorinating at 200°C, and in the dark, a mixture of n- and isopentanes (from petroleum) to the amyl chlorides which are hydrolysed with dilute sodium hydroxide solution plus a little sodium oleate for emulsification, to the amyl alcohols.

Seven isomeric amyl chlorides are theoretically possible, but in practices six are obtained. 2-Chloro-2-methylbutane (the only t-chloride produced) readily eliminates hydrogen chloride to form trimethylethylene. Pentasol (the mixture of six amyl alcohols) finds great use as a solvent in the lacquer industry.

Commercial 's-amyl alcohol' (80 per cent 2- and 20 per cent 3-pentanol) is made by hydrating 1-and 2-pentenes (from cracked petroleum).

A number of higher alcohols occur as esters in waxes, *e.g.*, cetyl (palmityl) alcohol (hexadecan-1-ol), $C_{16}H_{33}OH$, m.p. 49°C, occurs as the palmitate in spermaceti (obtained from the oil of the sperm whale); carnaubyl alcohol (tetracosan-1-ol), $C_{24}H_{49}OH$, m.p. 69°C, as esters in wool-grease; ceryl alcohol (hexacosan-1-ol), $C_{26}H_{53}OH$, m.p. 79°C, as the cerotate in chinese wax; myricyl (melissyl) alcohol (triacontan-1-ol), $C_{30}H_{61}OH$, m.p. 88°C, as esters in bees-wax.

A number of higher alcohols are now prepared industrially by the catalytic reduction of the ethyl (or glyceryl) esters of the higher monocarboxylic acids, particularly the alcohols lauryl (dodecan-1-ol), $C_{12}H_{25}OH$, m.p. 24°C; myristyl (tetradecan-1-ol), C14H29OH, m.p. 38°C; palmityl (hexadecan-1-ol), $C_{16}H_{33}OH$, m.p. 49°C; and stearyl (octaecan-1-ol), $C_{18}H_{37}OH$, m.p. 59°C.

These alcohols are used in the form of sodium alkyl sulphates, $ROSO_2ONa$, as detergents, emulsifying agents, wetting agents, insecticides, fungicides, etc. These sodium salts lather well, and are not affected by hard water, and hence can be used as a soap substitute.

DIHYDRIC ALCOHOLS OR GLYCOLS

Dihydric alcohols are compounds containing two hydroxyl groups. They are classified as α, β, γ . . . glycols, according to the relative positions of the two hydroxyl groups: α is the 1,2 glycol; β, 1,3; γ, 1,4.

Although, it is unusual to find a compound with two hydroxyl groups attached to the same carbon atom, ether derivatives of these, 1,1 glycols are stable, *e.g.*, acetals. The commonest glycols are the α-glycols.

Nomenclature

The common names of the α-glycols are derived from the corresponding alkene from which they may be prepared by direct hydroxylation, *e.g.*,

$CHCH_2CH_2OH$ ethylene glycol

$(CH_3)_2COHCH_2OH$ isobutene glycols,

β-, γ . . . Glycols are named as the corresponding *polymethlene* glycols, *e.g.*,

$HOCH_2CH_2CH_2OH$ trimethylene glycol

$HOCH_2CH_2CH_2CH_2CH_2OH$ pentamethylene glycol

According to the I.U.P.A.C. system of nomenclature, the class suffix is *-diol*, and numbers are used to indicate the positions of side-chains and two hydroxyl groups *e.g.*,

$$HOCH_2CH_2\underset{}{\overset{CH_3}{\overset{|}{C}}}HCH_2\overset{CH_3}{\overset{|}{C}}HCH_2OH$$ 2, 4-dimethylhexane-1,6-diol

Ethylene glycol, glycol (*ethane*-1,2-*diol*), is the simplest glycol, and may be prepared as follows:

(*i*) By passing ethylene into cold dilute alkaline permanganate solution.

(*ii*) By passing ethylene into hypochlorous acid, and then hydrolysing the ethylene chlorohydrin by boiling with aqueous sodium hydrogen carbonate:

$$CH{=}CH_2 + HOCl \longrightarrow ClCH_2CH_2OH \xrightarrow{\text{aq. NaHCO}_3} HOCH_2CH_2OH + NaCl + CO_2$$

(*iii*) By boiling ethylene dibromide with aqueous sodium carbonate:

$$BrCH_2CH_2Br + NaCO_3 + H_2O \longrightarrow HOCH_2CH_2OH + 2NaBr + CO_2\ (50\%)$$

The low yield in this reaction is due to the conversion of some ethylene dibromide into vinyl bromide:

$$BrCH_2CH_2Br + Na_2CO_2 \longrightarrow CH_2{=}\ CHBr + NaBr + NaHCO_3$$

If aqueous sodium hydroxide is used instead of sodium carbonate, vinyl bromide is again obtained as by-product. The best yield of glycol is obtained by heating ethylene dibromide with potassium acetate in glacial acetic acid, and subsequently hydrolysing the glycol diacetate with hydrogen chloride in methanolic solution:

$$BrCH_2CH_2Br + 2CH_3CO_2K \longrightarrow$$

$$CH_2(OCOCH_3)CH_2(OCOCH_3) + 2KBr \xrightarrow{HCl} HOCH_2CH_2OH$$
(90%) (83-84%)

(*iv*) By treating ethylene oxide with dilute hydrochloric acid:

$$\underset{CH_2}{\overset{CH_2}{|}} \!\!> O + H_2O \xrightarrow{HCl} HOCH_2CH_2OH$$

Glycol is prepared industrially by method (*iv*), and by the catalytic reduction of methyl glycollate which is produced synthetically.

Glycol is a colourless viscous liquid, p.b. 197°C, and has a sweet taste (the prefix *glyc*-indicates that the compound has a sweet taste: Greek *glukus*, sweet). It is miscible in all proportions with water and ethanol, but is insoluble in ether. It is widely used as a solvent and as an antifreeze agent.

The high boiling points and high solubility i water of polyhydric alcohols are due to hydrogen bonding involving all hydroxyl groups.

The chemical reactions of glycol are those which might have been expected of a monohydric primary alcohol. One hydroxyl group, however, is almost always completely attacked before the other reacts.

(*i*) Sodium at 50°C forms the monosodium salt, and the di-sodium salt only when the temperature is raised to 160°C.

(*ii*) Hydrogen chloride forms ethylene chlorohydrin at 160°C, whereas at 200°C, ethylene dichloride is formed.

Ethylene chlorohydrin (β-chloroethyl alcohol, 2-chloroethanol) is a colourless liquid, p.b. 128.8°C.

It is very useful in organic syntheses since it contains two different reactive groups; *e.g.*, by heating with aqueous sodium cyanide it may be converted into ethylene cyanohydrin, which, on hydrolysis, gives β-hydroxypropionic acid:

$$HOCH_2CH_2Cl + NaCH \longrightarrow HOCH_2CH_2CN + NaCl \ (79\text{-}80\%)$$

$$HOCH_2CH_2CN \xrightarrow{HCl} HOCH_2CH_2CO_2H$$

(*iii*) When glycol is treated with phosphorus trichloride or phosphorus tribromide, the corresponding ethylene dihalide is obtained:

$$\begin{matrix} CH_2OH \\ | \\ CH_2OH \end{matrix} \xrightarrow{PBr_3} \begin{matrix} CH_2Br \\ | \\ CH_2Br \end{matrix}$$

Phosphorus tri-iodide, however, produces ethylene. Ethylene di-iodide is formed as an intermediate, but readily eliminates iodine :

$$\begin{matrix} CH_2OH \\ | \\ CH_2OH \end{matrix} \xrightarrow{PI_3} \left[\begin{matrix} CH_2I \\ | \\ CH_2I \end{matrix}\right] \longrightarrow I_2 + \begin{matrix} CH_2 \\ \| \\ CH_2 \end{matrix}$$

(*iv*) When glycol is treated with organic acids or inorganic oxygen acids, the mono- or di-esters are obtained, depending of the relative amounts of glycol ad acid; *e.g.*, glycol diacetate is obtained by heating glycol with acetic acid in the presence of a small amount of sulphuric acid as catalyst:

$$\begin{matrix} CH_2OH \\ | \\ CH_2OH \end{matrix} + 2CH_3CO_2H \xrightarrow{PI_3} \begin{matrix} CH_2OCOCH_3 \\ | \\ CH_2OCOCH_3 \end{matrix} + 2H_2O$$

When glycol is heated with dicarboxylic acids, condensation polymers (polyesters) are obtained:

$$nHOCH_2CH_2OH + nHO_2C(CH_2)_xCO_2H \longrightarrow$$
$$HOCH_2(CH_2)_xCOOCH_2]_nCH_2OH$$

(*v*) Glycol condenses with aldehydes or ketones in he presence of *p*-toluenesulponic acid to yield respectively cyclic acetals or cyclic ketals (1,3-*dioxolans*):

$$\begin{matrix} CH_2OH \\ | \\ CH_2OH \end{matrix} + O{=}C\begin{matrix} R^1 \\ \\ R^2 \end{matrix} \longrightarrow \begin{matrix} CH_2O \\ | \\ CH_2O \end{matrix}\!\!>\!C\begin{matrix} R^1 \\ \\ R^2 \end{matrix} + H_2O$$

Dioxolan formation offers a means of protecting a carbonyl group in reactions carried out in alkaline media. The carbonyl compound may be regenerated by the addition of periodic acid to an aqueous dioxan solution of the dioxolan:

$$\begin{matrix} R^1 \\ \\ R^2 \end{matrix}\!\!>\!C\!<\!\!\begin{matrix} O-CH_2 \\ | \\ O-CH_2 \end{matrix} \xrightarrow{HIO_4} R^1COR^2 + 2CH_2O$$

(*vi*) When glycol is oxidised with nitric acid, glycollic and oxalic acids may be readily isolated. All theoretically possible oxidation products have been isolated, but since, except for glycollic and oxalic acids, they are more readily oxidised than glycol itself, they have only been obtained in very poor yields. Oxidation of glycols with acid, alkaline or neutral permanganate, acid dichromate or chromic acid (chromium trioxide in acetic acid), lead tetra-acetate or periodic acid results in fission of the carbon-carbon bond to give aldehydes, acids and/or ketones.

Oxidation of 1,2-glycols occurs more rapidly with *cis*-glycols than with the corresponding *trans*-isomers, and this has been interpreted as evidence for the formation of cyclic intermediates, *e.g.* The fact he *trans*-isomers *are* oxidise suggests that the reaction proceeds through a noncyclic intermediate for these compounds. This could be the non-cyclic ester formed in the first step. These oxidations are also subject to steric hindrance, *e.g.*, glycol is oxidised much faster than pinacol by periodic acid. According to Bunton *et al.* (1957), the rate-determining step for glycol oxidation is the fission of the complex, whereas the for pinacol is the formation of the complex.

(*vii*) Glycol is converted in acetaldehyde when heated with dehydrating agents, *e.g.*,

$$HOCH_2CH_2OH \xrightarrow{ZnCl_2} CH_3CHO + H_2O$$

On the other hand, when glycol is heated with a dehydrating agent such as phosphoric acid, *polyethylene glycols* are obtained, *e.g.*, diethylene glycol:

$$2HOCH_2CH_2OH \xrightarrow{-H_2O} HOCH_2CH_2OCH_2CH_2OH$$

By varying the amount of phosphoric acid and the temperature, the polyethylene glycols up to decaethylene glycol can be obtained. These condensation polymers contain both the alcohol and ether functional groups; they are soluble in water (alcohol function) and are very good solvents (ether function). They are soluble in water (alcohol function) and are very good solvents (ether function). They are widely used as solvents for gums, resins, cellulose esters, etc.

Ethylene oxide, b.p. 10.7°C. According to the I.U.P.A.C. system of nomenclature, an oxygen atom linked to two of the carbon atoms in a carbon chain is denoted by the prefix *epoxy* in all cases other than those in which a substance is named as a cyclic compound; thus ethylene oxide is *epoxyethane*. Epoxy-compounds contain the *oxiran* (*oxirane*) ring, and ethylene oxide is also known as *oxiran* (*oxirane*). Oxiran compounds are also referred to as *cyclic ethers* or *alkene oxides*. Epoxides may, in general, be prepared by epoxidation of alkenes with peroxy acids. Ethylene oxide may be prepared as follows:

$$HOCH_2CH_2Cl + KOH \longrightarrow \overset{\diagup O \diagdown}{CH_2—CH_2} + KCl + H_2O$$

The mechanism is possibly:

$$HO^- \; H-O(-CH_2-CH_2Cl) \xrightarrow{fast} H_2O + CH_2(-O^-)-CH_2 \; Cl$$

$$\xrightarrow{slow} CH_2-CH_2\,(epoxide,\ -O-) + Cl^-$$

This is an example of neighbouring group participation.

Ethylene oxide is manufactured by passing ethylene and air under pressure over a silver catalyst at 200-400°C:

$$C_2H_4 + 1/2O_2 \longrightarrow CH_2-CH_2\ (\text{bridged by O})$$

Ethylene oxide undergoes molecular rearrangement on heating of form acetaldehyde:

$$CH_2-CH(H)\ (\text{bridged by O}) \xrightarrow[\text{shift}]{H^-} CH_3CHO$$

Epoxides are reduced by lithium aluminium hydride to alcohols, unsymmetrical epoxides giving as main product the more highly substituted alcohol; thus (I) is the main product:

$$RCH-CH_2\ (\text{bridged by O}) \xrightarrow{LiAlH_4} \underset{(I)}{RCHOHCH_3} + \underset{(II)}{RCH_2CH_2OH}$$

On the other hand (II) is the main product if LAH—$AlCl_3$ is used (Eliel *et al.*, 1958. Brown *et al.* (1968) have shown that (II) is the major product (> 90%) when BH_3—BF_3 is the reducing reagent.

Epoxides are readily reduced to the corresponding alkenes by t-phosphines, *e.g.*,

$$RCH-CHR\ (\text{bridged by O}) + Ph_3P \longrightarrow RCH=CHR + Ph_3PO$$

They are also oxidised by dimethyl sulphoxide to α-hydroxyketones (acyloins).

$$RCH-CHR\ (\text{bridged by O}) + Me_2SO \xrightarrow{H^+} RCH(OH)-CR(H)(O-\overset{+}{S}Me_2) \longrightarrow RCHOHCOR + H^+ + Me_2S$$

Ethylene oxide is converted into ethylene glycol in *dilute* acid solution and into ethylene halogenohydrins with *concentrated* halogen acid solutions. Ethylene oxide also forms monoethers with alcohols in the presence of a small amount of acid as catalyst. The mechanisms of these reactions in acid media probably follow the same pattern, *e.g.*, with *dilute* acid (with alcohols, replace H_2O by ROH):

$$\text{(CH}_2\text{)}_2\text{O} + H_3O^+ \rightleftharpoons \text{(CH}_2\text{)}_2\overset{+}{O}H + H_2O$$

$$H_2O + \text{(CH}_2\text{)}_2\overset{+}{O}H \longrightarrow H_2\overset{+}{O}-CH_2-CH_2-OH \xrightarrow{-H^+} CH_2OH-CH_2OH$$

Attack occurs from the position remote from the ^+OH, and the product is the *trans* diol. This cannot be demonstrated with ethylene oxide itself but has been, *e.g.*, with 1,2-epoxycyclohexane. With *concentrated* halogen acid.

$$Cl^- + \text{(CH}_2\text{)}_2\overset{+}{O}H \longrightarrow CH_2Cl-CH_2OH$$

Ethylene oxide also reacts with alcohols under the influence of *basic* catalysts. A possible mechanism is:

$$EtO^- + \text{(CH}_2\text{)}_2O \longrightarrow EtO-CH_2-CH_2O^- \quad H-OEt \longrightarrow CH_2OEt-CH_2OEt + EtO^-$$

Of particular interest is the addition of hydrogen cyanide to form ethylene cyanohydrin:

$$\text{CH}_2\text{—CH}_2\text{(O)} + HCN \longrightarrow HOCH_2CH_2CN$$

The structure of ethylene oxide is uncertain; it may contain 'bent' bonds (see cyclopropane). When ethylene oxide is heated with methanol under pressure, the monomethyl ether of glycol is formed:

$$\text{CH}_2\text{—CH}_2\text{(O)} + CH_3OH \longrightarrow HOCH_2CH_2OCH_3$$

This is known as *methyl cellosolve*; the corresponding ethyl ether is known as *ethyl cellosolve*. Cellosolves are very useful as solvents since they contain both the alcohol and ether functional groups (*cf.* polyethylene glycols).

The further action of ethylene oxide on cellosolves produces *carbitols*, *e.g.*,

$$HOCH_2CH_2OCH_3 + \underset{CH_2-CH_2}{\overset{O}{\wedge}} \longrightarrow HOCH_2CH_2OCH_2CH_2OCH_3$$

When glycol is treated wit ethylene oxide, diethylene glycol is formed:

$$HOCH_2CH_2OH + \underset{CH_2-CH_2}{\overset{O}{\wedge}} \longrightarrow HOCH_2CH_2OCH_2CH_2OH$$

Diethylene glycol dimethyl ether (*diglyme*), p.b. 160°C, is used as solvent.

Ethylene oxide reacts with ammonia to form a mixture of three aminoalcohols which are usually referred to as the *ethanolamines*:

$$C_2H_4O + NH_3 \longrightarrow HOCH_2CH_2NH_2 \xrightarrow{C_2H_4O}$$

$$(HOCH_2CH_2-)_2NH \xrightarrow{C_2H_4O} (HOCH_2CH_2-)_3N$$

The ethanolamines are widely used as emulsifying agents.

The mechanism of these reactions is possibly:

$$H_3\ddot{N}\ \ \underset{CH_2}{\overset{CH_2}{|}}\!\!\!>O \longrightarrow \begin{matrix} H_3\overset{+}{N}-CH_2 \\ | \\ CH_2O^- \end{matrix} \longrightarrow \begin{matrix} H^+ + CH_2NH_2 \\ | \\ CH_2O^- \end{matrix} \longrightarrow \begin{matrix} CH_2NH_2 \\ | \\ CH_2OH \end{matrix}$$

In the presence of ethylene oxide, the reaction proceeds with the ethanolamine now being the amino reagent, etc.

The C—O group (stretch) in epoxides in he region 1 260–1 240cm^{-1} (s), and so these compounds may be readily distinguished from ethers.

The NMR spectra of epoxy compounds show a τ-value ~ 7·2-7·6 p.p.m. for methylene protons and 7·0 for methine protons. These ranges are sufficiently different from those of CH_2 and CH groups in other environments that epoxides may usually be recognised in this way.

Dixoan (1,4-*dioxan, diethylene dioxide*), p.b. 101·5°C, is manufactured:

(*i*) By distilling glycol with fuming sulphuric acid:

$$2HOCH_2CH_2OH \longrightarrow O\begin{matrix} CH_2-CH_2 \\ \\ CH_2-CH_2 \end{matrix}O + 2H_2O$$

(*ii*) Bydistilling ethylene oxide with dilute sulphuric acid:

$$2C_2H_4O \longrightarrow O\begin{matrix} CH_2-CH_2 \\ CH_2-CH_2 \end{matrix}O$$

Dioxan is a useful solvent for cryoscopic and ebullioscopic work.

Propene oxide, p.b. 35°C, is manufactured as follows:

$$(i)\ 2CH_3CH{=}CH2 \xrightarrow{2HOCl} CH_3CHOHCH_2Cl + CH_3CHClCH_2OH$$

$$\xrightarrow{Ca(OH)_2} 2CH_3\overset{O}{CH{-}CH_2}$$

$$(ii)\ CH_3CH{=}CH_2 \xrightarrow[\text{cat.}]{\text{t-BuOOH}} CH_3\overset{O}{CH{-}CH_2}$$

Its reactions are similar to those of ethylene oxide. When heated with dilute hydrochloric acid, propene glycol, p.b. 188°C, which is used as a solvent, as a substitute for glycerol, etc.

General Methods of Preparation of 1,2-Glycols Other Than Glycol Itself

1. By the reduction of ketones with magnesium amalgam, *e.g.*, pinacol.

$$2CH_3COCH_3 \xrightarrow[\text{(ii) } H_2O]{\text{(i) Mg--Hg}} (CH_3)_2C(OH)C(OH)C(OH)(CH_3)_2 (43\text{–}50\%)$$

2. By the action of a Grignard reagent of α-diketones, *e.g.*,

$$CH_3COCOCH_3 \xrightarrow{2RMgX} CH_3\overset{R}{\underset{XMgO}{C}}-\overset{R}{\underset{OMgX}{C}}CH_3 \xrightarrow{H_2O} CH_3\overset{R}{\underset{HO}{C}}-\overset{R}{\underset{OH}{C}}CH_3$$

3. By the action of a Grignard reagent on α-ketonic esters, *e.g.*,

$$CH_3COCO_2C_2H_5 \xrightarrow{3RMgX} CH_3\overset{R}{\underset{XMgO}{C}}-\overset{R}{\underset{OMgX}{C}}R \xrightarrow{H_2O} CH_3\overset{R}{\underset{HO}{C}}-\overset{R}{\underset{OH}{C}}R$$

4. By the catalytic reduction of acyloins:

$$RCH(OH)COR \xrightarrow[Ni]{H_2} RCH(OH)CH(OH)R$$

5. By hydroxylation of unsaturated compounds.

Pinacol (*tetramethylethylene glycol, 2,3-dimethylbutane-2, 3-diol*) is most conveniently prepared by reducing acetone with magnesium amalgam (see above). It crystallises out of solution as the hexahydrate. The most important reaction of pinacol is the rearrangement it undergoes with distilled with dilute sulphuric acid:

$$(CH_3)_2C(OH)C(OH)(CH_3)_2 \xrightarrow{H_2SO_4} CH_3COC(CH_3)_3 + H_2O (65\text{–}72\%)$$

The *pinacol-pinacolone rearrangement* is general for pinacols (ditertiary alcohols). Some diene is formed as by-product:

$$CH_3-\underset{OH}{\overset{CH_3}{\underset{|}{\overset{|}{C}}}}-\underset{OH}{\overset{CH_3}{\underset{|}{\overset{|}{C}}}}-CH_3 \xrightarrow{H_2SO_4} CH_2=\overset{CH_3}{\overset{|}{C}}-\overset{CH_3}{\overset{|}{C}}=CH_2 + 2H_2O$$

POLYMETHYLENE GLYCOLS

A general method of preparing polymethylene glycols is to reduce α,ω-dicarboxylic esters with sodium and ethanol or lithium aluminium hydride, or catalytically. Individual polymethylene glycols are usually prepared by special methods.

Trimethylene glycol (propane-1,3-diol), b.p. 214°C (with decomposition), may be prepared by the hydrolysis of trimethylene dibromide.

Tetramethylene glycol (butane-1, 4-diol), b.p. 230°C, and *hexamethylene glycol (hexane-1, 6-diol)*, m.p. 42°C, are most conveniently prepared by reducing the corresponding α,ω-dicarboxylic esters (succinic and adipic, respectively). Tetramethylene glycol is prepared industrially by hydrogenating butynediol. It is used for preparing butadiene, γ-butyrolactone and tetrahydrofuran.

Pentamethylene glycol (pentane-1,5-diol), b.p. 239°C, can be prepared from pentamethylene dibromide, which is obtained from piperidine by the *von Braun reaction*, by heating with potassium acetate in glacial acetic acid, and subsequently hydrolysing the diacetate with alkali (cf. glycol):

$$BrCH_2(CH_2)_3CH_2Br \xrightarrow{CH_3CO_2K} CH_3COOCH_2(CH_2)_3CH_2OOCCH_3$$

$$\xrightarrow{NaOH} HOCH_2(CH_2)_3CH_2OH \quad \text{(v.g.)}$$

Pentamethylene glycol may also be prepared by the catalytic hydrogenation (copper chromite) of tetrahydrofurfuryl alcohol under pressure

(40 – 47 per cent yield). The polymethylene glycols are readily converted into the corresponding mono- or di-halogen derivative by halogen acid, according to the amount of halogen acid used. These polymethylene halides are useful reagents in organic syntheses since they contain two reactive halogen atoms, one or both of which may be made to undergo reaction.

If the synthesis requires reaction with one halogen atom only, the most satisfactory procedure is to 'protect' the other halogen atom by ether formation and subsequently decompose the ether with concentrated hydrobromic acid, *e.g.*,

$$C_2H_5ONa + BrCH_2CH_2CH_2Br \longrightarrow NaBr + C_2H_5OCH_2CH_2CH_2Br$$

$$\xrightarrow{AgNO_2} C_2H_5OCH_2CH_2CH_2NO_2 \xrightarrow{HBr} BrCH_2CH_2CH_2NO_2$$

Then *e.g.*,

$$BrCH_2CH_2CH_2NO_2 \xrightarrow{KCN} NCCH_2CH_2CH_2NO_2$$

TRIHYDRIC ALCOHOLS

The only important trihydric alcohol is *glycerol (propane-1, 2, 3-triol),* b.p. 290°C. It occurs in almost all animal and vegetable oils and fats as the glyceryl esters of mainly palmitic, stearic and oleic acids.

Glycerol is obtained in large quantities as a by-product in the manufacture of soap, and this is still a commercial source of glycerol (see below). It is prepared synthetically as follows:

$$CH_3CH{=}CH_2 \xrightarrow[480-500°C]{Cl_2} ClCH_2CH{=}CH_2 \xrightarrow[150°C\ 12\ atm]{aq.\ Na_2CO_3}$$

$$\underset{\text{allyl alcohol}}{HOCH_2CH{=}CH_2} \xrightarrow{HOCl} \underset{\text{glycerol }\beta\text{-monochlorohydrin}}{HOCH_2CHClCH_2OH}$$

$$\xrightarrow{NaOH} HOCH_2CHOHCH_2OH$$

An alternative route that is used is:

$$ClCH_2CH{=}CH_2 \xrightarrow{HOCl} \left.\begin{matrix} ClCH_2CHOHCH_2Cl \\ + \\ ClCH_2CHClCH_2OH \end{matrix}\right\} \xrightarrow{CaO}$$

$$\underset{\text{epichlorohydrin}}{\overset{\;\;O}{CH_2{-}CHCH_2Cl}} \xrightarrow{NaOH} HOCH_2CHOHCH_2OH$$

Other methods of manufacture are:

$$(i)\ CH_2CH{=}CHO \xrightarrow[\text{cat.}]{H_2} CH_2{=}CHCH_2OH \xrightarrow{H_2O_2} HOCH_2CHOHCH_2OH$$

$$(ii)\ CH_2{=}CHCH_3 + O_2 \xrightarrow[300-400^\circ C]{Cu_2O} CH_2{=}CHCHO$$

$$CH_2{=}CHCH_3 + H_2O \xrightarrow[WO_3-ZnO]{220-250^\circ} CH_3CHOHCH_3$$

$$CH_2{=}CHCHO + (CH_3)_2CHOH \xrightarrow[400^\circ C]{MgO-ZnO} (CH_3)_2CO + CH_2{=}CHCH_2OH$$

$$(CH_3)_2CHOH + O_2 \xrightarrow{100-140^\circ C} (CH_3)_2CO + H_2O_2$$

$$CH_2CHCH_2OH + H_2O_2 \xrightarrow[WO_3]{60-70^\circ C} HOCH_2CHOHCH_2OH$$

Acetone is obtained as a by-product.

Glycerol is used as an antifreeze, for making explosives, and, because of its hygroscopic properties, as a moistening agent for tobacco, shaving soaps, etc. Glycerol contains one secondary and two primary alcoholic groups, and it undergoes many of the reactions to be expected of these types of alcohols. The carbon atoms in glycerol are indicated as shown:

$$\overset{\alpha}{CH_2}\,OH\overset{\beta}{C}HO\,H\overset{\alpha'\ or\ \gamma}{CH_2}\,OH\ \cdot$$

(*i*) When glycerol is treated with sodium, one α-hydroxyl group is readily attacked, and the other α-group less readily; the β-hydroxyl groups is not attacked at all:

$$HOCH_2CHOHCH_2OH \xrightarrow{Na} NaOCH_2CHOHCH_2OH \xrightarrow{Na} NaOCH_2CHOHCH_2ONa$$

(*ii*) On passing hydrogen chloride into glycerol at 110°C until there is the theoretical increase in weight corresponding to the esterification of one hydroxyl group, both α- and β-glycerol monochlorohydrin are found, the former predominating (66 per cent).

Continued action of hydrogen chloride at 110°C, using 25 per cent of acid in excess required by theory for the esterification of two hydroxyl groups, produces glycerol α, α′-dichlorohydrin (α-dichlorohydrin) and glycerol α,β-dichlorohydrin (β-dichlorohydrin), the former predominating (55 – 57 per cent); some other products are also formed:

$$HOCH_2CHOHCH_2OH \xrightarrow{HCl} ClCH_2CHOHCH_2OH + HOCH_2CHClCH_2OH$$
$$\xrightarrow{HCl} ClCH_2CHOHCH_2Cl + ClCH_2CHClCH_2OH$$

When either of these dichlorohydrins or glycerol itself is treated with phosphorus pentachloride, glycerol trichlorohydrin (1, 2, 3-trichloropropane) is obtained. This is a liquid, b.p. 156 – 158°C.

When glycerol α, α'-dichlorohydrin is oxidised with sodium dichromate-sulphuric acid mixture, 1,3-dichloroacetone (1,3-dichloropropan-2-one), m.p. 45°C, is obtained:

$$ClCH_2CHOHCH_2Cl \xrightarrow{[O]} ClCH_2COCH_2Cl \quad (68-75\%)$$

When glycerol α, α'-dichlorohydrin is treated with powdered sodium hydroxide in ether solution, epichlorohydrin (3-chloro-1,2-epoxypropane), b.p. 117°C, is obtained.

$$ClCH_2CHOHCH_2Cl + NaOH \longrightarrow ClCH_2\overset{O}{CH-CH_2} + NaCl + H_2O \quad (76-81\%)$$

Epichlorohydrin is also obtained by distilling an alkaline solution of the α-dichlorohydrin under reduced pressure (yield: 90 per cent). It is used for the manufacture of epoxy resins, plasticisers, etc.

Hydrogen bromide and the phosphorus bromides react with glycerol in the same way as the corresponding chlorine compounds, but the analogous iodine compounds behave differently, the products depending on the amount of reagent used. When glycerol is heated with a *small* amount of hydrogen iodide or phosphorus tri-iodide, *allyl iodide* is the main product:

$$HOCH_2CHOHCH_2OH \xrightarrow{PI_3} [ICH_2CHICH_2I] \longrightarrow I_2 + CH_2{=}CHCH_2I$$

When a *large* amount of phosphorus tri-iodide is used, the main product is *isopropyl iodide:*

$$CH_2{=}CHCH_2I \xrightarrow{HI} [CH_3CHICH_2I] \xrightarrow{-I_2} CH_3CH{=}CH_2$$
$$\xrightarrow{HI} CH_3CHICH_3 \quad (80\%)$$

(*iii*) When glycerol is treated with monocarboxylic acids, esters are obtained which may be mono-, di- or tri-esters, according to the amount of acid used; high temperature and an excess of acid favour he formation of the tri-ester (see the glycerides, below).

Nitroglycerine is manufactured by adding glycerol in a thin stream to a cold mixture of concentrated nitric and sulphuric acids:

$$\begin{array}{l} CH_2OH \\ | \\ CHOH + 3HNO_3 \\ | \\ CH_2OH \end{array} \longrightarrow \begin{array}{l} CH_2ONO_2 \\ | \\ CHONO_2 + 3H_2O \\ | \\ CH_2ONO_2 \end{array}$$

Nitroglycerine is an ester, not a nitro-compound; it is *glyceryl trinitrate*. The incorrect name appears to have been introduced due to the use of 'mixed acid' which is normally used for nitration.

Nitroglycerine is a poisonous, colourless, oily liquid, insoluble in water. It usually burns quietly when ignited, but when heated, rapidly struck or detonated, it explodes violently. Nobel (1867) found that nitroglycerine could be stabilised by absorbing it in kieselguhr.

This was *dynamite*, which is now, however, usually manufactured by using wood pulp as the absorbent, and adding solid ammonium nitrate. *Blasting gelatin* or *gelignite* is made by mixing nitroglycerine with gun-cotton (cellulose nitrate). The smokeless powder, *cordite*, is a mixture of nitroglycerine, gun-cotton and vaseline. Nitroglycerine is also used in the treatment of angina pectoris.

When heated with formic acid or oxalic acid at 260°C, glycerol is converted into allyl alcohol. With dibasic acids, glycerol forms condensation polymers known as *alkyd resins*, the commonest of which is *glyptal*, formed by heating glycerol with phthalic anhydride.

When boric acid is added to an aqueous solution of glycerol, a complex is produced which has a higher electrical conductivity than boric acid itself, *i.e.*, is a stronger acid. This complex is believed to be a borospiranic acid in which the two rings attached to the boron atom are perpendicular to each other (hence name spiran): This borospiranic acid is formed from α-glycols in which the two hydroxyl groups are in the *cis*-configuration. Ethylene glycol apparently does not form this complex; the reason is unknown.

Dilute nitric acid converts glycerol into glyceric and tartronic acids; concentrated nitric acid oxidises it to mainly glyceric acid (80 per cent); bismuth nitrate produces mainly mesoxalic acid. Bromine water, sodium hypobromite and Fenton's reagent (hydrogen peroxide and ferrous sulphate) oxidise glycerol to a mixture of glyceraldehyde (predominantly) and dihydroxyacetone; this mixture is known as *glycerose*. These two compounds are interconvertible in the presence of anhydrous pyridine, this being known as the *Lobry de Bruyn-van Ekenstein rearrangement.*

$$\begin{array}{l} CHO \\ | \\ CHOH \\ | \\ CH_2OH \end{array} \rightleftharpoons \left[\begin{array}{l} CHOH \\ \| \\ COH \\ | \\ CH_2OH \end{array}\right] \rightleftharpoons \begin{array}{l} CH_2OH \\ | \\ CO \\ | \\ CH_2OH \end{array}$$

(*v*) When heated with potassium hydrogen sulphate, glycerol is dehydrated to acraldehyde:

$$HOCH_2CHOHCH_2OH \xrightarrow{KHSO_2} CH_2{=}CHCHO + 2H_2O$$

(*vi*) The mixed ethers of glycerol may be conveniently prepared by the action of sodium alkoxide on a glycerol chlorohydrin, *e.g.*, *triethylin*, the triethyl ether of glycerol, is prepared by heating 1,2,3-trichloropropane with sodium ethoxide:

$$\begin{array}{l} CH_2Cl \\ | \\ CHCl \\ | \\ CH_2Cl \end{array} + 3C_2H_5ONa \longrightarrow \begin{array}{l} CH_2OC_2H_5 \\ | \\ CHOC_2H_5 \\ | \\ CH_2OC_2H_5 \end{array} + 3NaCl$$

Helferich and co-workers (1923) found that triphenylmethyl chloride (*trityl chloride*), $(C_6H_5)_3CCl$, usually formed ethers only with primary alcoholic groups. Thus, the α-mono- and the α,α′-ditriphenylmethyl (trityl) ethers of glycerol can be prepared, the latter offering a means of preparing β-esters glycerol, *e.g.*,

$$\begin{array}{l} CH_2OH \\ | \\ CHOH \\ | \\ CH_2OH \end{array} \xrightarrow{(C_6H_5)_3CCl} \begin{array}{l} CH_2OC(C_6H_5)_3 \\ | \\ CHOH \\ | \\ CH_2OC(C_6H_5)_3 \end{array} \xrightarrow{RCOCl} \begin{array}{l} CH_2OC(C_6H_5)_3 \\ | \\ CHOCOR \\ | \\ CH_2OC(C_6H_5)_3 \end{array} \xrightarrow{HBr} \begin{array}{l} CH_2OH \\ | \\ CHOCOR \\ | \\ CH_2OH \end{array}$$

A better means of obtaining the β-ester is to protect the α,α′-hydroxyl groups by cyclic ether formation with benzaldehyde (Bergmann and Carter, 1930). This cyclic acetal is formed when glycerol and benzaldehyde are heated together, or the cool mixture treated with hydrogen chloride:

$$\begin{array}{l} CH_2OH \\ | \\ CHOH + C_6H_5CHO \\ | \\ CH_2OH \end{array} \xrightarrow{HCl} \begin{array}{l} CH_2O \diagdown \\ | \qquad\qquad CHC_6H_5 + H_2O \\ CHOH \\ | \qquad\qquad \diagup \\ CH_2O \end{array}$$

(*vii*) Glycerol can be fermented to produce a variety of compounds, *e.g.*, propionic acid, succinic acid, acetic acid, n-butanol, trimethylene glycol, lactic acid, n-butyric acid, etc. Fermentation by means of a particular micro-organism usually produces more than one compound, *e.g.*, propionic acid bacteria produce propionic acid, succinic acid and acetic acid.

POLYHYDRIC ALCOHOLS

Tetrahydric alcohols. *Erythritol,* $HOCH_2CHOHCHOHCH_2OH$, exists in three forms: dextrorotatory erythritol, m.p. 88°; laevorotatory erythritol, m.p. 88°; *meso*-erythritol, m.p. 121.5°C. The *meso*-form occurs in certain lichens and seaweeds. All three forms may be oxidised to tartaric acid,

$$HO_2CCHOHCHOHCO_2H$$

Pentaerythritol, $C(CH_2OH)_4$, is prepared by the condensation of formaldehyde with acetaldehyde.

Pentahydric alcohols (*pentitols*). There are four pentitols with the structure $HOCH_2(CHOH)_3CH_2OH$, which may be prepared by reducing the corresponding aldopentoses. Two are optically active, forming a pair of enantiomers —D- and L-arabitol—and the other two are optically inactive, existing as *meso* forms—adonitol and xylitol.

Aldopentose	*Pentitol*
D- and L-ribose	adonitol (ribitol)
D- and L-xylose	xylitol
D-arabinose	
D-lyxose	D-arabitol (D-lyxitol)
L-arabinose	
L-lyxose	L-arabitol (L-lyxitol)

Hexahydric alcohols (*hexitols*). There are ten hexitols with the structure $HOCH_2(CHOH)_4CH_2OH$, which may be prepared by reducing the corresponding aldohexoses. Eight exist as four pairs of enantiomers, and the remaining two as meso forms.

Aldohexose	*Pentitol*
D-glucose and L-glucose	D-sorbitol (D-glucitol)
L-glucose and D-glucose	L-sorbitol (L-glucitol)
D-mannose	D-mannitol
L-mannose	L-mannitol
D-idose	D-iditol
L-idose	L-iditol
D-talose and D-altrose	D-talitol
L-talose and L-altrose	L-talitol
D- and L-galactose	dulcitol
D- and L-allose	allodulcitol (allitol)

D-*Sorbitol, D-mannitol* and *dulcitol* occur naturally.

LIPIDS

The term 'lipids' embraces a variety of naturally occurring compounds which have in common the property of being soluble in organic solvents, but sparingly soluble in water. Also, all lipids yield mono-carboxylic acids (saturated and unsaturated) on hydrolysis. The lipids include oils, fats, waxes and phospholipids.

OILS AND FATS

This group of lipids is often referred to as the 'simple' lipids. They are compounds of glycerol and various organic acids, *i.e.*, they are glyceryl esters of *glycerides*. Oils, which are liquids at ordinary temperatures, contain a larger proportion of unsaturated acids than do the fats, which are solids at ordinary temperatures.

The acids present in the glycerides are almost exclusively straight-chain acids, and almost always contain an even number of carbon atoms. The chief saturated acids are lauric, myristic, palmitic and stearic. The chief unsaturated acids are oleic, linoelic and linoenic, and these usually occur in the *cis*-form.

Palmitic acid is the most abundant of the saturated acids, and acids containing less than eighteen carbon atoms are usually present only as minor constituents, but sometimes they are present in appreciable amounts in insect waxes and marine fats. Glycerides are named according to the nature of the acids present, the suffix – *ic* of the common name of the acid being changed to –*in*. Glycerides are said to be 'simple' when all the acids are the same, and 'mixed' when the acids are different:

$$\begin{array}{l} CH_2OCOC_{17}H_{35} \\ | \\ CHOCOC_{17}H_{35} \\ | \\ CH_2OCOC_{17}H_{35} \end{array} \qquad \begin{array}{l} CH_2OCOC_{15}H_{31} \\ | \\ CHOCOC_{17}H_{33} \\ | \\ CH_2OCOC_{17}H_{33} \end{array}$$

tristearin (simple glyceride) — α-palmito-α′, β-diolein (mixed glyceride)

It is still not certain whether simple glycerides occur naturally; if they do, they are definitely not as common as the mixed glycerides.

According to Desnuelle *et al.* (1959), the structure of natural glycerides is not random; the position of the acid residue depends on the chain-length of the unsaturated acids and on the degree of their unsaturation. In vegetable oils the saturated acids occur mainly at the 1- and 3-positions and unsaturated acids at the 2-position. In animal fats the distribution is not so rigid.

SYNTHESIS OF GLYCERIDES

Simple triglycerides are readily prepared from glycerol and an excess of acyl chloride, or from 1,2,3-tribromopropane and the silver or potassium salts of the acids. Mixed triglycerides are far more difficult to prepare. A number of methods have been developed, *e.g.*,

(*i*) The sodium salt of an acid is heated with glycerol monochlorohydrin, and the monoester so formed is acylated:

$$\begin{array}{l} CH_2Cl \\ | \\ CHOH \\ | \\ CH_2OH \end{array} \xrightarrow{R^1CO_2Na} \begin{array}{l} CH_2OCOR^1 \\ | \\ CHOH \\ | \\ CH_2OH \end{array} \xrightarrow{R^2COCl} \begin{array}{l} CH_2OCOR^1 \\ | \\ CHOCOR^2 \\ | \\ CH_2OCOR^2 \end{array}$$

The preparation of 1-monoglycerides has been improved by Hartman (1960). The 2- and 3-positions in glycerol are protected by the formation of the isopropylidene derivative, and the protecting group is removed by boric acid in hot 2-methoxyethanol.

(*ii*) 1,3-benzylidene-glycerol is treated with an acid chloride, the benzaldehyde residue is removed by hydrolysis, and the monoester so formed is then further acylated:

$$\begin{array}{l} CH_2O \\ | \\ CHOH \quad CHC_6H_5 \\ | \\ CH_2O \end{array} \xrightarrow{R^1COCl} \begin{array}{l} CH_2O \\ | \\ CHOCOR^1 \; CHC_6H_5 \\ | \\ CH_2O \end{array} \xrightarrow{HCl} \begin{array}{l} CH_2OH \\ | \\ CHOCOR^1 \\ | \\ CH_2OH \end{array}$$

$$\xrightarrow{R^2COCl} \begin{array}{l} CH_2OCOR^2 \\ | \\ CHOCOR^1 \\ | \\ CH_2OCOR^2 \end{array}$$

In both methods (*i*) and (*ii*) there is, however, still difficulty in knowing with certainty the position of the acyl group introduced first, since the acyl group tends to migrate in the monoester, producing the following equilibrium:

$$\begin{array}{l} CH_2OCOR \\ | \\ CHOH \\ | \\ CH_2OH \end{array} \rightleftharpoons \begin{array}{l} CH_2OH \\ | \\ CHOCOR \\ | \\ CH_2OH \end{array}$$

This isomerisation has been shown to be intramolecular.

Analysis of Oils and Fats

Oils and fats are characterised by means of physical as well as chemical tests. The usual physical constants that are determined are melting point, solidifying point, density and refractive index. The chemical tests give an indication of the type of fatty acids present in the oil or fat.

The **acid value**, which is the number of milligrams of potassium hydroxide required to neutralise 1 gram of the oil or fat, indicates the amount of free acid present.

The **saponification value** is the number of milligrams of potassium hydroxide required to neutralise the acids resulting from the complete hydrolysis of 1 gram of the oil or fat.

The **iodine value**, which is the number of grams of iodine that combine with 100 grams of oil or fat, gives the degree of unsaturation of the acids in the substance.

Several methods are used for determining the iodine value. In *Hubl's method*, a carbon tetrachloride solution of the substance is treated with a solution of iodine and mercuric chloride in ethanol; in *Wijs' method*, iodine monochloride in glacial acetic acid is used. Another method is to use a solution of glacial acetic acid containing pyridine, bromine and concentrated sulphuric acid (*Dam's solution*).

The **Reichert-Meissl value**, which is the number of ml of 0.1 N-potassium hydroxide required to neutralise the distillate of 5 grams of hydrolysed fat, indicates the amount of steam-volatile fatty acids (*i.e.*, acids up to lauric) present in the substance.

The **acetyl value**, which is the number of milligrams of potassium hydroxide required to neutralise the acetic acid obtained when 1 gram of an acetylated oil or fat is hydrolysed, indicates the number of free hydroxyl groups present in the substance.

PREPARATION OF GLYCEROL FROM OILS AND FATS

Glycerol and some acids are prepared for industrial use by the hydrolysis of oils and fats with water under pressure at 220°C. The glycerol is recovered from the aqueous solution; the free fatty acids are used in the manufacture of candles.

$$\begin{array}{l} CH_2OCOR \\ | \\ CHOCOR + 3H_2O \\ | \\ CH_2OCOR \end{array} \longrightarrow \begin{array}{l} CH_2OH \\ | \\ CHOH + 3RCO_2R \\ | \\ CH_2OH \end{array}$$

If an oil is saponified, *i.e.*, hydrolysed with alkali, soaps are obtained. Any metallic salt of a fatty acid is a soap, but the term soap is usually applied to the water-soluble salts since only these have detergent properties.

The saturated fats give hard soaps whereas the unsaturated fats, *i.e.*, the oils, give soft soaps. Ordinary soap is a mixture of the sodium salts of the even fatty acids from octanoic to stearic. The sodium salts of a given oil or fat are harder and less soluble than the corresponding potassium salts. Thus soft soaps are usually the potassium salts, particularly when they are derived from oils.

Hardening of Oils

Glycerides of the unsaturated acids are liquid at room temperature and so are unsuitable for edible fats. By converting the unsaturated acids into saturated acids, oils are changed into fats. This introduction of hydrogen is known as the *hardening of oils*. The oil is heated to 150 – 200°C and hydrogen is passed in, under pressure, in the presence of a finely divided nickel catalyst. In the hydrogenation process only a proportion of the unsaturated acids are converted into the saturated acids, otherwise a fat as hard as tallow would be obtained; the hydrogenation is carried out until a fat of the desired consistency is obtained.

Phospholipids

These differ from the simple lipids in that they also contain phosphorus and nitrogen; they are glycerides in which one organic acid residue, the α- or β-, is replaced by a group containing phosphoric acid and a base. One subgroup of the phospholipids is the **phosphatides**. When the base is **cholamine**, $CH_2OCH_2NH_2$, the phosphatide is known as a **kephalin**; when the base is **choline**, $CH_2OHCH_2\overset{+}{N}(CH_3)_3\}O\overset{-}{H}$, the phosphatide is known as a **lecithin**. These compounds have betaine structures, and it appears unlikely that β-kephalins or β-lecithins occur in nature.

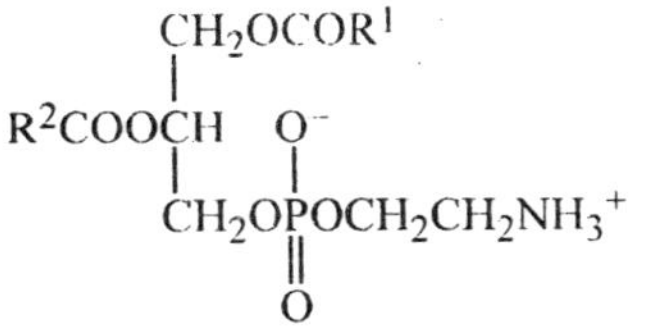

```
        CH2OCOR1                         CH2OCOR1
        |                                |
R2COOCH    O-                    R2COOCH    O-
        |   |                            |   |          +
        CH2OPOCH2CH2NH3+                 CH2OPOCH2CH2N(CH3)3
            ||                               ||
            O                                O
```

α-kephalin 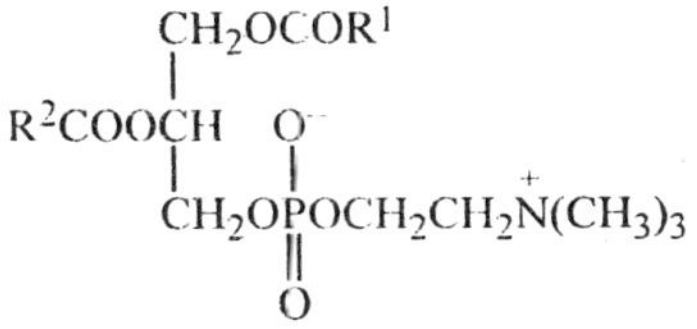 α-lecithin

Kephalins may also contain L-serine instead of aminoethanol. These three types of phospholipids have been called *phosphoglycerides*, in order to distinguish them from another group of phospholipids, the *acetal phospholipids* (*phasmalogens*). The structures of these compounds appear to be uncertain; one proposal is (for an ethanolamine plasmalogen):

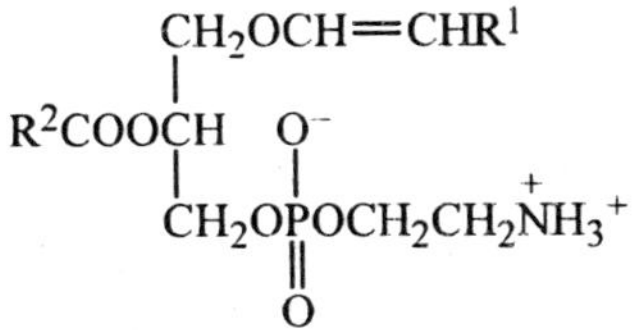

```
        CH2OCH=CHR1
        |
R2COOCH    O-
        |   |            +
        CH2OPOCH2CH2NH3+
            ||
            O
```

All of these phosphoglycerides occur in the brain and the spinal chord.

Only a few monocarboxylic acids have been isolated from phosphatides: stearic, oleic, linoelic and arachidonic from kephalins, and plamitic, stearic, oleic, linoeic, linoenic and arachidonic from lecithins. Enzymic studies of phospholipids have shown that saturated acids predominate at the 1-position and unsaturated acids at the 2-position (Hanahan *et al.*, 1960).

Another group of phospholipids is the **sphingolipids**. These are *not* glycerides, and yield, on hydrolysis, fatty acids, phosphoric acid, choline and sphingosine (which has the *trans*-configuration about the double bond). Sphingolipids are also known as **sphingomyelins**, and their structure is believed to be:

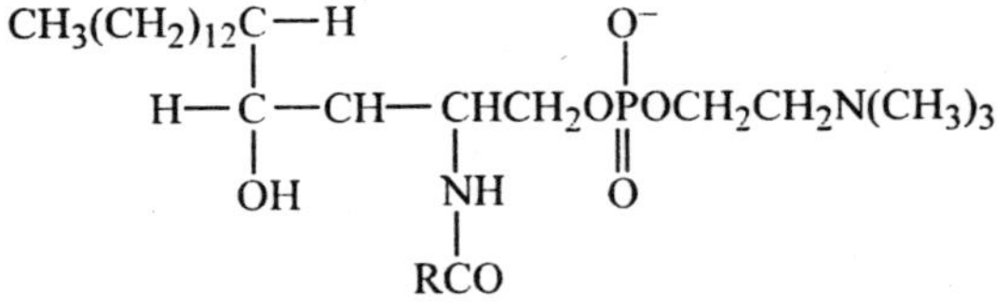

```
CH3(CH2)12C—H                          O⁻
          |                            |
       H—C—CH—  CHCH2OPOCH2CH2N(CH3)3
          |       |       ||
         OH      NH       O
                  |
                 RCO
```

Glycolipids (cerebrosides) also contain sphingosine, but instead of the phosphocholine group, they have $\cdots$CHCH$_2$OCH(CHOH)$_3$CHCH$_2$OH,

```
···CHCH2OCH(CHOH)3CHCH2OH,
   |     |_____O_____|
 —NH
```

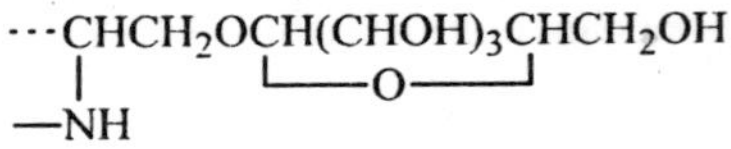

i.e., they are glycosides, and apparently only two sugars have been isolated so far, glucose and galactose.

Drying Oils

These are oils which, on exposure to air, change into hard solids, *e.g.*, linseed oil. All drying oils contain a large proportion of the unsaturated acids linoleic and linolenic, and it is the 'drying' property which makes these oils valuable in the paint industry. The mechanism of drying appears to be a complicated process involving oxidation, polymerisation and colloidal gel formation, and it has been found to be catalysed by various metallic oxides, particularly lead monoxide.

These are esters of the higher homologues of both the fatty acids and monohydric alcohols, *e.g.*,

beeswax	myricyl palmitate	$C_{15}H_{31}CO_2C_{30}H_{61}$
spermaceti	cetyl palmitate	$C_{15}H_{31}CO_2C_{16}H_{33}$
carnauba wax	myricyl cerotate	$C_{25}H_{51}CO_2C_{30}H_{61}$

Some waxes are also esters of cholesterol, *e.g.*, cholesteryl esters occur in wool-wax. Also present in waxes are free acids, long-chain alcohols, ketones and alkanes.

CROTYL ALCOHOL

The crotyl alcohol (*crotonyl alcohol, but-2-en-l-ol*), b.p. 118°C, may be prepared by the reduction of crotonaldehyde with aluminium isopropoxide:

$$CH_3CH = CHCHO \xrightarrow{[(CH_3)_2CHO]_3Al} CH_3CH = CHCH_2OH \text{ (85–90\%)}$$

When treated with hydrogen bromide, crotyl alcohol produces a mixture ofcrotyl and methyl-vinylcarbinyl bromide:

$$CH_3CH = CHCH_2OH \xrightarrow{HBr} CH_3CH = CHCH_2Br + CH_3CHBrCH = CH_2$$

The formation of the rearranged proiduct is an example of anionotropy, and the most widely studied example is the three-carbon system, *i.e.*, the allylic system. Nucleophilic substitutior reactions of allylic halides may occur by the S_N2 mechanism, and when this is operating substitution proceeds normally. When, however, nucleophilic substitution reactions are carried out under conditions which favour the unimolecular mechanism, then the product may be a mixture of two isomers the rearranged product being an example of the *allylic rearrangement*. The mechanism by which this rearrangement occurs is known as SN)', and is believed to occur via a resonance hybrid (ambident) cation:

$$MeC = CHCH_2OH \underset{}{\overset{H^+}{\rightleftharpoons}} MeCH = CHCH_2O\,H_2^+ \underset{+H_2O}{\overset{-H_2O}{\rightleftharpoons}}$$

$$MeCH{=}CHCH_2^+ \longleftrightarrow Me\overset{+}{C}HCH{=}CH_2$$

$$\downarrow Br^- \qquad\qquad \downarrow Br^-$$

$$\underset{S_N1}{MeCH{=}CHCH_2Br} \qquad \underset{S_N1'}{MeCHBrCH{=}CH_2}$$

There is much evidence to support this S_N1' mechanism, *e.g.*, Hughes *et al.* (1948) examine the conversion of α-methylallyl and crotyl chloride into α-methylallyl and crotyl ether by reaction with ethanolic sodium ethoxide. The concentration of the sodium ethoxide was made so small as to give first-order kinetics, or so large as to give second-order kinetics. The result were:

$$MeCHClCH{=}CH_2 \xrightarrow{\text{1st order}} \downarrow\ \downarrow \xleftarrow{\text{1st order}} MeCH{=}CHCH_2Cl$$

$$MeCHClCH{=}CH_2 \xrightarrow{\text{2nd order}} MeCH(OEt)CH{=}CH_2 \text{ (100\%)}$$

$$MeCH{=}CHCH_2Cl \xrightarrow{\text{2nd order}} MeCH{=}CHCH_2OEt \text{ (100\%)}$$

$$\text{1st order: } \left\{ \begin{array}{l} MeCH(OEt)CH{=}CH_2 \text{ (13\%)} \\ + \\ MeCH{=}CHCH_2OEt \text{ (87\%)} \end{array} \right.$$

With second-order substitution, each of chloride gives its own unrearranged ether (thus the S_N2 mechanism). With first-order substitution, however, each chloride gives the *same* mixture isomeric ethers. This latter result may be explained by assuming that *two* mechanisms a operating, S_N1 (without rearrangement) and S_N1' (with rearrangement), both proceeding through a common intermediate ambident cation.

In addition to the S_N1' mechanism, there is also the S_N2', *i.e.*, a bimolecular nucleophilic substitution with allylic rearrangement. This has been difficult to demonstrate, but a good example of it has been given by de la Mare *et al.* (1953). These authors have suppressed the S_N2 reaction by steric factors and reduced the S,1 reaction by means of a reagent of high nucleophilic activity and a solvent of low ionising power. They found that the reaction between α, α-dimethylallyl chloride and sodium thiophenoxide in ethanol gives 62 per cent of rearranged product by a second-order reaction (*i.e.*, S_N2'):

$$Me_2\overset{Cl}{\overset{|}{C}}—CH=CH_2 \quad SPh \longrightarrow Me_2C=CHCH_2SPh + NaCl$$

One other problem that will be discussed here is the case of ion pairs in the allylic rearrangement (see also). Young *et al.* (1951) have shown that α, α-dimethylallyl chloride in acetic acid undergoes acetolysis accompanied by rearrangement to γ, g-dimethylallyl chloride:

$$Me_2CClCH = CH_2 \xrightarrow{MeCO_2H} \begin{matrix}\text{acetolysis}\\ \text{products}\end{matrix} + Me_2C = CHCH_2Cl$$

The rate of the isomerisation was shown to be proportional to the concentration of the starting chloride only. Thus a possible mechanism is the S_N1'. If this were so, then the rate of isomerisation would be affected by addition of chloride ions (since the first step is ionisation).

Experiment showed that the rate was unaffected by added chloride ions, and it was also shown that when an excess of radioactive chloride ion was added, the rate of isomerisation was far greater than the incorporation of Cl^-.

It thus appears that the chloride ion is set free during the isomerisation and is mainly the 'first ion' to recombine, even though the solution contains a large excess of Cl^{*-} ions. Young *et al.*, explained this result by assuming that the starting chloride undergoes ionisation to form an ion-pair, and this is followed by internal return:

$$Me_2CClCH{=}CH_2 \rightleftharpoons \underset{\text{ion pair}}{[Me_2C{\cdots}CH{\cdots}CH_2]^+Cl}$$
$$\rightleftharpoons Me_2C{=}CHCH_2Cl$$

Nevertheless, some ion pairs escape from the solvent cage, and these Tree' carbonium ions are capable of undergoing acetolysis and combination with Cl^{*-}.

A number of unsaturated alcohols occur in essential oils. The essential oils are volatile oils with pleasant odours, and are obtained from various plants to which they impart their characteristic odours. They are widely used in the perfume industry.

The simplest acetylenic alcohol is *propargyl alcohol* (prop-2-yn-l-ol), b.p. 114°C. This may be prepared from 1,2,3-tribromopropane:

$$BrCH_2CHBrCH_2Br \xrightarrow[\text{KOH}]{\text{ethanolic}} CH_2 = CBrCH_2Br$$
$$\xrightarrow[\text{KOH}]{\text{aqueous}} CH_2 = CBrCH_2OH \xrightarrow[\text{KOH}]{\text{ethanolic}} CH \equiv CCH_2OH$$

Propargyl alcohol (together with butynediol) is prepared by the interaction of acetylene and formaldehyde in the presence of silver or cuprous acetylide as catalyst:

$$C_2H_2 + HCHO \xrightarrow{Ag_2C_2} CH \equiv CCH_2OH$$
$$\xrightarrow[\text{HCHO}]{Ag_2C_2} HOCH_2C \equiv CCH_2OH$$

Propargyl alcohol behaves in many ways like acetylene, *e.g.*, it adds on two or four bromine atoms, and forms the silver or cuprous compounds when treated with an ammoniacal solution of silver nitrate or cuprous chloride. It may be catalytically reduced to allyl and n-propyl alcohols.

MONOCARBOXYLIC ACIDS

The monocarboxylic acids are also referred to as the fatty acids, a name which arose from the fact that some of the higher members, particularly palmitic and stearic acids, occur in natural fats. Their general formula is $C_nH_{2n}O_2$ but, because their functional group is the carboxyl group, —CO_2H, they are more conveniently expressed as $C_nH_{2n+1}CO_2H$, since these show the nature of the functional group. The structure of the carboxyl group is written as $-C{\overset{\displaystyle O}{\diagup\!\!\diagup}}\!\!\diagdown_{OH}$ (or —COOH), but, as we shall later, it does not represent accurately the behaviour of this group. Also, since only the carboxyl group hydrogen atom is replaceable by a metal, the monocarboxylic acids are *monobasic*.

Nomenclature

The monocarboxylic acids are commonly known by the trivial names, which have been derived from the source of the particular acid; *e.g.*, formic acid, HCO_2H, was so named because it was first obtained by the distillation of ants; the Latin word for ant is *formica*. Acetic acid, CH_3CO_2H, is the chief constituent of vinegar, the Latin word for which is *acetum*; etc.

Another system of nomenclature considers the acids, except formic acid, as derivatives of acetic acid, *e.g.*,

$CH_3CH_2CO_2H$	methylacetic acid
$(CH_3)_3CCO_2H$	trimethylacetic acid
$(CH_3)_2CHCH_2CO_2H$	isopropylacetic acid

In the above two systems the positions of substituents in the chain are indicated by Greek letters, the α-carbon atom being the one joined to the caboxyl group, *e.g.*,

$CH_3CH(OH)CH_2CH_2CO_2H$	γ-hydroxyvaleric acid
$CH_3CHClCH(CH_3)CO_2H$	β-chloro α-methylbutyric acid.

According to the I.U.P.A.C. system of nomenclature, the suffix of the monocarboxylic acids is *-oic*, which is added to the name of the alkane corresponding to the longest carbon chain containing the carboxyl group, *e.g.*,

HCO_2H	methanoic acid
$CH_3CH_2CO_2H$	propanoic acid

The positions of side-chains (or substituents) are indicated by numbers,. *the carboxyl group always being given number* 1:

$CH_3CH(CH_3)CH(CH_3)CH_2CO_2H$ 3,4-dimethylpentanoic acid

Alternatively, the carboxyl group is regarded as a substituent, and is denoted by the suffix carboxylic acid, *e.g.*,

$\overset{4}{C}H_3\overset{3}{C}H_2\overset{2}{C}H(CH_3)\overset{1}{C}H_2CO_2H$ 2-methylbutane-1-carboxylic acid

General Methods of Preparation

1. Oxidation of alcohols, aldehydes or ketones with acid dichromate yields acids:

$$RCH_2OH \xrightarrow{[O]} RCHO \xrightarrow{[O]} RCO_2H \ (g.\text{–}v.g.)$$

$$R_2CHOH \xrightarrow{[O]} R_2CO \xrightarrow{[O]} R^1CO^2H + R^2CO_2H \ (g.)$$

Many *primary* alcohols, on oxidation with acid dichromate, form esters in addition to acids, *e.g.*, n-butanol gives n-butyric acid and n-butyl n-butyrate (41–47 per cent). Several mechanisms have been proposed, but the generally accepted one involves hemiacetal formation.

$$RCH_2OH \xrightarrow{[O]} RCHO \xrightarrow{RCH_2OH} RCH(OH)OCH_2R \xrightarrow{[O]} RCO_2CH_2R$$

2. A very good synthetic method is the hydrolysis of cyanides with or alkali:

$$RC{\equiv}N \xrightarrow{H_2O} \left[R{-}C{\overset{OH}{\underset{NH}{\lessgtr}}} \right] \longrightarrow R{-}C{\overset{O}{\underset{NH_2}{\lessgtr}}} \xrightarrow{H_2O}$$

$$RCO_2H + NH_3 \quad (g.\text{-}r.g.)$$

The amide, $RCONH_2$, may be isolated if suitable precautions are taken.

3. By the reaction between a Grignard reagent and carbon dioxide:

$$RMgX + CO_2 \longrightarrow RCO_2MgX \xrightarrow{H_2O} RCO_2H$$

4. Many monocarboxylic acids may be conveniently synthesised from alkyl halides and ethyl malonate or acetoacetic ester. These syntheses will be discussed in detail later.

5. A number of higher acids may be obtained by the hydrolysis of natural fats or may be prehalf-ester of a dicarboxylic acid (Linstead *et al.*, 1950; *cf.*).

$$RCO_2H + HO_2C(CH_2)_nCO_2CH_3 \longrightarrow H_2 + 2CO_2 + R(CH_2)_nCO_2CH_3$$

The chain of a monocarboxylic acid can be extended by five or six carbon atoms in a synthesis involving enamines.

6. Monocarboxylic acids may be obtained by heating a sodium alkoxide with carbon monoxide under pressure.

$$R\bar{O}\overset{+}{N}a + CO \longrightarrow RCO_2^-\overset{+}{N}a$$

Acids may also be obtained by heating the alcohol with carbon monoxide under pressure in the presence of a catalyst.

The sodium salts of acetic and propionic acids are formed by the interaction of methylsodium or ethylsodium, respectively, and carbon dioxide, *e.g.*,

$$C_2\bar{H}_5\overset{+}{N}a + CO_2 \longrightarrow C_2H_5CO_2Na$$

Monocarboxylic acids may be prepared by heating an alkene with carbon monoxide and steam under pressure at 300 – 400°C in the presence of a catalyst, *e.g.*, phosphoric acid:

$$CH_2{=}CH_2 + CO + H_2O \longrightarrow CH_3CH_2CO_2H$$

$$CH_3CH{=}CH_2 + CO + H_2O \longrightarrow (CH_2)_2CHCO_2H$$

General Reactions of the Monocarboxylic Acids

1. The acids are acted upon by the strongly electropositive metals with the liberation of hydrogen and formation of a salt:

   $$RCO_2H + Na \longrightarrow RCO_2^-Na^+ + \frac{1}{2}H_2$$

   Salts are also formed by the reaction between an acid and an alkali:

   $$RCO_2H + NaOH \longrightarrow RCO_2^-Na^+ + H_2O$$

2. Monocarboxylic acids react with alcohols to form *esters*:

   $$R^1CO_2H + R^2OH \rightleftharpoons R^1CO_2R^2 + H_2O$$

3. Phosphorus trichloride, phosphorus pentachloride or thionyl chloride act upon the monocarboxylic acids to from *acid chlorides*:

   $$3RCO_2H + PCl_3 \longrightarrow 3RCOCl + H_3PO_3$$

   $$RCO_2H + PCl_5 \longrightarrow RCOCl + HCl + POCl_3$$

   $$RCO_2H + SOCl_2 \longrightarrow RCOCl + HCl + SO_3$$

   The acid series may be 'stepped up' via the acid chloride, but since the process involves many steps, the yield of the higher acid homologue is usually only *f.-f.g.*:

$$(a)\ RCO_2H \xrightarrow{PCl_5} RCOCl \xrightarrow[\text{catalyst}]{H_2} RCHO \xrightarrow[\text{catalyst}]{H_2}$$

$$RCH_2OH \xrightarrow{PBr_3} RCH_2Br \xrightarrow{KCN} RCH_2CN \xrightarrow{H^+} RCH_2CO_2H$$

$$(b)\ RCO_2H \xrightarrow{PCl_5} RCOCl \xrightarrow{KCN} RCOCN$$

$$\xrightarrow{H^+} RCOCO_2H \xrightarrow[\text{reduction}]{\text{Clemmensen}} RCH_2CO_2H$$

Although, method (*b*) involves fewer steps than (*a*), the final yield of acid by route (*a*) is higher than that by (*b*), since the yields in each step of (*b*) are generally low. Furthermore, the yield from (*a*) can be improved by direct reduction of acid to alcohol by means of various metallic hydrides.

4. When the ammonium salts of the monocarboxylic acids are strongly heated, the *acid amide* is formed by the elimination of water:

$$RCO_2NH_4 \longrightarrow RCONH_2 + H_2O$$

5. When the anhydrous sodium salt of a monocarboxylic acid is heated with soda-lime, an alkane and other products are formed.

The mechanism of this decarboxylation is uncertain, but there is much evidence to show that *salts* decompose by an S_E1 reaction :

$$R{-}C(=O)O^- \longrightarrow \bar{R}: + CO_2 \xrightarrow{H^+} R{-}H$$

The thermal decarboxylation of *free* acids may be:

$$R{-}C(=O){-}O{-}H \longrightarrow \overset{+}{H} + R{-}C(=O)O^- \longrightarrow \overset{+}{H} + \bar{R}: + CO_2 \longrightarrow R{-}H + CO_2$$

Monocarboxylic acids may be 'stepped down' by heating the calcium salt with calcium acetate, and the methyl ketone produced is oxidised with acid dichromate:

$$RCH_2CO_2Ca/_2 + CH_3CO_2Ca/_2 \longrightarrow RCH_2COCH_3$$

$$\xrightarrow{[O]} RCO_2H + CH_3CO_2H$$

It should be noted that the second step is based on the general rule that when an unsymmetrical ketone is oxidised, the carbonyl group remains chiefly with the smaller alkyl group. Since monocarboxylic acids containing an odd number of carbon atoms occur only in small amounts, this method of descending the series affords a useful means of obtaining 'odd' acids from 'even'. When a mixture of the calcium salt (or any of the others mentioned above) and calcium formate is heated, an aldehyde is obtained. The pyrolysis of iron salts gives very good yields of symmetrical ketones. The pyrolysis of iron slats gives very good yields of symmetrical ketones. Metallic salts, particularly the silver slats, are converted by bromine into the alkyl bromide. This reaction may also be used to 'step down' the acid series.

6. When a concentrated aqueous solution of the sodium or potassium salt of a monocarboxylic acid is electrolysed, an alkane is obtained.

$$2RCO_2K + 2H_2O \longrightarrow R{-}R + 2CO_2 + 2KOH + H_2$$

7. Monocarboxylic acids react slowly with chlorine or bromine in the cold, but at higher temperatures and in the presence of a small amount of phosphorus, reaction proceeds smoothly to give a-halogeno-acids; *e.g.*,

$$RCH_2CO_2H \xrightarrow{P/Br_2} RCHBrCO_2H \xrightarrow{P/Br_2} RCBr_2CO_2H$$

8. All the acids, except formic acids, are extremely resistant to oxidation, but prolonged heating with oxidising agents ultimately produces carbon dioxide and water.

9. The reduction products of monocarboxylic acids depend on the nature of the reducing agent. Thus, heating with hydrogen iodide-red phosphorous under pressure, or with hydrogen under pressure at elevated temperature in the presence of a nickel catalyst, produces an alkane.

$$RCO_2H + 3H_2 \xrightarrow{Ni} RCH_3 + 2H_2O$$

If the catalyst is Ru—C, the alcohol is formed. Alcohols are also produced by reduction with various metallic hydrides.

10. The acids undergo the *Schmidt reaction* to form a primary amine:

$$RCO_2H + HN_3 \xrightarrow{H_2SO_4} RNH_2 + CO_2 + N_2$$

Schmidt's reaction with acids is a modification of the Curtius reaction. The following mechanism has been proposed :

$$R{-}C({=}O){-}OH \xrightarrow{H^+} R{-}\overset{+}{C}(OH)(OH) \xrightarrow{HN_3} R{-}C(OH)(OH){-}N(H){-}\overset{+}{N}{\equiv}N \xrightarrow{-H_2O} R{-}C(OH){-}NH{-}\overset{+}{N}{\equiv}N \xrightarrow{-N_2}$$

$$O{=}C(R){-}\overset{+}{N}{-}H \xrightarrow{-H^+} O{=}C{=}NR \xrightarrow{H_2O} RNH_2 + CO_2$$

The rearrangement (1,2-shift) has been shown to be intramolecular, *e.g.*, Keynon *et al.* (1939) showed retention of optical activity when α-phenylpropionic acid underwent the Schmidt reaction:

$$C_6H_5CH(CH_3)CO_2H \longrightarrow C_6H_5CH(CH_3)NH_2$$

THE HYDROGEN BOND

Compounds containing OH or NH groups often exhibit unexpected properties such as relatively high boiling points, and it was soon felt necessary to assume that the elements oxygen or nitrogen were linked by means of hydrogen, thereby producing the *hydrogen bond.* Detailed study has shown that the unexpected properties were exhibited only when the atoms participating in the bond had high electronegativity—fluorine, oxygen and nitrogen (decreasing in this order), and to a less extent, chlorine and sulphur. Thus the hydrogen bond explained, for example, the association of hydroxylic compounds such as water, alcohols, etc., and the association of ammonia.

The exact nature of the hydrogen bond has been the subject of much discussion. The difficulty lies mainly in the fact that the energy of a hydrogen bond varies between that of the 'van der Waals forces' (4.184 kJ mol^{-1}) and that of a chemical bond. Some values obtained are: H—F....H, 41.84 kJ mol^{-1}; H—O....H, 29.29 kJ; H—N....H, 8.37 kJ. These are 'weak' hydrogen bonds, and for these the geometry of Z—H and Y is little changed when the hydrogen bond produces the complex Z—H....Y. Hence, the bond length of Z—H is almost the same in both Z—H and Z—H.....Y. It is accepted that a number of factors contribute, but it appears that the most important one is electrostatic. In bond Z—H, if Z has high electronegativity, there will be a relatively large amount of polarity, *i.e.*, the state of affairs will be $\overset{\delta-}{Z}—\overset{\delta+}{H}$, where δ+ is relatively large. Since the hydrogen atom has a tiny volume, the $\overset{\delta+}{H}$ will exert a large electrostatic force and so can attract atoms with a relatively large δ– charge, providing these atoms have a small atomic radius. Fluorine, oxygen and nitrogen are of this character. If the atom has a greater radius the electrostatic forces are weaker; thus chlorine, although it has about the same electronegativity as nitrogen, forms very weak hydrogen bonds since its atomic radius is greater. C—H bonds do not normally form hydrogen bonds, but when the electronegativity of carbon is increased by sp and sp^2 hybridisation, then apparently this group can form hydrogen bonds.

Hydrogen bond formation *intramolecularly*, *i.e.*, involving one molecule only, gives rise to *ring formation* or *chelation*, and this usually when the formation of a 5-, 6- or 7-membered ring is possible. Hydrogen bonding intermolecularly, *i.e.*, between two or more molecules, gives rise to association. Many examples of hydrogen bonding will be found in the text, and this is represented by a dotted line between the hydrogen and other atom involved (as shown above).

Hydrogen bonding affects all physico-chemical properties such as m.p., b.p., solubility, spectra, etc., *e.g.*, association produces a higher boiling point than expected (*e.g.*, from the molecular weight of the compound). On the other hand, chelation usually produces a lower boiling point than expected, *e.g.*, a nitre-compound usually has a higher boiling point than its parent compound, but if chelation is possible in the nitro-compound, the boiling point is lowered.

Atomic and Molecular Orbitals

So far, we have discussed the structure of molecules in terms of valency bonds. There is an alternative method of investigating the structure of molecules, and to appreciate this approach—and to extend the other—it is necessary to consider the structure of matter from the point of view of wave-mechanics. Classical physics (*i.e.*, the laws of mechanics, etc.) is satisfactory when dealing with large masses. These laws are approximations, but deviations become significant only when dealing with very small particles such as electrons and nuclei. The behaviour of these small particles, however, may be satisfactorily studied by wave (quantum) mechanics.

This uses the idea of the particle-wave duality of matter. It has already been pointed out that the electron may be regarded as a tiny mass carrying a negative charge. In 1923, de Broglie proposed that every moving particle has wave properties associated with it. This was first experimentally verified in the case of the electron (Davison and Germer, 1927; G. P. Thomson, 1928). Thus an electron has a dual nature, particle and wave, but it behaves as one or the other according to the nature of the experiment; *it cannot at the same time behave as both.* According to wave-mechanics, a moving particle is represented by a wave function ψ such that $\psi^2 d\tau$ is the probability of finding the particle in the element of volume $d\tau$. The greater the value of ψ^2, the greater is the probability of finding the electron in that volume $d\tau$.

Theoretically, ψ has a finite value at a large distance (compared with atomic dimensions) from the nucleus, but in practice there is very little probability of finding the electron beyond a distance of 2.3Å. This spatial behaviour of an electron, in terms of probability, is known as an orbital. When associated with one nucleus, the electron is said to be in an atomic orbital (A.O.), and when with a number of nuclei, the electron is said to be in a molecular orbital (M.O.). The behaviour of an electron may be represented by a charge cloud, the density of the cloud at any point being equal to the value of ψ^2 at that point.

In 1926, Schrodinger developed the wave-equation, which connected the wave function ψ of an electron with its energy. E. This equation has an infinite number of solutions, but very few of these solutions describe the known behaviour of electrons. Thus only certain values for E are permissible, since certain conditions must be satisfied. The permitted solutions for ψ are called the eigen functions, and the corresponding values of E are called the eigenvalues. A number of *eigenfunctions* exist, the simplest being those which possess spherical symmetry (ψ_s function), and the next simplest being those which possess an axis of symmetry (ψ_p function). Atomic orbitals are thus classified as s, p,d,f,. . . orbitals, and the energy of any A.O. is the eigenvalue (energy value) corresponding to that wave function ψ. We shall here be concerned with only s and p orbitals.

In addition to its wave function ψ, an electron also has spin. Two electrons can have the same wave function, *i.e.*, can occupy the same orbital, *provided their spins are opposite* (Pauli exclusion principle). In this case the electrons are said to be paired.

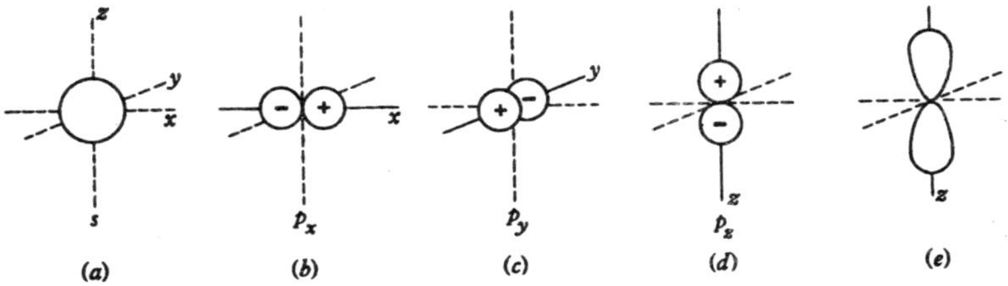

(a) s (b) p_x (c) p_y (d) p_z (e)

Fig. 15

The wave function ψ is a function of the co-ordinates of the electron and consists of two parts, the distance of the electron from the nucleus, *i.e.*, the *radial distribution function*, and the orientation of the electron with respect to the nucleus, *i.e.*, the *angular distribution function.* The two common ways of representing orbitals omit the radial distribution function and draw polar graphs in three dimensions of the angular function or its square, and the orbitals are depicted as sections through these polar (three-dimensional) graphs. The plot of the angular function for an s-orbital consists of a sphere, and the orbital is thus represented as a circle.

This is also the case for the plot of the square of the angular function (Fig. 4.6(a)). On the other hand, the plot of the angular function for p_z-orbital consists of two spheres in contact at the origin, and hence this orbital is represented as two circles in contact (Figs. 4.6 b, c, d). If, however, the polar plot is of the square of the angular function, then the p-orbital consists of two lobes in contact at the origin, and hence is represented, *e.g.*, the p_z-orbital, as shown in Fig. 4.6e. It-can be seen in Fig. 4.6 that p-orbitals have a marked directional character, each orbital having an axis at right angles to those of the other two, and hence they are known as the p_x,p_y,p_z orbitals, respectively. These orbitals are entirely equivalent except for their directional property. It should also be noted that one circle of a p-orbital is marked with a positive sign and the other with a negative sign.

When the plot of the square of the angular function is used, the resultant values are always positive and so both lobes are positive. In the same way, an s-orbital can be + or – for the plot of the angular function (but not its square). These signs arise from the fact that a wave function can have positive and negative regions, but the signs have no physical significance. Thus, it does not matter which circle in a p-orbital is given the positive sign; the other is then given the – sign. Where ψ is zero, there is a node, *i.e.*, there is no likelihood of finding an electron in a nodal plane. These relative signs are useful when considering the overlap of atomic orbitals, and lobes are often used instead of circles.

The order of orbital energies is Is < 2s < 2p < 3s < 3p . . . . When an electron absorbs energy, it is driven into an orbital of higher energy. The atom is then said to be *'excited'* and is more reactive. On returning to its normal orbital, the electron loses this extra energy. When all the electrons in an atom are in their normal orbitals, *i.e.*, orbitals of lowest energy, the atom is said to be in the 'ground' state. So far, we have dealt only with atoms, *i.e.*, with electrons associated with one nucleus. The wave-equations for molecules cannot be solved without making some approximations. Two types of approximations have been made, one set giving rise to the *valence-bond* method (V.B.); and the other set to the *molecular orbital* (M.O.). The V.B. method—due mainly to the work of Heitler, London, Slater and Pauling—considers the molecule as being made up of atoms with electrons in atomic orbitals on each atom. Thus a molecule is treated as if it were composed of atoms which, to some extent, retain their individual character when linked to other atoms.

The M.O. method—due mainly to the work of Hund, Lennard-Jones and Mulliken—treats a molecule in the same way as an atom, except that in the molecule an electron moves in the field of more than one nucleus, *i.e.*, molecular orbitals axe polycentric. Thus each electron in a molecule is described by a certain wave function, the molecular orbital, for which various shapes can be drawn as for A.O.s, but differing in that the former are polycentric and the latter monocentric. In general, the greater the freedom (*i.e.*, the larger the region for movement) allowed to an electron, the lower will be its energy. Hence atoms combine to form a molecule because, owing to the overlap of the A.O.s when the atoms are brought together, the electrons acquire a greater freedom, and the energy of the system is lowered below that of the separate atoms. Energy would therefore have to be supplied to separate the atoms in the molecule, and the greater the amount of energy necessary, the stronger are the bonds formed between the various atoms.

Let us now consider the case of the hydrogen molecule. A hydrogen atom has one Is electron. When the bond is formed between two hydrogen atoms to form the hydrogen molecule, these two 1s electrons become *paired* to form *molecular electrons*, *i.e.*, both occupy the same M.O., a state of affairs which is possible provided their spins are antiparallel. A very important principle for obtaining the M.O. is that the bond energy is greatest when the component A.O.s overlap one another as much as possible. To get the maximum amount of overlap of orbitals, the orbitals should be in the same plane. Thus the M.O. is considered as being a *linear combination of atomic orbitals with maximum overlap* (L.C.A.O.). Furthermore, according to L.C.A.O. theory, "the binding energy is greater the more nearly equal are the energies of the component A.O.s. If these energies differ very much, then there will be no significant combination between the two atoms concerned.

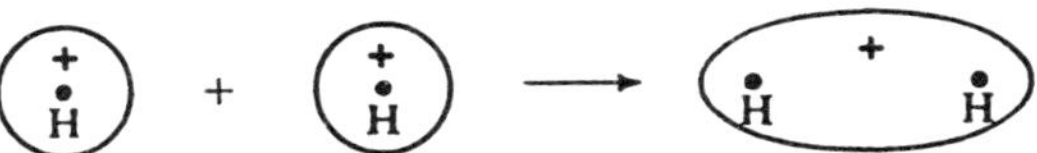

Fig. 16

Since the hydrogen molecule is composed of two identical atoms, the probability of finding both electrons simultaneously near the same nucleus is very small. Hence one might expect the M.O. to be symmetrical with respect to the two hydrogen nuclei, *i.e.*, the M.O. in the hydrogen molecule will be 'plum-shaped'.

Although the probability of finding the two electrons simultaneously near the same nucleus is very small, nevertheless this probability exists, and gives rise to the two ionic structures H^+H^- and $H^-\ H^+$. Thus the hydrogen molecule will be a resonance hybrid of three resonating structures, one purely covalent (*i.e.*, the two electrons are equally shared), and two ionic (*i.e.*, the pair of electrons are associated with one nucleus all the time):

$$H{:}H \leftrightarrow \overset{+}{H}{:}\overset{-}{H} \leftrightarrow \overset{-}{H}{:}\overset{+}{H}$$

Calculation has shown that the ionic structures contribute very little to the actual state of the hydrogen molecule, and the bond between the two hydrogen atoms is described as a *covalent* bond with partial ionic character. It should here be noted that when the single bond is formed between the two hydrogen atoms, the probability of finding the electrons is greatest in the region between the two nuclei. It is this concentration of the negatively charged electrons between the two positive hydrogen nuclei that binds the nuclei together. Since electrons are negatively charged, they will repel each other and so tend to keep out of the region between the two nuclei. On the other hand, since the spins of the two electrons are antiparallel, this produces attraction between the two electrons, thereby tending to concentrate them in the internuclear region. The net result is that the electron density for paired electrons is greatest between the two nuclei. Such a bond is said to be a localised M.O., and preserves the idea of a bond connecting the two atoms. This localisation (in a covalent bond) gives rise to the properties of bond lengths, dipole moments, polarisability and force constants.

FORMIC ACID

Formic acid (*methanoic acid*) is prepared industrially by heating sodium hydroxide with carbon monoxide at 210°C, and at a pressure of 6–10 atmospheres:

$$NaOH + CO \longrightarrow HCO_2Na$$

Anhydrous formic acid is obtained from the aqueous solution (70–77 per cent) by the addition of butyl formate followed by distillation.

The first fraction is an azeotrope of ester and water, and then the excess of ester is removed from the formic acid by fractionation. The most convenient laboratory preparation of formic acid is to heat glycerol with oxalic acid at 100–100 C. Glyceryl monoxalate is produced, and decomposes into glyceryl monoformate (monoformin) and carbon dioxide. When the evolution of carbon dioxide ceases, more oxalic acid is added, whereupon formic acid is produced:

$$\begin{array}{l} CH_2OH \\ | \\ CHOH \\ | \\ CH_2OH \end{array} + \begin{array}{l} COOH \\ | \\ COOH \end{array} \longrightarrow \begin{array}{l} CH_2OCOCOOH \\ | \\ CHOH \\ | \\ CH_2OH \end{array} + H_2O \xrightarrow{100\text{-}110^\circ C}$$

$$CO_2 + \begin{array}{l} CH_2OCHO \\ | \\ CHOH \\ | \\ CH_2OH \end{array} \xrightarrow{(COOH)_2} \begin{array}{l} CH_2OCOCOOH \\ | \\ CHOH \\ | \\ CH_2OH \end{array} + HCO_2H \xrightarrow{100\text{-}110^\circ C}$$

The distillate contains formic acid and water. The aqueous formic acid solution cannot be fractionated to give anhydrous formic acid because the boiling point of the acid is 100.5°C. The procedure adopted is to neutralise the aqueous acid solution with lead carbonate, and concentrate the solution until lead formate crystallises out. The precipitate is then recrystallised, dried, and heated at 100°C in a current of hydrogen sulphide:

$$(HCO_2)_2Pb + H_2S \longrightarrow 2HCO_2H + PbS$$

The anhydrous formic acid which distils over contains a small amount of hydrogen sulphide, and may be freed from the latter by adding some dry lead formate and redistilling. This procedure for obtaining the anhydrous acid from its aqueous solution can only be used for volatile acids.

Formic acid is a pungent corrosive liquid, m.p. 8.4°C, b.p. 100.5°C, miscible in all proportions with water, ethanol and ether. It forms salts which, except for the lead and silver salts, are readily soluble in water. Formic acid is a stronger acid than any of its homologues.

Formic acid is dehydrated to carbon monoxide by concentrated sulphuric acid and when heated under pressure at 160°C, is decomposed into carbon dioxide and hydrogen:

$$CO + H_2O \xleftarrow{H_2SO_4} HCO_2H \xrightarrow{\text{heat}} CO_2 + H_2$$

The same decomposition takes place at room temperature in the presence of a catalyst such as iridium, rhodium, etc.

When metallic formates are heated with an alkali, hydrogen is evolve.

$$HCO_2Na + NaOH \longrightarrow H_2 + Na_2CO_3$$

When calcium or zinc formate is strongly heated, formaldehyde is produced:

$$(HCO_2)_2Ca \longrightarrow HCHO + CaCO_3$$

When sodium or potassium formate is rapidly heated to 360°C, hydrogen is evolved and the oxalate is formed.

$$2HCO_2Na \longrightarrow (COONa)_2 + H_2$$

Formic acid forms esters, but since it is a relatively strong acid it is not necessary to use a catalyst: refluxing 90 per cent formic acid with the alcohol is usually sufficient. Formic acid differs from the rest of the members of the monocarboxylic acid series in being a powerful reducing agent; it reduces ammoniacal silver nitrate and the salts of many of the heavy metals, *e.g.*, it converts mercuric chloride into mercurous chloride.

Acetic acid (*ethanoic acid*) is prepared industrially by various methods, *e.g.*, (*i*) the air oxidation of acetaldehyde in acetic acid in the presence of manganous ion as catalyst: (*ii*) the air oxidation of n-butane in the liquid phase under pressure and at 130–230°C in the presence of a suitable catalyst, *e.g.*, manganous ion; (*iii*) by reaction between carbon monoxide and methanol under pressure in the presence of cobalt octacarbonyl at about 210°C:

$$CO + CH_3OH \longrightarrow CH_3CO_2H.$$

One of the earliest methods for preparing acetic acid was by the destructive distillation of wood to give pyroligneous acid. This contains about 10 per cent acetic acid, and was originally treated by neutralising with lime and then distilling off the volatile compounds (these are mainly methanol and acetone). On distillation with dilute sulphuric acid, the residue gives dilute acetic acid. More recently, the acetic acid is extracted by means of solvents, *e.g.*, isopropyl ether.

Vinegar, which is 6–10 per cent aqueous solution of acetic acid, may be made in several ways. Malt vinegar is prepared by the oxidation of wort by means of the bacteria *Mycoderma aceti*:

$$CH_3CH_2OH + O_2 \longrightarrow CH_3CO_2H + H_2O$$

In the 'quick' vinegar process', beech shavings, contained in barrels, are moistened with strong vinegar containing the bacteria. A 10 per cent aqueous solution of ethanol containing phosphates and inorganic salts (which are necessary for the fermentation) is then poured through the barrels, and the ethanol is thereby oxidised to acetic acid. A plentiful supply of air is necessary, otherwise the oxidation is incomplete and acetaldehyde is produced.

Acetic acid is a pungent corrosive liquid, m.p. 16.6°C, b.p. 118°C, miscible in all proportions with water, ethanol ad ether It is stable towards oxidising agents, and so is a useful solvent for chromium trioxide oxidations. Acetic acid is commonly used as a solvent, and in the preparation of acetates, acetone, acetic anhydride, etc.

Strengths of the monocarboxylic acids. If the structure of the carboxyl group were simply (I), then, because of the –I effect of the carbonyl group (as shown in (II), proton release will be facilitated as compared with alcohols.

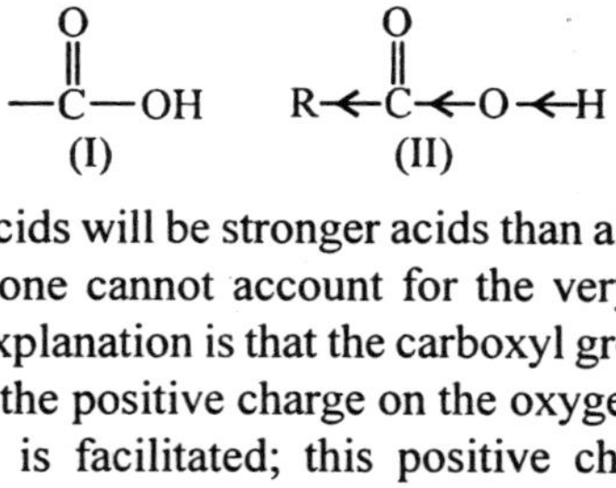

$$\underset{\text{(I)}}{-\overset{\overset{\text{O}}{\|}}{\text{C}}-\text{OH}} \qquad \underset{\text{(II)}}{\text{R}\leftarrow\overset{\overset{\text{O}}{\|}}{\text{C}}\leftarrow\text{O}\leftarrow\text{H}}$$

Thus carboxylic acids will be stronger acids than alcohols. However, the inductive effect alone cannot account for the very large difference in acid strength. One explanation is that the carboxyl group is a resonance hybrid and because of the positive charge on the oxygen atom of the OH group, proton release is facilitated; this positive charge is absent in alcohols since they are not resonance hybrids.

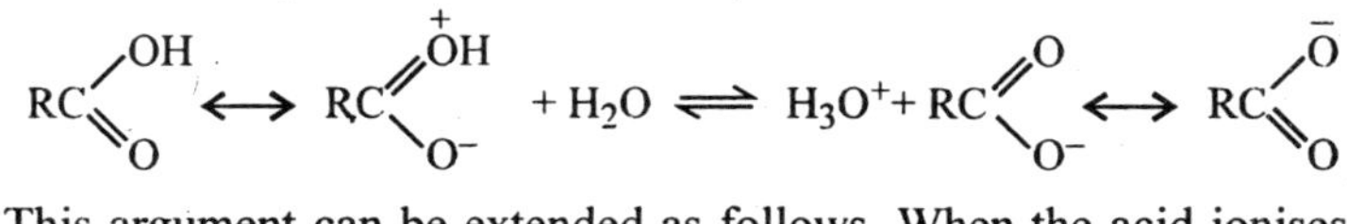

$$\text{RC}\begin{matrix}\diagup\text{OH}\\ \diagdown\!\!\diagdown\text{O}\end{matrix} \longleftrightarrow \text{RC}\begin{matrix}\diagup\!\!\diagup\overset{+}{\text{O}}\text{H}\\ \diagdown\text{O}^-\end{matrix} + H_2O \rightleftharpoons H_3O^+ + \text{RC}\begin{matrix}\diagup\!\!\diagup\text{O}\\ \diagdown\text{O}^-\end{matrix} \longleftrightarrow \text{RC}\begin{matrix}\diagup\bar{\text{O}}\\ \diagdown\!\!\diagdown\text{O}\end{matrix}$$

This argument can be extended as follows. When the acid ionises, *i.e.*, donates its proton to water (which is the solvent), the conjugate base of the acid. *i.e.*, the carboxylate ion, is a resonance hybrid of two *equivalent* resonating structures. At the same time, since the negative charge is spread, the resonance energy of this anion will be greater than that of the un-ionised acid, which is a resonance hybrid of two different resonating structures, one with separation of *unlike* charges. Therefore the internal energy of the anion is lower than that of the un-ionised acid. In the case of alcohols, thee is no resonance stabilisation in the alkoxide ion. It therefore follows that $\Delta G^\ominus$ for the ionisation of monocarboyxlic acids will be more negative than that of alcohols. Hence, the equilibrium constant of the former will be greater than that of the latter, *i.e.*, the pK_a values of the acids will be lower than those of the alcohols. Thus alcohols are weaker acids than the monocaboxylic acids.

X-ray and electron-diffraction measurements of the carboxylate ion have shown that the two C—O bonds are of equal length.

Peracetic (peroxyacetic) acid, $CH_3C(=O)O{-}OH$ may be prepared by treating acetic anhydride with concentrated hydrogen peroxide, and then distilling under reduced pressure.

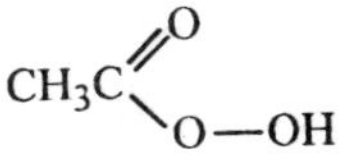

Swern *et al.* (1962) have prepared per-acids (85–97 per cent yield) by treating a solution or slurry of an aliphatic (or aromatic) acid in methanesulphonic acid with 90–95 per cent hydrogen peroxide. This appears to be the best method so far. Peracetic acid is an unpleasant-smelling liquid, f.p. + 0.1°C, soluble in water, ethanol and ether. It explodes violently when heated above 110°C. It is a powerful oxidising agent; it oxidises the olefinic bond to the oxide.

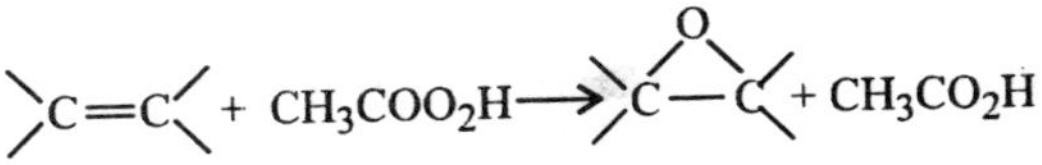

$$>C=C< + CH_3COO_2H \longrightarrow >C\overset{O}{—}C< + CH_3CO_2H$$

It also oxidises primary aromatic amines to nitroso-compounds, *e.g.*, aniline is converted into nitrosobenzene:

$$C_6H_5NH_2 + 2CH_3COO_2H \longrightarrow C_6H_5NO + 2CH_2CO_2H + H_2O$$

Infra-red absorption spectra measurements of per-acids show that these acids exist in solution very largely in the monomeric, intra-molecularly hydrogen-bonded form (*inter alia*, Minkoff, 1954).

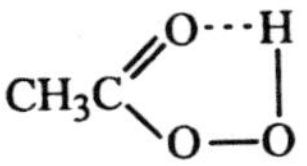

$$CH_3C(=O\cdots H)-O-O \text{ (ring: } O\cdots H-O-O\text{)}$$

Propionic acid (*propanoic acid*) is prepared industrially by the oxidation of n-propanol:

$$CH_3CH_2CH_2OH \xrightarrow{[O]} CH_3CH_2CO_2H$$

It is a colourless liquid with an acrid odour, m.p. – 22°C, b.p. 141°C, miscible with water, ethanol and ether in all proportions.

Butyric acids. There are two isomers possible, and both are known.

n-Butyric acid occurs as the glyceryl ester in butter, and as the free acid in perspiration. It is prepared industrially by the oxidation of n-butanol.

n-Butyric acid is a viscous unpleasant-smelling liquid, m.p. – 4.7°C, b.p. 162°C, miscible with water, ethanol and ether. It is the liberation of free n-butyric acid that gives stale butter its rancid odour.

Isobutyric acid occurs in the free state and as its esters in many plants. It is prepared industrially by the oxidation of isobutanol and may be prepared synthetically as follows:

$$(CH_3)_2CHOH \xrightarrow{PBr_3} (CH_3)_2CHBr \xrightarrow{KCN}$$

$$(CH_3)_2CHCN \xrightarrow[acid]{H_2O} (CH_3)_2CHO_2H$$

Isobutyric acid is a liquid, m.p. – 47°C, b.p. 154°C. Its calcium salt is more soluble in hot water than in cold, whereas calcium butyrate is more soluble in cold water than in hot.

Valeric acids. There are four isomers possible, and all are known:

n-Valeric acid, $CH_3CH_2CH_2CH_2CO_2H$, m.p. – 34.5°C, b.p. 187°C.

Isovaleric acid, $(CH_3)_2CHCH_2CO_2H$, m.p. – 51°, b.p. 175°C.

Ethylmethylacetic acid or *active valeric acid,* CH_3CH_2CH-$(CH_3)CO_2H$, p.b. 175°C.

Trimethylacetic acid or *pivalic acid,* $(CH_3)_3CCO_2H$, m.p. 35.5°, p.b. 164°C.

The higher monocarboxylic acids which occur in nature usually have straight chains, and usually contain an *even* number of carbon atoms. **Caproic** (*hexanoic*), $C_6H_{12}O_2$ (m.p. – 9.5°, b.p. 205°C), **caprylic** (*octanoic*), $C_8H_{16}O_2$ (m.p. 16°, p.b. 237°C) and *capric* (*decanoic*) acid, $C_{10}H_{20}O_2$ (m.p. 31.5° p.b. 270°C), are present as glyceryl esters in goats' butter. *Lauric* (*dodecanoic*), $C_{12}H_{24}O_2$ (m.p. 44°C) and **myristic** (*tetradecanoic*) acid, $C_{14}H_{28}O_2$ (m.p. 58°C), occur as their glyceryl esters in certain vegetable oils.

The most important higher acids are *palmitic* (*hexadecanoic*), $C_{16}H_{32}O_2$ (m.p. 64°C), and **stearic** (*octadecanoic*), $C_{18}H_{36}O_2$ (m.p. 72°C), which are very widely distributed as their glyceryl esters (together with oleic acid) in most animal ad vegetable oils and fats. The sodium and potassium salts of palmitic and stearic acids are the constituents of ordinary soaps.

Some still higher acids are found in waxes: **arachidic** (*eicosanoic*), $C_{20}H_{40}O_2$ (m.p. 77°C), **behenic** (*docosanoic*), $C_{22}H_{44}O_2$ (m.p. 82°C), **lignoceric** (*tetracosanoic*), $C_{24}H_{48}O_2$ (m.p. 83.5°C), **cerotic** (*hexacosanoic*), $C_{26}H_{52}O_2$ (m.p. 87.7°C), and **melissic** acid (*triacontanoic*), $C_{30}H_{60}O_2$ (m.p. 90°C).

The odd fatty acids may be obtained by degrading an even acid (see below). Two odd acids which may be prepared readily are *n-heptanoic* or *oenanthylic* acid, $C_7H_{14}O_2$ (m.p. – 10°, p.b. 223.5°C), and *nonanoic* or *pelargonic* acid $C_9H_{18}O_2$ (m.p. 12°, p.b. 254°C). n-Heptanoic acid is prepared by the oxidation of n-heptaldehyde, which is obtained by destructively distilling castor oil which contains ricinoleic acid (*q.v.*):

$$CH_3(CH_2)_5CHO + [O] \xrightarrow[H_2SO_4]{KMnO_4} CH_3(CH_2)_5CO_2H \ (68\text{–}70\%)$$

Nonanoic acid is obtained by the oxidation of oleic acid (*q.v.*).

Margaric acid (*heptadecanoic acid*), $C_{16}H_{33}CO_2H$, m.p. 61°C, is prepared by heating a mixture of the calcium salts of stearic and acetic acids, and then oxidising the heptadecyl methyl ketone so produced.

$$C_{17}H_{35}CO_2ca + CH_3CO_2ca \longrightarrow CaCO_3 + C_{17}H_{35}COCH_3 \xrightarrow{[O]} C_{16}H_{33}CO_2H$$

Odd-numbered monocarboxylic acids occur naturally in very small amounts up to $n = 23$, and their main source is butter fat. Branched-chain acids have also been isolated from butter fat.

General Properties of the Monocarboxylic Acids

The first three acids are colourless, pungent-smelling liquids; the acids from butyric, $C_4H_8O_2$, to nonanoic, $C_9H_{18}O_2$, are oils which smell like goats' butter; and those higher than decanoic acid, $C_{10}H_{20}O_2$, are odourless solids. The lower members are far less volatile than is to be expected from their molecular weight. This can be explained by hydrogen bonding, and electron diffraction studies (Pauling, 1934) have shown that an eight-membered ring is present, *i.e.*, the acids exist as cyclic dimers. The lower members exist as dimers in the vapour phase and in aqueous solution, but in the liquid phase they exist as 'polymers'.

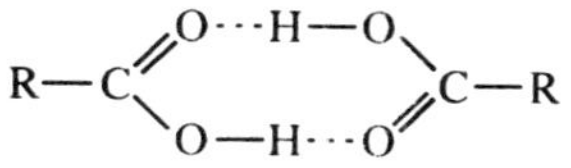

$$R-C\overset{\diagup O\cdots H-O\diagdown}{\underset{\diagdown O-H\cdots O\diagup}{}}C-R$$

The melting point of the n-monocarboyxlic acids show alternation or oscillation from one member to the next, the melting point of an 'even' acid being higher than that of the 'odd' acid immediately below and above it in the series (The physical constants of the individuals).

A number of homologous series follow this oscillation or 'saw-tooth' rule. The first four members are very soluble in water, and the solubility decreases as the molecular weight increases. This solubility is due to the acids being capable of forming hydrogen bonds with water.

The infra-red absorption region of the carboxyl group contains a number of bands. The carbonyl group (stretch) absorbs in the same region as that of acyclic ketones. 1725–1700 cm^{-1}(s), but the (free) hydroxyl group (stretch) absorbs at 3560–3500 cm^{-1}(m), which is in the lower frequency region than for a alcohols (3670–3580 cm^{-1}). The infra-red spectrum of stearic acid (as a 1 per cent solution in carbon tetrachloride).

It should be noted that the band for the bonded O—H (str.) is hidden by the strong C—H (str.)) On the other hand the spectrum of stearic acid as a Nujol mull. In this case, the stearic acid is in the solid (dispersed) state, and it can be seen that there is an extremely large difference in the region 1300–1180 cm^{-1}.

Mass spectra. Acids, esters, and amides undergo similar fragmentation patterns, *e.g.*, α-cleavage for lower members (Z=OH, OR, NH_2):

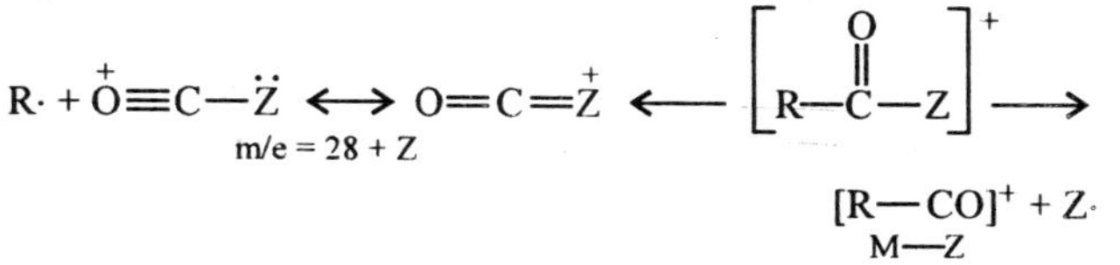

$$R\cdot + \underset{m/e\,=\,28\,+\,Z}{\overset{+}{O}\equiv C-\ddot{Z}} \longleftrightarrow O=C=\overset{+}{Z} \longleftarrow \left[R-\overset{O}{\overset{\|}{C}}-Z\right]^{+} \longrightarrow \underset{M-Z}{[R-CO]^{+}} + Z\cdot$$

Also, R and Z may carry the positive charge. Hence, peaks are obtained at $M-17$ and *m/e* 45 (Z = OH), $M-31$, M–45, and *m/e* 59, 73 (Z = OMe and OEt, respectively), and $M-16$ and *m/e* 44 (Z = NH_2).

Since they are more volatile than their corresponding acids, esters are used preferentially. All can undergo the McLafferty rearrangement if they contain a γ-hydrogen atom:

$$\begin{matrix} CH_2\text{—}H \\ | \\ CH_2 \\ \diagdown CH_2 \diagup \end{matrix} \quad \overset{\cdot +}{O}=CZ \longrightarrow \begin{matrix} CH_2 \\ \| \\ CH_2 \end{matrix} + \left[\begin{matrix} HO \\ \diagdown \\ C\text{—}Z \\ \| \\ CH_2 \end{matrix} \right]^{\dot{+}}$$

m/e 60 (Z = OH); 74 (Z = OMe); 88 (Z = OEt); 59 ($Z = NH_2$)

When the alkyl group in O of esters is higher than methyl, esters can also undergo the McLafferty rearrangement to give alkene and acid, *e.g.*,

$$\left[\begin{matrix} CH_3CH\text{—}H & O \\ | & \| \\ CH_2 & CR \\ \diagdown O \diagup \end{matrix} \right] \longrightarrow \left[\begin{matrix} CH_3CH \\ \| \\ CH_2 \end{matrix} \right]^{+} + RCO_2H$$

General Reactions of the Monocarboxylic Acids

Acids and their derivatives may be represented as follows, where Z = OH, OR, X, NH_2, etc.

$$RC\begin{matrix} \diagup\!\!\diagup O \\ \diagdown \ddot{Z} \end{matrix} \longleftrightarrow R\overset{+}{C}\begin{matrix} \diagup\!\!\diagup O^- \\ \diagdown \ddot{Z} \end{matrix} \longleftrightarrow RC\begin{matrix} \diagup O^- \\ \diagdown\!\!\diagdown Z^+ \end{matrix}$$

Thus, all are resonance hybrids and the behaviour of the 'carbonyl' group depends on the nature of Z. Z may have a – I and a + R effect and consequently the actual weights of the canonical forms will depend on the relative contributions of these two electronic effects, *e.g.*, for Z = NH_2 (acid amide), + R > – I, whereas for Z = Cl (acid chloride), – I > + R. Hence, the rates of reaction of the carbonyl group in these compounds can be expected to be different. Many of the reactions of the acids and their derivatives may be generalised as:

(i) $$R\text{—}\underset{Z}{\overset{O}{\overset{\|}{\underset{|}{C}}}} + Y{:} \underset{}{\overset{r/d}{\rightleftharpoons}} R\text{—}\underset{Z}{\overset{O^-}{\overset{|}{\underset{|}{C}}}}\text{+}Y \longrightarrow R\text{—}\overset{O}{\overset{\|}{C}}\text{—}Y + Z{:}$$

(ii) $$R\text{—}\underset{Z}{\overset{O}{\overset{\|}{\underset{|}{C}}}} + H^+ \rightleftharpoons R\text{—}\underset{Z}{\overset{\overset{+}{O}H}{\overset{\|}{\underset{|}{C}}}} \longrightarrow R\text{—}\underset{Z}{\overset{OH}{\overset{|}{\underset{|}{C^+}}}} \xrightarrow[r/d]{Y:} R\text{—}\underset{Z}{\overset{OH}{\overset{|}{\underset{|}{C}}}}\text{—}Y$$

$$\longrightarrow R\text{—}\overset{O}{\overset{\|}{C}}\text{—}Y + HZ$$

These reactions are typical of the behaviour of the carbonyl group (in aldehydes and ketones), and the greater the positive charge on the carbonyl atom, the more easily is this carbon atom attacked by a nucleophile (in the r/d step). Since the + R effect tends to neutralise the positive charge, the greater the + R effect of Z, the slower will be the reaction. Experimental work has shown the rates of hydrolysis are in the order $RCONH_2 < RCO_2R < RCOCl$, and this is the reverse order of the + R effects of the groups, *i.e.*, $NH_2 > OR > Cl$.

ESTERS

Esters are compounds which are formed when the hydroxylic hydrogen atom in oxygen acids is replaced by an alkyl group; the acid may be organic or inorganic. The most important esters are derived from the caboxylic acids. The general formula of the carboxylic esters is $C_2H_{2n}O_2$, which is the same as that of the carboxylic acids, and they are named as the alkyl salts of the acid. *e.g.*,

$CH_3COOC_2H_5$ ethyl acetate

$(CH_3)_2CHOOCH(CH_3)_2$ isopropyl isobutyrate

Carboxylic esters are formed by the action of the acid on an alcohol: acid + alcohol:

$$\text{acid} + \text{alcohol} \rightleftharpoons \text{ester} + \text{water}$$

The reaction is reversible, the *forward* reaction being known as *esterification*, and the *backward* reaction as *hydrolysis* (for mechanisms).

General Methods of Preparation of the Carboxylic Esters

1. The usual method is *esterification*. The reaction is always slow, but is speeded up by the presence of small amounts of inorganic acids as catalysts, *e.g.*, when the acid is refluxed with the alcohol in the presence of 5–10 per cent concentrated sulphuric acid:

$$R^1COOH + R^2OH \xrightarrow{H_2SO_4} R^1COOR^2 + H_2O \ (v.g.)$$

Alternatively, hydrogen chloride is passed into the mixture of alcohol and acid until there is a 3 per cent increase in weight, and the mixture is refluxed (the yields are very good). This is known as the *Fische–Speier method* (1895), and is more satisfactory for secondary and tertiary alcohols than the sulphuric acid method, which tends to dehydrate the alcohol to alkene. Esterification without the use of catalysts, and starting with one mole of acid and one mole of alcohol, gives rise to about 2/3 mole of ester. The yield of ester may be increased by using excess of acid or alcohol, the cheaper usually being the one in excess.

Increased yields may also be effected by dehydrating agents, *e.g.*, concentrated sulphuric acid behaves both as a catalyst and a dehydrating agent. The same effect may be obtained by removing the water or ester from the reaction mixture by distillation, which is particularly useful for high-boiling acids and alcohols.

On the other hand, the water may be removed from the reaction mixture by the addition of benzene or carbon tetrachloride, each of which forms a binary mixture with water (and may form a ternary mixture with water and the alcohol), the azeotropic mixtures boiling at a lower temperature than any of the components.

2. Acid chlorides or anhydrides react rapidly with alochols to form esters:

$$R^1COCl + R_2OH \longrightarrow R^1CO_2R^2 + HCl \ (\textit{v.g.–ex.})$$

$$(R^1CO)_2O + R^2OH \longrightarrow R^1CO_2R^2 + R^1CO_2H \ (\textit{v.g.–ex.})$$

The reaction with tertiary alcohols is very show and is often accompanied by side reactions (alkene or alkyl halide formation), but by using the appropriate conditions, esters are produced. Esters of tertiary alochols may also be conveniently prepared by means of a Grignard reagent.

3. Esters may be prepared by refluxing the silver salt of an acid with an alkyl halide in ethanolic solution:

$$R^1CO_2Ag + R^2Br \longrightarrow R^1CO_2R^2 + AgBr \ (\textit{v.g. –ex.})$$

This method is useful where direct esterification is difficult.

4. *Methyl* esters are very conveniently prepared by treating an acid with an ethereal solution of diazomethane (*q.v.*):

$$RCO_2H + CH_2N_2 \longrightarrow RCO_2CH_3 + N_2 \ (\textit{ex.})$$

According to Morrison *et al.* (1961), the BF_3—MeOH reagent is better than diazomethane for the methylation of carboxylic acids.

5. The interaction between an ether and carbon monoxide at 125–180°C, under 500 atmospheres pressure, in the presence of boron trifluoride plus a little water, produces esters:

$$R_2O + CO \longrightarrow RCO_2R$$

Esters may also be prepared by the Tischenko reaction (with aldehydes); ethyl acetate is made commercially this way.

General Properties of Esters

The carboxylic esters are pleasant-smelling liquids or solids. The boiling points of the straight chain isomers are higher than those of the branched-chain isomers. The boiling points of the methyl and ethyl esters are lower than those of the corresponding acid, and this is due to the fact that the esters are not associated since they cannot form intermolecular hydrogen bonds. The esters of low molecular weight are fairly soluble in water–hydrogen bonding between ester and water is possible–and the solubility decreases as the series is ascended; all esters are soluble in most organic solvents.

General Reactions of Esters

1. Esters are hydrolysed by acids or alkalis:

$$R^1CO_2H + R^2OH \xleftarrow{H_3O^+} R^1CO_2R^2 \xrightarrow{OH^-} R^1CO_2^- + R^2OH$$

When hydrolysis is carried out with alkali, the salt of the acid is obtained, and since the alkali salts of the higher acids are soaps, alkaline hydrolysis is known as *saponification* (derived from Latin word meaning soap); saponification is far more rapid than acid hydrolysis.

Mechanisms of carboxyl esterification and hydrolysis of carboxylic esters. Mechanistic studies have shown that in alkaline or neutral hydrolysis, it is the neutral ester molecule which undergoes reaction, whereas in acid-catalysed hydrolysis, it is the conjugate acid of the ester which undergoes reaction.

$$\underset{\text{neutral ester}}{R^1\!-\!\overset{\overset{\displaystyle O}{\|}}{C}\!-\!OH^2} \qquad R^1\!-\!\overset{\overset{\displaystyle \overset{+}{O}H}{\|}}{C}\!-\!OH^2 \quad \text{rather than} \quad \underset{\text{conjugate acid}}{R^1\!-\!\overset{\overset{\displaystyle O}{\|}}{C}\!-\!\overset{\overset{\displaystyle H}{|}}{O^+}\!-\!R^2}$$

The protonated carbonyl oxygen form of the cA is preferred, and this is based on the work of Fraenkel (1961), who examined the NMR spectrum of methyl formate in, *e.g.*, 100 per cent sulphuric acid and concluded from his results that the ester is protonated chiefly on the carbonyl oxygen.

This formulation is also used for *acids*, partly from analogy with esters and partly from the fact that of the two possible conjugate acids, (I) and (II) , (I) may be written as a resonance hybrid to which two identical canonical forms contribute, whereas for (II) the two contributing structures are different and one has fewer bonds and carries unlike charges.

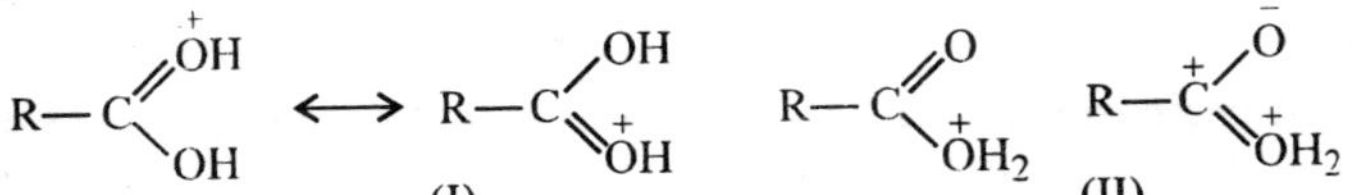

$$\mathrm{R{-}C({=}\overset{+}{O}H)(OH)} \longleftrightarrow \mathrm{R{-}C(OH)({=}\overset{+}{O}H)} \quad \text{(I)} \qquad \mathrm{R{-}C({=}O)(\overset{+}{O}H_2)} \quad \mathrm{R{-}\overset{+}{C}(\overset{-}{O})({=}\overset{+}{O}H_2)} \quad \text{(II)}$$

The hydrolysis of carboxylic esters may be formulated in two ways:

$$\mathrm{R^1{-}\overset{O}{\overset{\|}{C}}{-}OR^2} \qquad \mathrm{R^1{-}\overset{O}{\overset{\|}{C}}{-}O{-}R^2}$$

acyl-oxygen heterolysis — alkyl-oxygen heterolysis

Both modes of fission have been demonstrated experimentally, and kinetic studies have shown that two types of mechanism may operate for each mode, *unimolecular* and *bimolecular*. Also, since in principle, the *acid-catalysed* reactions (esterification and hydrolysis) are reversible, but not *alkaline* hydrolysis, then *four* mechanisms are possible for esterification.

A represents the cA of the ester or acid (reaction in *acid* solution), *B* represents the unprotonated ester (reaction in *basic* or *neutral* solution), subscripts AC and AL, respectively, denote acyl- and alkyl-heterolysis, and numbers 1 and 2 represent the molecularity of the rate-determining step (Ingold, 1941):

Most of the experimental work has been carried out on hydrolysis because this process can be studied in acidic, basic, or neutral media, whereas esterification can be studied only in acidic media. It might appear at first sight that esterification could also be studied in neutral media, but since carboxylic acids are acids, reaction in the absence of added acid (as catalyst) is still effectively in an acid medium. Because of this, mechanisms of esterification are partly based on the principle of microscopic reversibility.

Bimolecular basic hydrolysis with acyl-oxygen heterolysis (B_{AC} 2 mechanism). It has long been known that the alkaline hydrolysis of esters is a second-order reaction, *i.e.*, rate = k [ester] [OH^-]. Evidence for acyl-oxygen heterolysis has been obtained in several ways, *e.g.*,

Polanyl *et al.*, (1934) showed that the alkaline hydrolysis of n-pentyl acetate in water enriched with ^{18}O gave n-pentyl alcohol containing *no* ^{18}O. Therefore acyl-oxygen heterolysis must have occurred:

$$\text{CH5CO—OC5H11} + \dot{\text{O}}\text{H}^- \longrightarrow CH_3CO\dot{O}H + C_5H_{11}O^- \longrightarrow$$

$$CH_3CO\dot{O}^- + C_5H_{11}OH$$

Another method uses optical activity as a diagnostic test for demonstrating acyl-oxygen heterolysis. If the alcohol group is optically active, and if the $B_{AC}2$ mechanism operates, then the ion RO^- will be liberated and will retain its optical activity since the bond R—O is never broken. Various examples of retention are known when the reaction is bimolecular.

A mechanism consistent with the above facts (and with the work of Bender (1951), who used esters labelled with ^{18}O in the carbonyl group) is:

$$HO^- + R^1C(=O)-OR^2 \underset{\text{fast}}{\overset{\text{slow}}{\rightleftharpoons}} HO-C(O^-)(R^1)-OR^2 \underset{\text{slow}}{\overset{\text{fast}}{\rightleftharpoons}}$$

$$HO-C(=O)-R^1 + {}^-OR^2 \xrightarrow{\text{fast}} {}^-O-C(=O)-R^1 + R^2OH$$

Any R^1 group with – I effect should accelerate hydrolysis by creating a positive charge on the attached carbon atom, thereby facilitating attack by the hydroxide ion. On the other hand, because the r/d step is bimolecular, the larger R^1 is the larger can be expected to be the steric retardation, since the hybridisation of the carbon atom of the carbonyl group changes from sp^2 to sp^3 (with consequent increased crowding). These anticipated results have been observed in practice, *e.g.*,

– I effect

Ester	$MeCO_2Me$	$ClCH_2CO_2Me$	$CHCl_2CO_2Me$
Relative rates	1	761	16000

Steric effect

Ester	$MeCO_2Et$	$EtCO_2Et$	$isoPrCO_2Et$
Relative rates	1.00	0.47	0.10

Bimolecular acid-catalysed hydrolysis and esterification with acyl-oxygen heterolysis ($A_{AC}2$ mechanism). Acid-catalysed hydrolysis and acid-catalysed esterification are the reverse of each other, and so according to the principle of microscopic reversibility the mechanism of acid-catalysed hydrolysis will be that of acid-catalysed esterification, but in reverse.

The rate law for acid-catalysed hydrolysis has been shown to be second-order, first-order in both ester and hydrogen ion concentration, *i.e.*, rate = k [ester] $[H_3O^+]$. Also, evidence for acyl=oxygen heterolysis has been obtained in several ways, *e.g.*, Ingold *et al.* (1939) showed that the acid-catalysed hydrolysis of methyl hydrogen succinate in water enriched with ^{18}O gave methanol with no extra ^{18}O. Thus, acyl-oxygen heterolysis must have occurred:

$$HO_2CCH_2CH_2CO\text{—}OMe + H_2\dot{O} \xrightarrow{H^+} HO_2CCH_2CH_2CO\dot{O}H + MeOH$$

Furthermore, Robers *et al.* (1938) esterified benzoic acid with methanol enriched with 18O and obtained water *not* enriched with 18O. Thus, acyl-oxygen heterolysis must have occurred:

$$PhCO\text{—}OH + Me\dot{O}H \longrightarrow PhCO\dot{O}me + H_2O$$

It should be noted that the r/d step for esterification is the addition of alcohol, whereas that for hydrolysis is the addition of water. In both cases, the r/d step involves change of hybridisation of the carbon atom of the carbonyl group from sp_2 to sp_3, and consequently steric retardation can be expected to increase as R1 increase as size (*cf.* the BAC_2 mechanism), *e.g.* (esterification with methanol):

Acid	$MeCO_2H$	n-$PrCO_2H$	$MeCCO_2H$	Et_3CCO_2H
Relative rates	1.0	0.51	0.037	0.00016

Unimolecular acid-catalysed hydrolysis and esterification with acyl-oxygen heterolysis ($A_{AC}1$ mechanism). Let us consider the following *observed* results (*M* represents the mesityl group, 1,3,5-trimethylphenyl, in *mesitoic acid*, $Me_3C_6H_2CO_2H$):.

$PhCO_2H$ + MeOH $\xrightarrow{HCl}$ $PhCO_2Me$ *occurs readily*

$PhCO_2ME$ + OH^- $-\!\!\longrightarrow$ $PhCO_2^-$ + MeOH *occurs readily*

MCO_2H + MeOH $\xrightarrow{HCl}$ *No reaction*

$$MCO_2Me + OH^- \longrightarrow MCO_2^- + MeOH \text{ very slow}$$

Hammett *et al.* (1937) showed that when methyl benzoate is dissolved in concentrated sulphuric acid and the solution then poured into ice/water, almost all of the ester is recovered. On the other hand, methyl mestioate, on the same treatment, gave a quantitative yield of mestioic acid. Then Newman (1941) showed that when a solution of mesitoic acid in concentrated sulphuric acid is poured into a cold methanol, methyl mesitoate is formed. This different behaviour of mesitoyl derivatives from benzoyl derivatives suggests that different mechanisms are operating in esterification and hydrolysis for these two sets of compounds. Now, Hammett (1937) had shown that when mesitoic acid is dissolved in concentrated sulphuric acid, the van't Hoff factor *i* is 4 (measured by depression of freezing point). Under the same conditions, benzoic acid gives *i* = 2. Hammett therefore proposed that the mesitoyl reactions proceed through an intermediate *acylium* (*oxocarbonium*) ion (*mesitoylium* ion in this case) formed by acyl-oxygen heterolysis:

$$\text{(i)}\ MCO_2H + H_2SO_4 \overset{\text{fast}}{\rightleftharpoons} M\overset{+}{C}(OH)_2 + HSO_4^-$$

$$\text{(ii)}\ M\overset{+}{C}(OH)_2 + H_2SO_4 \overset{\text{slow}}{\rightleftharpoons} M\overset{+}{C}\ O + H_3O^+ + HSO_4^-$$

The overall equation may be written:

$$MCO_2H + 2H_2SO_4 \rightleftharpoons M\overset{+}{C}\ O + H_2O^+ + 2HSO_4^-$$

This accounts for *i* = 4 for mesitoic acid, and for *i* = 2 for benzoic acid if only reaction (*i*) is involved. Thus, the reactions of mesitoic acid and its ester may be formulated as:

$$\left.\begin{array}{l} MCO_2H \\ MCO_2H \end{array}\right\} \xrightarrow[\text{slow}]{H_2SO_4} MC{=}O \xrightarrow{\text{fast}} \left\{\begin{array}{l} \xrightarrow{MeOH} MCO_2Me \\ \xrightarrow{H_2O} MCO_2H \end{array}\right.$$

This type of reaction occurs with highly sterically hindered acids and esters, and so steric hindrance must be an important factor. Further support for this mechanism has been provided by Olah *et al.* (1967), who showed that acids (and esters), when dissolved in $FSO_3H—SbF_5—SO_2$ (a very strongly acidic medium), at 60°C, lose a molecule of water at 0°C to form the acylium ion:

$$RC(=O)OH \xrightarrow[-60^\circ C]{H^+} RC(=\overset{+}{O}H)OH \longleftrightarrow RC(OH)=\overset{+}{O}H \xrightarrow[0^\circ C]{-H_2O} R\overset{+}{C}{=}O$$

Unimolecular basic hydrolysis with alkyl-oxygen heterolysis ($B_{AL}1$ mechanism). This mechanism has been shown to occur when the alkyl group of the alcohol is capable of forming a relatively stable carbonium ion, the solvent has high ionising power (thereby encouraging alkyl-oxygen heterolysis), and the hydroxide ion concentration is low (thereby suppressing the bimolecular mechanism). An example of the $B_{AL}1$ mechanism is the hydrolysis of optically active α-phenylethyl hydrogen phthalate in *faintly* alkaline solution. Kenyon *et al.* (1936) showed that the product was recemissed α-phenylethanol (phenylmethylmethanol); this is in keeping with the formation of an intermediate carbonium ion (sp^2 hybridised) [R = *o*–$HO_2CC_6H_4$—]:

$$RCO{-}O{-}CHMePh \underset{}{\overset{slow}{\rightleftharpoons}} RCO_2^- + PhMeCH^+ \xrightarrow[fast]{H_2O} PhMeCHOH_2^+ \xrightarrow[fast]{RCO_2^-} (\pm)\text{-}PhMeCHOH + RCO_2H$$

Bimolecular basic hydrolysis with alkyl-oxygen heterolysis ($B_{AL}2$ mechanism). This mechanism is rare. One example is the reaction between methyl benzoate and sodium methoxide in methanol (Bunnett *et al.*, 1950). The products were dimethyl ether and sodium benzoate; these results are readily explained on the basis of the $B_{AL}2$ mechanism:

$$MeO^- \;\; Me{-}O{-}OCPh \longrightarrow Me_2O + PhCO_2^-$$

Unimolecular acid-catalysed hydrolysis and esterification with alkyl-oxygen heterolysis ($A_{AL}1$ mechanism). Definite evidence for this mechanism was obtained by Hughes, Ingold *et al.* (1939), who showed that the esterification of acetic acid with optically active octan-2-ol in the presence of sulphuric acid gave a large amount of racemised ester. Racemisation is to be expected if an intermediate carbonium ion is formed, and so the reaction may be formulated as $A_{AL}1$ (R = n-$C_6H_{11}CHMe$—):

$$ROH + H^+ \underset{fast}{\overset{fast}{\rightleftharpoons}} ROH_2^+ \underset{fast}{\overset{slow}{\rightleftharpoons}} H_2O + R^+$$

$$MeC(=O)OH \;\; R^+ \underset{slow}{\overset{fast}{\rightleftharpoons}} MeC(=\overset{+}{O}R)OH \underset{fast}{\overset{fast}{\rightleftharpoons}} MeC(OR)=O + H^+$$

Since all steps are reversible, the $A_{AL}1$ mechanism should, in principle, be possible for ester hydrolysis, Evidence for its occurrence has been obtained by, *e.g.*, Bunton *et al.* (1951), who carried out the acid-catalysed hydrolysis of t-butyl acetate in water enriched with ^{18}O and obtained t-butanol containing ^{18}O.

$$\mathrm{Me\overset{\overset{\displaystyle O}{\|}}{C}{-}O{-}Bu^t + H^+ \underset{fast}{\overset{fast}{\rightleftharpoons}} Me{-}\overset{\overset{\displaystyle \overset{+}{O}H}{\|}}{C}{-}O{-}Bu^t \underset{fast}{\overset{slow}{\rightleftharpoons}} MeC\begin{matrix} \diagup OH \\ \diagdown\!\!\diagdown O \end{matrix} + \overset{+}{B}u^t}$$

$$\mathrm{\overset{+}{B}u^t + H_2\dot{O} \underset{fast}{\overset{fast}{\rightleftharpoons}} Bu^t\dot{O}H_2{}^+ \underset{slow}{\overset{fast}{\rightleftharpoons}} Bu^t\dot{O}H + H^+}$$

The $A_{AL}1$ mechanism is common for t-alcohols; these from relatively stable carbonium ions (*cf.* the $B_{AL}1$ mechanism).

2. Esters are converted into alcohols by the Bouveault-Blanc reduction, catalytic hydrogenation, and by metallic hydrides.

3. Esters react with ammonia to form amides. This reaction is example of *ammonolysis* (which means, literally, splitting by ammonia):

$$\mathrm{R^1COOR^2 + NH_3 \longrightarrow R^1CONH_2 + R^2OH}\ (g.)$$

With hydrazine, esters form acid hydrazides:

$$\mathrm{R^1COOR^2 + H_2NNH_2 \longrightarrow R^1CONHNH_2 + R^2OH}$$

Esters react with phosphorus pentachloride or thionyl chloride to form acyl chlorides, *e.g.*,

$$\mathrm{R^1CO_2R^2 + PCl_5 \longrightarrow R^1COCl + POCl_3 + R^2Cl}$$

4. By means of *alcoholysis* (splitting by alcohol), an alcohol residue in an ester can be replaced by another alcohol residue. Alcoholysis is carried out by refluxing the ester with a large excess of alcohol, preferably in the presence of a small amount of acid or sodium alkoxide as catalyst.

$$\mathrm{CH_3CO_2C_4H_9 + C_2H_5OH \overset{C_2H_5ONa}{\rightleftharpoons} CH_3CO_2C_2H_5 + C_4H_9OH}$$

Alcoholysis of esters is also known as *transesterification.*

The Mechanism of alcoholysis is uncertain; a possibility is (in basic media):

$$\mathrm{R^1\overset{\overset{\displaystyle O}{\|}}{\underset{\underset{\displaystyle OR^2}{|}}{C}} + EtO^- \rightleftharpoons R^1\overset{\overset{\displaystyle O^-}{\|}}{\underset{\underset{\displaystyle OR^2}{|}}{C}}{-}OEt \rightleftharpoons R^1\overset{\overset{\displaystyle O}{\|}}{C}{-}OEt + R^2O^-}$$

$$\mathrm{R^2O^- + EtOH \rightleftharpoons R^2OH + EtO^-}$$

In *acidolysis*, the acid residue is displaced from its ester by another acid residue, *e.g.*,

$$\mathrm{CH_3CO_2C_2H_5 + C_5H_{11}CO_2H \rightleftharpoons C_5H_{11}CO_2C_2H_5 + CH_3CO_2H}$$

Acidolysis is a useful reaction for converting the natural ester of a dibasic acid into its acid ester.

5. When an ester – preferably the methyl or ethyl ester – is treated with sodium in an inert solvent, *e.g.*, ether, benzene or toluene, and subsequently with acid, an *acyloin* is formed (50–70 per cent yield). It is important that the reaction be carried out *in the absence of any free alcohol*. Acyloins are α,β-keto-alcohols, and according to Kharasch and his co-workers (1939), their formation takes place via a free-redical mechanism, *e.g.*, propionin from ethyl propionate:

Acyloins are reduced by the Clemmensen method to alkanes.

6. Carboxylic esters which contain α-hydrogen atoms react with sodamide in liquid ammonia solution to form the acid amide and a condensation product involving two molecules of the ester, *e.g.*, ethyl acetate gives acetamide and acetoacetic ester :

$$CH_3CO_2C_2H_5 + NaNH_2 \longrightarrow CH_3CONH_2 + C_2H_5ONa$$

$$CH_3CO_2C_2H_5 + NaNH_2 \longrightarrow NH_3 + NH_3 + [\ddot{C}H_2CO_2C_2H_5]^- Na^+$$

$$\xrightarrow{CH_3CO_2C_2H_5} CH_3COCH_2CO_2C_2H_5 + C2H_5OH$$

Of particular interest is the *carbonation* of esters, *i.e.*, the introduction of the carboxyl group.

$$R^1CH_2CO_2R^2 + NaNH_2 \xrightarrow{\text{liquid } NH_3}$$

$$NH_3 + [R^1CHCO_2R^2]^- Na^+ \xrightarrow[\text{ether}]{CO_2} \underset{\displaystyle CO_2Na}{R^1\overset{|}{C}HCO_2R^2}$$

The yields of these *malonic acid derivatives* are 54–60 per cent with acetates, and becomes progressively lower as the molecular weight of he acid increases.

7. The pyrolysis of esters, particularly acetates, gives alkenes.

Esters are used as solvents for cellulose, oils, gums, resins, etc., and as plasticisers. They are also used for making artificial flavours and essences, *e.g.*, isoamyl acetate – banana oil; amylbutyrate–apricot; isoamyl isovalerate–apple; methyl butyrate–banana oil; amyl butyrate–apicot; isoamyl isovalerate–apple; methyl butyrate–pineapple; etc.

Ortho-esters are compounds of the type $R^1C(OR^2)_3$. They are derived from the ortho-acids, $RC(OH)_3$, which have not yet been isolated, but which are possibly present in aqueous solution:

$$R{-}C{\overset{\displaystyle O}{\diagup\!\!\diagup}}\!\!\diagdown_{OH} + H_2O \quad RC(OH)_3$$

The most important ortho-esters are the orthoformic esters, particularly ethyl orthoformate. This may be prepared by running ethanol and chloroform into sodium covered with ether:

$$2CHCl_3 + 6C_2H_5OH + 6Na \longrightarrow 2HC(OC_2H_5)_3 + 6NaCl + 3H_3 \text{ (70\%)}$$

A possible mechanism is via the formation of dichloromethylene:

$$CHCl_3 \xrightarrow{EtO^-} :CCl_2 \xrightarrow{EtOH} EtOCHCl_2 \xrightarrow{2EtO^-}$$
$$HC(OEt)_3 + 2Cl^-$$

Ethyl orthoformate may be used for preparing ketals and for preparing aldehydes by means of a Grignard reagent.

Atomic and Molecular Orbitals

So far, we have discussed the structure of molecules in terms of valency bonds. There is an alternative method of investigating the structure of molecules, and to appreciate this approach—and to extend the other—it is necessary to consider the structure of matter from the point of view of wave-mechanics. Classical physics (*i.e.*, the laws of mechanics, etc.) is satisfactory when dealing with large masses. These laws are approximations, but deviations become significant only when dealing with very small particles such as electrons and nuclei. The behaviour of these small particles, however, may be satisfactorily studied by wave (quantum) mechanics.

This uses the idea of the particle-wave duality of matter. It has already been pointed out that the electron may be regarded as a tiny mass carrying a negative charge. In 1923, de Broglie proposed that every moving particle has wave properties associated with it. This was first experimentally verified in the case of the electron (Davison and Germer, 1927; G. P. Thomson, 1928). Thus an electron has a dual nature, particle and wave, but it behaves as one or the other according to the nature of the experiment; *it cannot at the same time behave as both*. According to wave-mechanics, a moving particle is represented by a wave function ψ such that $\psi^2 d\tau$ is the probability of finding the particle in the element of volume $d\tau$. The greater the value of ψ^2, the greater is the probability of finding the electron in that volume $d\tau$.

Theoretically, ψ has a finite value at a large distance (compared with atomic dimensions) from the nucleus, but in practice there is very little probability of finding the electron beyond a distance of 2.3Å. This spatial behaviour of an electron, in terms of probability, is known as an orbital. When associated with one nucleus, the electron is said to be in an atomic orbital (A.O.), and when with a number of nuclei, the electron is said to be in a molecular orbital (M.O.). The behaviour of an electron may be represented by a charge cloud, the density of the cloud at any point being equal to the value of ψ^2 at that point.

In 1926, Schrodinger developed the wave-equation, which connected the wave function ψ of an electron with its energy. E. This equation has an infinite number of solutions, but very few of these solutions describe the known behaviour of electrons. Thus only certain values for E are permissible, since certain conditions must be satisfied. The permitted solutions for ψ are called the eigen functions, and the corresponding values of E are called the eigenvalues. A number of *eigenfunctions* exist, the simplest being those which possess spherical symmetry (ψ_s function), and the next simplest being those which possess an axis of symmetry (ψ_p function). Atomic orbitals are thus classified as s, p,d,f,. . . orbitals, and the energy of any A.O. is the eigenvalue (energy value) corresponding to that wave function ψ. We shall here be concerned with only s and p orbitals.

ESTERS OF THE INORGANIC ACIDS

Alkyl sulphates, $(RO)_2SO_2$. Only methanol and ethanol give a good yield of the alkyl sulphate by reaction with concentrated sulphuric acid; the higher alochols give mainly alkenes and ethers. On the other hand, all the alcohols give a fair yield of alkyl hydrogen sulphate when a mixture of alcohol and concentrated sulphuric acid is heated on a steam bath.

Methyl sulphate is prepared industrially:

(*i*) By heating methanol with concentrated sulphuric acid, and then distilling the methyl hydrogen sulphate under reduced pressure:

$$CH_3OH + H_2SO_4 \longrightarrow H_2O + CH_3OSO_3H \xrightarrow{2} (CH_3O)_2SO_2 + H_2SO_4$$

(*ii*) By treating methanol with sulphur trioxide at low temperatures:

$$2SO_3 + 2CH_3OH \longrightarrow (CH_3)_2SO_4 + H_2SO_4$$

Ethyl sulphate may be prepared by the same methods as methyl sulphate, but in addition, there is the industrial preparation by passing ethylene in excess into cold concentrated sulphuric acid:

$$2C_2H_4 + H_2SO_4 \longrightarrow (C_2H_5)_2SO_4$$

Sulphates of the higher alcohols are conveniently prepared by the oxidation of sulphites as follows:

$$2ROH_4 + SOCl_2 \longrightarrow 2HCl + (RO)_2SO \xrightarrow{KMnO_4} (RO)_2SO_2$$

Methyl sulphate, b.p. 188°C, and ethyl sulphate, b.p. 208°C, are heavy poisonous liquids. They are largely used as alkylating agents since the alkyl group will replace the hydrogen atom of the groups —OH, —NH— or —SH. The alkylation may be carried out by treating the compound with the alkyl sulphate in sodium hydroxide solution. Usually only one of the alkyl groups takes part in the reaction, *e.g.*, methylation of a primary amine:

$$RNH_2 + (CH_3)_2SO_4 + NaOH \longrightarrow RNHCH_3 + CH_3NaSO_4 + H_2O$$

The methyl and ethyl esters of the carboxylic acids maybe conveniently prepared by treating the sodium salt of the acid with respectively methyl or ethyl sulphate:

$$RCO_2Na + (C_2H_5)_2SO_4 \longrightarrow RCO_2C_2H_5 + C_2H_5NaSO_4$$

The sodium alkyl sulphates of the higher alcohols are used as detergents, *e.g.*, sodium lauryl sulphate:

Alkyl nitrates, $RONO_2$. The only important alkyl nitrate is ethyl nitrate, $C_2H_5ONO_2$. This may be prepared by heating ethyl iodide with silver nitrate in ethanolic solution:

$$C2H_5I + AgNO3 \longrightarrow C_2H_5ONO_2 + AgI$$

When concentrated nitric acid is added to ethanol, the reaction is usually violent; part of the ethanol is oxidised, and some of the nitric acid is reduced to nitrous acid.

Apparently it is the presence of the nitrous acid which produces the violent oxidation of the alcohol by the nitric acid. This danger may be avoided by first boiling the nitric acid with urea, which destroys any nitrous acid present, and then adding this mixture to cool ethanol, any nitrous acid produced being immediately destroyed by the urea:

$$CO(NH_2)_2 + 2HNO_2 \longrightarrow CO_2 + 2N_2 + 3H_2O$$

Ethyl nitrate is a pleasant-smelling liquid, b.p. 87.5°C. When reduced with tin and hydrochloric acid, it forms hydroxylamine and ethanol:

$$C_2H_2ONO_2 \xrightarrow{Sn/HCl} C_2H_5OH + NH_2OH + H_2O$$

Alkyl nitrites, RONO. Alkyl nitrites are isomeric with the nitroalkanes. The only important alkyl nitrites are the ethyl and amyl nitrites; the latter is actually mainly isoamyl nitrite, since the amyl alcohol used is the isopentanol from fusel oil. Ethyl and amyl nitrites are prepared by adding concentrated hydrochloric acid or sulphuric acid to aqueous sodium nitrite and the alcohol:

$$R{-}O{-}H + NO^+ \longrightarrow R{-}O\begin{smallmatrix}\diagup H\\ \diagdown NO\end{smallmatrix} \xrightarrow{-H^+} RONO$$

Ethyl nitrite, p.b. 17°C, and amyl nitrite, p.b. 99°C, are pleasant-smelling liquids, and are used as a means of preparing nitrous acid in anhydrous media; *e.g.*, an ethanolic solution of nitrous acid may be prepared by passing dry hydrogen chloride into amyl nitrite dissolved in ethanol.

MONOHYDRIC ALCOHOLS

The alcohol is a compound that contains one or more *hydroxyl* groups, *i.e.*, alcohols are hydroxy-derviatives of the alkanes. They are classified according to the number of hydroxyl groups present. Monohydric alcohols contain one hydroxyl group; dihydric, two; trihydric, three; etc. When the alcohols contain four or more hydroxyl groups, they are usually called polyhydric alcohols. The monohydric alcohols from an homologous series with the general formula $C_nH_{2n-2}O$, but, since their functional group is the hydroxyl group, their general formula is more satisfactorily written as $C_nH_{2n+1}OH$ or ROH.

Nomenclature

The simpler alcohols are commonly known by their trivial names, which are obtained by naming the alcohol as a derivative of the alkyl group attached to the hydroxyl group, *e.g.*, CH_3OH, methyl alcohol; $CH_3CH_2CH_2OH$. n-propyl alcohol; $CH_3CH(OH)CH_3$, isopropyl alcohol; $(CH_3)_3COH$, t-butyl alcohol. Another system of nomenclature considers the alcohols as derivatives of methyl alcohol, which is named *carbinol* or *methanol*, *e.g.*, $CH_3CH_2CHOHCH_3$, ethylmethylcarbinol or ethylmethylmethanol.

In the I.U.P.A.C. system of nomenclature, the longest carbon chain containing the hydroxyl group is chosen as the parent hydrocarbon. The class suffix is-*ol*, and the positions of side-chains and the hydroxyl group are indicated by numbers, the lowest possible number being given to the hydroxyl group, *e.g.*, CH_3OH, methanol; C_2H_5OH, ethanol; CH_3CH_2-CH_2OH, Propan-1-ol; $(CH_3)_2CHCHOHCH_3$, 3-methylbuttan-2-ol. According to this system the alcohols are referred to as the *alkanols*.

Monohydric alcohols are subdivided into primary, secondary and tertiary alcohols according as the alkyl group attached to the hydroxyl group is a primary, secondary or tertiary group, respectively. Primary alcohols contain the *primary alcoholic group ---CH_2OH*, *e.g.*, ethanol, CH_3CH_2OH; secondary alcohols, the *secondary alcoholic group* CH(OH), *e.g.*, isoprepanol, $(CH_3)_2CHOH$; and tertiary alcohols the *tertiary alcoholic group ≡C(OH)*, *e.g.*, t-butanol, $(CH_3)_3COH$.

GENERAL METHODS OF PREPARATION

1. By the hydrolysis of an alkyl halide with aqueous alkali or silver oxide suspended in water:

$$RX + \text{'AgOH'} \longrightarrow ROH + AgX$$

2. By the hydrolysis of esters with alkali:

$$R^1CO_2R^2 + KOH \longrightarrow R^1CO_2K + R^2OH$$

This method is important industrially for preparing certain alcohols that occur naturally as esters. Since tertiary alkyl halides give large amounts of alkene on hydrolysis, they are best converted into alcohols as follows:

$$R_3CX + CH_3CO_2Ag \xrightarrow[\text{heat}]{\text{EtOH}} AgX + CH_3CO_2\dot{C}R_3$$

$$\xrightarrow{\text{NaOH}} CH_3CO_2Na + R_3COH$$

3. By heating ethers with dilute sulphuric acid under pressure, *e.g.*, diethyl ether forms ethanol:

$$(C_2H_5)_2O + H_2O \xrightarrow{H_2SO_4} 2C_2H_5OH$$

This method is important industrially, since ethers are formed as by-products in the preparation of certain alcohols (see ethanol and propanols).

4. By the reduction of aldehydes, ketones or esters by means of excess of sodium and ethanol or n-butanol as the reducing agent (*Bouveault-Blanc reduction*, 1903), *e.g.*,

(*i*) *Aldehydes:* $RCHO \xrightarrow{e;\ H^+} RCH_2OH$ (g.-v.g.)

(*ii*) *Esters:* $R^1CO_2R^2 \xrightarrow{e;\ H^+} R^1CH_2OH + R^2OH$ (g.)

(*iii*) *Ketones:* $R_2CO \xrightarrow{e;\ H^+} R_2CHOH$ (g.)

The Bouvealult-Blanc reduction is believed to occur in steps involving the transfer of one electron at a time; *e.g.*,

$$\underset{\underset{O\cdot}{\|}}{RC}—OEt \xrightarrow{Na} \underset{\underset{O\cdot}{|}}{R\ddot{C}}—OEt \xrightarrow{EtOH} \underset{\underset{O\cdot}{|}}{RCH}—OEt \xrightarrow{Na} \underset{\underset{O^-}{|}}{RCH}—OEt \longrightarrow$$

$$OEt^- + RCH{=}O \xrightarrow{Na} R\ddot{C}H—O\cdot \xrightarrow{EtOH} RCH_2—O\cdot \xrightarrow{Na}$$

$$RCH_2—O^- \xrightarrow{EtOH} RCH_2OH$$

REDUCTION WITH METALLIC HYDRIDES

Many complex metallic hydrides reduce various functional groups, and their umber is increasing rapidly. The most versatile reagent is **lithium aluminium hydride** (LAH). This reduces most functional groups, Table 6.1, but does not normally reduce the olefinic bond. An unusual feature of this regent is its reduction of the *carboxyl group* to primary alcohol.

Reductions with lithium aluminium hydride are usually carried out in ethereal solutions, the compound in ether being added to the lithium aluminium hydride solution. In certain cases the reverse addition is necessary, *i.e.*, the hydride solution is added to the solution of the compound to be reduced.

Sodium borohydride, which is insoluble in ether, is used in *ethanolic* solution to reduce *carbonyl* compounds, the only important exception being the carboxyl group. It also does not normally reduce esters, but reduction to primary alcohol can often be effected by use of a large excess of reagent in methanol. The reduction of the carbonyl group by LAH or sodium borohydride occurs in a stepwise manner, each step involving hydride ion transfer, *e.g.*,

$$\overset{\overset{O}{\|}}{R_2C}\ H—AlH_3 \longrightarrow \left[\overset{\overset{O}{|}}{R_2CH} + AlH_3\right] \longrightarrow R_2CH—O—AlH_3 \xrightarrow{R_2CO}$$

$$(R_2CHO)_2AlH_2 \xrightarrow{R_2CO} (R_2CHO)_3AlH \xrightarrow{R_2CO} (R_2CHO)_4Al^- \xrightarrow{H^+} 4R_2CHOH$$

Experimental work has shown that the rate of reduction is decreased as the size of R increases. One contributing factor is steric hindrance, and this is made use of by employing reducing complexes containing large groups, *e.g.*, when the substrate contain is two carbonyl groups ad one is sterically hindered, lithium tri-t-butoxyaluminium hydride, reduce only the nonhindered carbonyl group. This reagent is generally used in tetrahydrofuran (THF) solution. In addition to steric hindrance, there is also the inductive effect in the anion, *e.g.*,

$$(R-O\leftarrow)_3Al\leftarrow H \quad >C=O$$

The electronegative oxygen atom exerts a strong – I effect and consequently makes hydride ion transfer more difficult. Hence, the reactivity of lithium tri-t-butoxyaluminium hydride is lower than that of LAH. Many other metallic hydrides are also reducing reagents, and because of their different reactivities, they offer a means of *selective* reduction.

Lithium borohydride (in ether or THF) is more reactive than sodium borohydride but less than LAH.

Lithium tri-t-butoxyaluminium hydride (in THF) useful for reducing acid chlorides to aldehydes at –78 C.

Sodium aluminium hydride (in THF) is similar to LAH, but it reduces esters to aldehydes.

Diborane (in THF) reduces many functional groups.

Aluminium hydride (in ether or THF) is similar to LAH, but is better for reducing α, β-unsaturated carbonyl compounds to unsaturated alcohols.

The reducing power of a *given* metallic hydride is affected by the nature of the solvent and the presence of certain other compounds, *e.g.*, **LAH and aluminium chloride** in ether form AlH_3, AlH_2Cl, or $AlHCl_2$ according to the proportion of reagents used:

$$3LiAlH_4 + AlCl_3 \longrightarrow 4AlH_3 + 3LiCl$$

$$LiAlH_4 + AlCl_3 \longrightarrow 2AlH_2Cl + LiCl$$

$$LiAlH_4 + 3AlCl_3 \longrightarrow 4AlHCl_2 + LiCl$$

Hence, the reducing power of the reagent will depend on which one is actually present; all are milder reducing reagents than LAH, *e.g.*, LAH-

-$AlCl_3$ does not reduce alkyl halides. It has also been shown that a solution of LAH in pyridine which has been allowed to stand will reduce a carbonyl group but not a carboxyl or carbalkoxyl group. Thus, LAH has been converted into a milder reducing reagent, the structure of which appears to be uncertain. *Sodium borohydride-aluminium chloride* (in diethylene glycol dimethyl ether, *i.e.*, diglyme is more reactive than sodium borohydride itself.

OPTICAL ACTIVITY OF BIPHENYL COMPOUNDS

Christie and Kenner's work has been extended by other workers, who showed that compounds in which at least three of the four ortho-positions in biphenyl are occupied by certain groups could be resolved. It was then soon found that two conditions were necessary for biphenyl compounds to exhibit optical activity

(i) Neither ring must have a vertical plane of symmetry. Thus (I) is not resolvable, but (II) is.

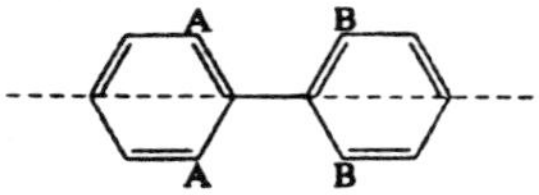

(I) 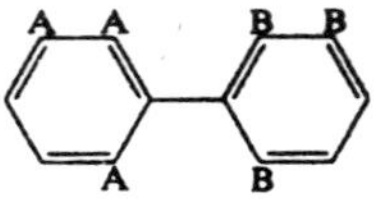 (II)

(ii) The substituents in the ortho-positions must have a large size, *e.g*, the following compounds were resolved: 6-nitrodiphenic acid, 6,6′-dinitrodiphenic acid, 6,6′-dichlorodiphenic acid, 2,2′diamino-6,6′-dimethylbiphenyl.

The earlier work showed that three groups had to be present in the ortho-positions. This gave rise to the theory that the groups in these positions impinged on one another when free rotation was attempted, *i.e.*, the steric effect prevented free rotation. This theory of restricted rotation about the single bond joining the two benzene rings (in the co-axial formula) was suggested simultaneously in 1926 by Turner and Le Fevre, Bell and Kenyon, and Mills. Consider molecule (III) and its mirror image (IV). Provided that the groups A, B and C are large enough to 'interfere mechanically', *i.e.*, to behave as 'obstacles', then free rotation about the single bond is restricted.

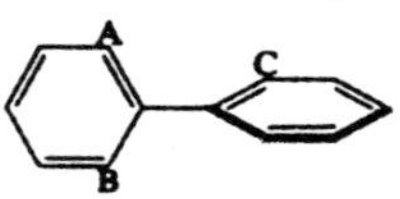

(III) 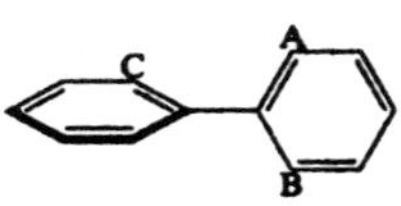 (IV)

Thus, the two benzene rings cannot be coplanar and consequently (IV) is not superimposable on (III), *i.e.,* (III) and (IV) are enantiomers. In molecule (III) there is no chiral centre; it is the molecule as a whole which is chiral, due to the restricted rotation.

In biphenyl the two benzene rings are co-axial, and in optically active biphenyl derivatives the rings are inclined to each other due to the steric and repulsive effects of the groups in the orthopositions. The actual angle of inclination of the two rings depends on the nature of the substituent groups, but it appears to be usually in the vicinity of 90°, *i.e.,* the rings tend to be approximately perpendicular to each other. Thus, in order to exhibit optical activity, the substituent groups in the ortho-positions must be large enough to prevent the two rings from becoming coplanar, in which case the molecule would possess a plane or a centre of symmetry, *e.g,* diphenic acid is not optically active. In configuration (V) the molecule has a plane of symmetry, and in configuration (VI) a centre of symmetry; of these two, (VI) is the more likely because of the repulsion between the two carboxyl groups.

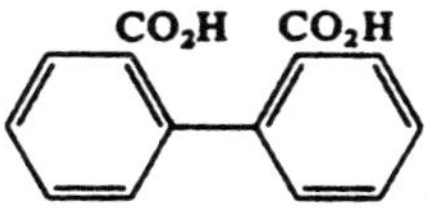

CO_2H CO_2H

(V)

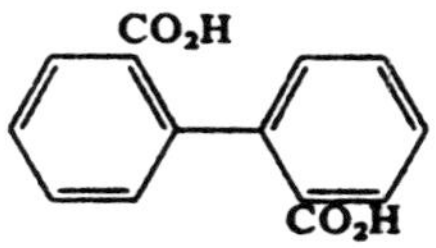

CO_2H

CO_2H

(VI)

If restricted rotation in biphenyl compounds is due entirely to the spatial effect, then theoretically we have only to calculate the size of the group in order to ascertain whether the groups will impinge and thereby give rise to optical activity. The 'size' of the group, however, must be calculated from van der Waals radii the results are in good agreement with experiment.

Later work has shown that if the substituent groups are large enough, then only two in the o- and o′-positions will produce restricted rotation, *e.g,* Lesslie and Turner (1932) resolved biphenyl-2,2'disulphonic acid (VII). In this molecule the sulphonic acid group is large enough to be impeded by the ortho-hydrogen atoms. This molecule was readily racemised on heating, but Lesslie *et al.,* (1962) prepared the enantiomers of 2,2′-di-t-butylbiphenyl from the corresponding optically active 6,6′-di-t-butylbiphenyl-3,3′-dicarboxylic acids. These were found to be highly optically stable (the t-butyl group is very large). Lesslie and Turner (1933) have also shown that the (+)-camphorsulphonate of 3′-

bromobiphenyl-2-trimethylarsonium iodide (VIII) undergoes mutarotation. The trimethylarsonium group is large enough to be impeded by the ortho-hydrogen atoms (the bromine atom in the meta-position gives asymmetry to this ring). Attempts to isolate the active biphenyl compound failed because it racemised rapidly. This mutarotation indicates that the biphenyl is optically active and that the two enantiomers are readily interconvertible.

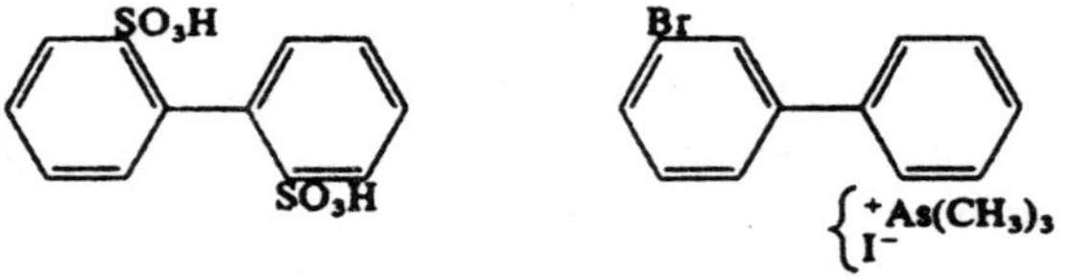

SO_3H SO_3H

(VII)

Br $\{^{+}As(CH_3)_3$ I^-

(VIII)

Since one phenyl group can rotate with respect to the other (about the interannular bond), the various positions would correspond to different conformations. In acyclic compounds, although there are preferred conformations the energy barriers are very low, and so the various conformations are interconvertible.

In the case of the biphenyls, however, because of steric hindrance, the molecules have large energy barriers separating the two forms, and these barriers are large enough (75-105 kJ mol^{-1}) to produce separable rotational isomers. Such isomers are called atropisomers, and the two conditions for atropisomerism have been given above. It has already been pointed out that diphenic acid is not optically active, and that its configuration is most probably (VI).

Now calculation shows that the effective diameter of the carboxyl group is large enough to prevent configuration (V) from being planar, and consequently, if the two rings could be held more or less in this configuration, the molecule would not be coplanar and hence would be resolvable. Such a compound, (IX), was prepared and resolved by Adams and Kornblum (1941). The two benzene rings are not coplanar and are held fairly rigid by the large methylene bridge.

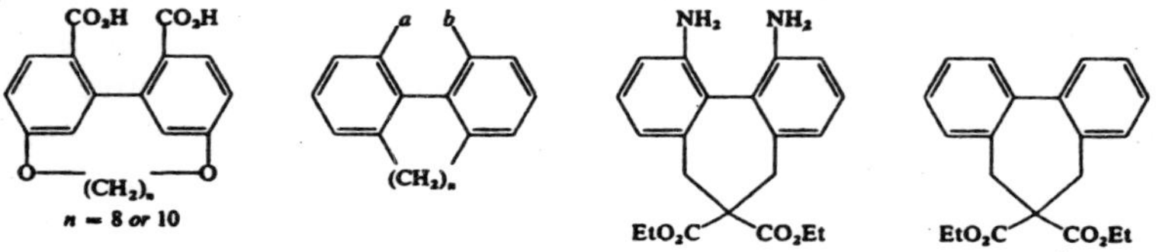

CO_2H CO_2H O $(CH_2)_n$ O $n = 8$ or 10

(IX)

a b $(CH_2)_n$

(X)

NH_2 NH_2 EtO_2C CO_2Et

(XI)

EtO_2C CO_2Et

(XII)

Many biphenyls have been studied from the point of view of the effect of a 2,2′-bridge on the optical activity of molecules of the type (X). When n = 1, the molecule is a disubstituted fluorene. Since this molecule is flat, it is not resolvable. When n = 2, the molecule is a disubstituted 9,10-dihydrophenanthrene. Such compounds have been resolved, *e.g,* (XVI) and (XVII) (see below). When n = 3, the molecules are resolvable and are highly optically stable. Iffland *et al.,* (1956) prepared the optically active biphenyl (XI) which has two amino-groups in the 6,6′-positions. On the other hand, these authors have also prepared (XII) in optically active forms. Mislow (1957) has also obtained the dibenzocyclo-octadiene acids (XIII) in optically active forms; both forms were highly optically labile. Similar to (XIII) is (XIV) which has been resolved by Bell (1952). Mislow *et al.,* (1961) have also resolved the biphenyl derivative (XV), and in 1962, prepared the (–)-form of (XVI). Turner *et al.,* (1955) had prepared (XVII), and Mislow, on comparing the optical stability of (XVI) with (XVII), found that the latter was the more stable one. In (XVI), the two methyl groups can slip past each other comparatively easily by bending out of the ring plane, but in (XVI) the second benzene ring in each naphthalene nucleus behaves as a large group and is also rigid, and consequently bending is very difficult.

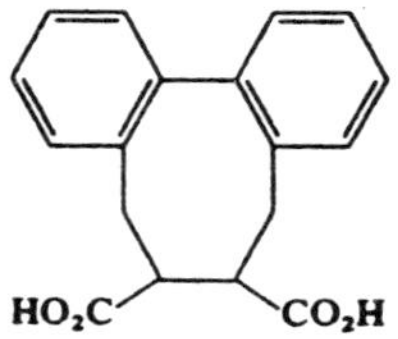

(XIII) 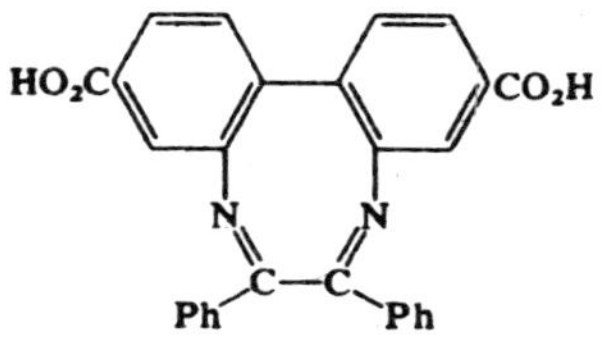 (XIV)

Fig. 17

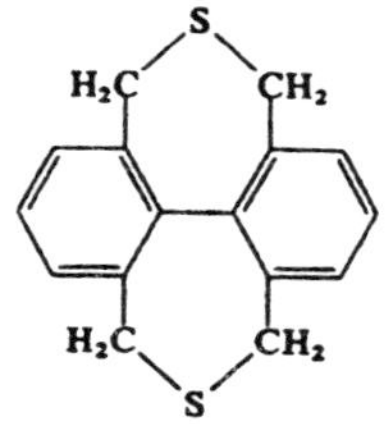

(XV) 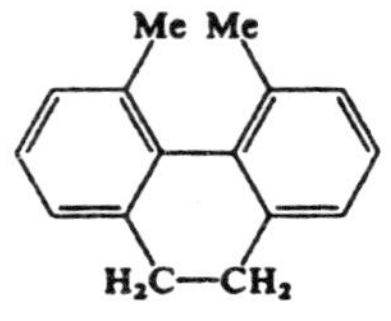 (XVI) 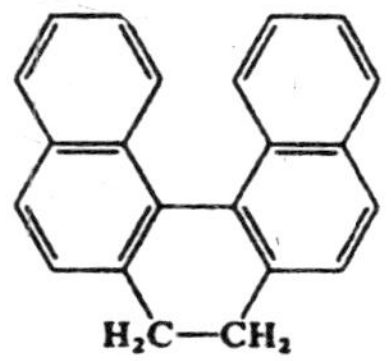 (XVII)

Fig. 18

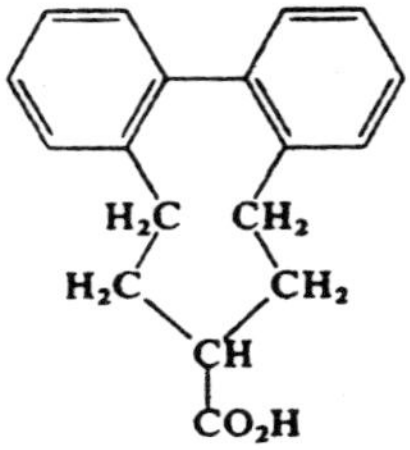

H₂C CH₂ H₂C CH₂ CH CO₂H

(XVIII)

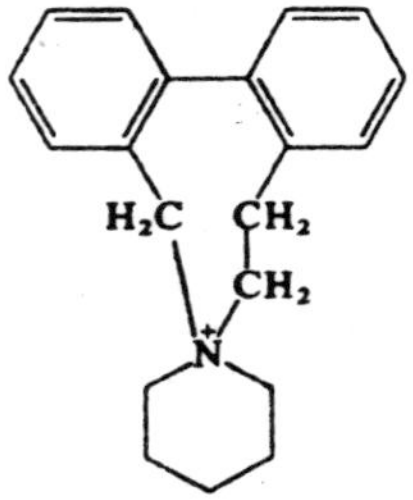

H₂C CH₂ CH₂ N

(XIX)

Fig. 19

Mislow *et al.*, (1961) have also prepared the (–)-form of dibenzocyclononadienecarboxylic acid (XVIII), and Hall *et al.*, (1959) prepared the piperidinium salt (XIX) as the picrate and found it was optically labile. Apart from the resolution of 2,2'-biphenyls, the stereochemistry of these compounds has been studied by u.v. and NMR spectroscopy and by X-ray analysis. In this way, a more detailed threedimensional structure of these molecules has been obtained, *e.g,* Wahl, jun. *et al.*, (1972) have carried out an X-ray diffraction study of (XX) and have shown that this molecule exists in the pseudochair form (XXI). (XX) had been prepared and resolved chromatographically (on cellulose acetate) by Liittringhaus *et al.*, (1967). These latter authors had, on the basis of consideration of bond angles predicted that the pseudo-tub conformation (XXII) was the more likely one.

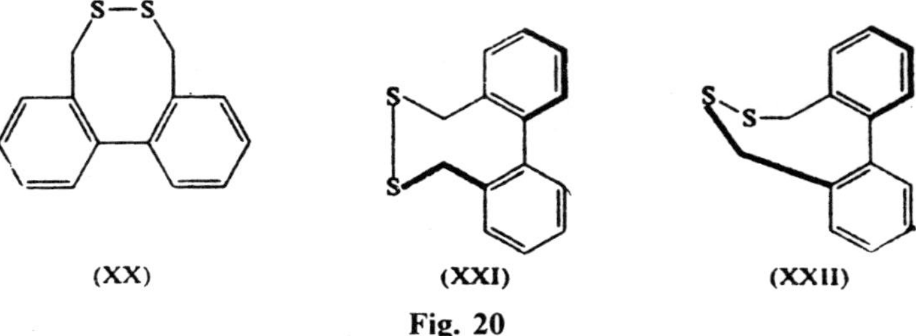

Fig. 20

A point of interest in connection with optically active biphenyls is thatSchmidt *et al.*, (1957) have shown that 4,4′,5,5′,6,6′ hexahydroxy-diphenic acid occurs naturally in an optically active form.

Absolute Configurations of Biphenyls

Since biphenyls owe their asymmetry to the molecule being asymmetric, the methods used for correlating configurations of compounds containing asymmetric carbon atoms cannot be applied. However, Mislow *et al.*, (1957) used asymmetric synthesis as a means of establishing the absolute configuration of 6,6′-dinitro-2,2′-diphenic acid.

Their method was chemical; assignment of absolute configuration has been obtained from a consideration of the transition states in the Meerwein-Ponndorf-Verley reduction of a dissymmetric diphenylic ketone by asymmetric alcohols of known absolute configuration. The (+)- and (–)-ketones (I) were partially reduced with (S)-(+)-methyl-t-butylmethanol in the presence of aluminium t-butoxide. The products were unchanged ketone with the (+)-form predominating, and the (+)- and (–)-alcohols, (II), with the (–)-form predominating.

Thus, as far as the unchanged ketone is concerned, this reaction is a kinetic resolution, since the ketone is now enriched in the (+)-form, and at the same time, the alcohol has become enriched in the (-)-form. Examination of models showed that, in the single conformation possible for the (S)-enantiomer (of the ketone), hydride transfer to either side of the carbonyl group is hindered by steric repulsion between the t-butyl group and a phenyl group, whereas for the (R)enantiomer, the repulsion is only between the methyl group and a phenyl group.

(*S*)-(+) (I) (*R*)-(–) + t-Bu–C(OH)(H)–Me (*S*)-(+)

$\xrightarrow{(t\text{-BuO})_3Al}$ (*S*)-(+) (II) (*R*)-(–)

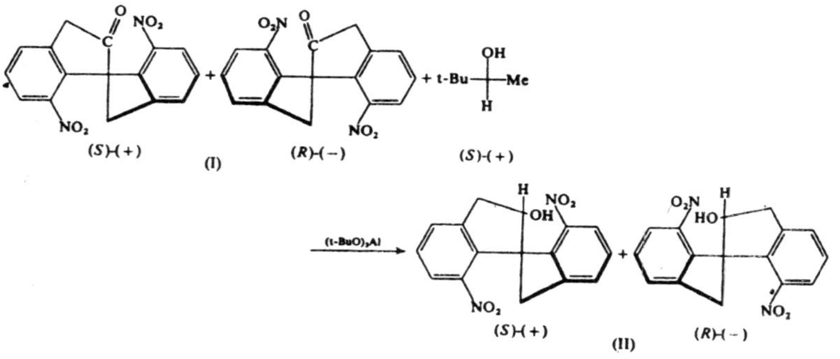

Fig. 21

Thus the (+)-ketone the (+)-ketone is (S)-(+) and the (–)-alcohol is (R)-(–). Furthermore, the (S)-(+)-ketone had been prepared from (–)-6,6′-dinitro-2,2′-diphenic acid (via the dimethyl ester), and so this (–)-enantiomer is the (S)-(–)-acid. This assignment of absolute configuration has been confirmed by X-ray diffraction (Akimoto *et al.*, 1968).

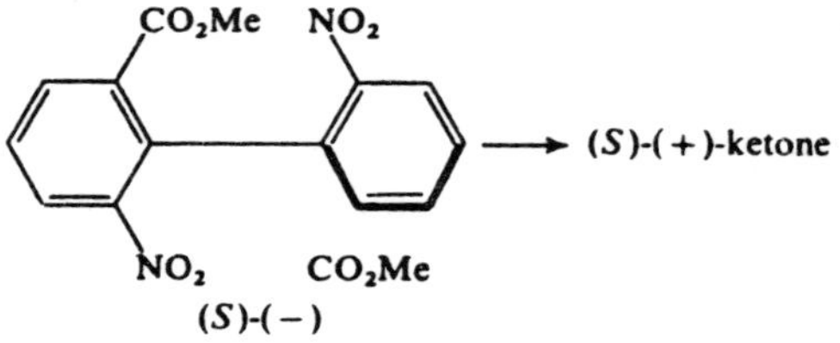
CO₂Me
NO₂
(S)-(+)-ketone
NO₂
CO₂Me
(S)-(−)

Fig. 22

The method of chemical correlation is reliable only if there is no change in configuration during the transformation or if the change in configuration occurs in a predictable manner. Thus, using the (S)-(–)-acid as absolute standard, Mislow *et al.*, carried out a number of chemical correlations, *e.g*,

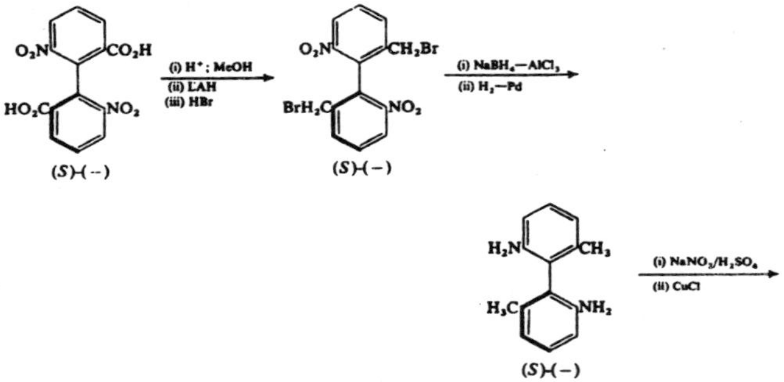
O₂N
CO₂H
HO₂C
NO₂
(S)-(−)
(i) H⁺; MeOH
(ii) LAH
(iii) HBr
O₂N
CH₂Br
BrH₂C
NO₂
(S)-(−)
(i) NaBH₄—AlCl₃
(ii) H₂—Pd
H₂N
CH₃
H₃C
NH₂
(S)-(−)
(i) NaNO₂/H₂SO₄
(ii) CuCl
Cl
CH₃
H₃C
Cl
(S)-(−)

Fig. 23

Mislow *et al.*, (1958), using the (S)-(–)-acid as the absolute standard, also correlated configurations in the biphenyl series by the quasi-racemate method. In this way these authors determined the configurations of 6,6′-dichloro- and 6,6′-dimethyl-2,2′-diphenic acid. Mislow *et al.*, (1960) have also confirmed absolute configurations in the biphenyl series by the rotatory dispersion method. The authors showed that the shapes of the ORD curves depended on the configuration and conformation of the biphenyl compounds; these compounds contain inherently dissymmetric chromophores.

The specification of absolute configuration of biphenyl compounds is carried out as follows. Since biphenyls do not owe their asymmetry to the presence of asymmetric carbon atoms (which are chiral centres), the criterion now is the presence of a chiral axis (*i.e.*, an axis of asymmetry).

This chiral axis may be derived from a chiral centre Z, (III), by 'extending' the point into a line AB which now passes through an elongated tetrahedron, (IV). For Z to be a chiral centre in (III), a, a′, b, and b′ must all be different, but for AB to be a chiral axis in (IV), it is sufficient that a and b be different, and a′ and b′ be different; it is not necessary that a should be different from a′, or b from b′. To apply the sequence rule to axial chirality (or axial asymmetry), it is necessary to use an additional rule, viz. with respect to an external point on the chiral axis, groups at the near end of the axis are given precedence over groups at the far end.

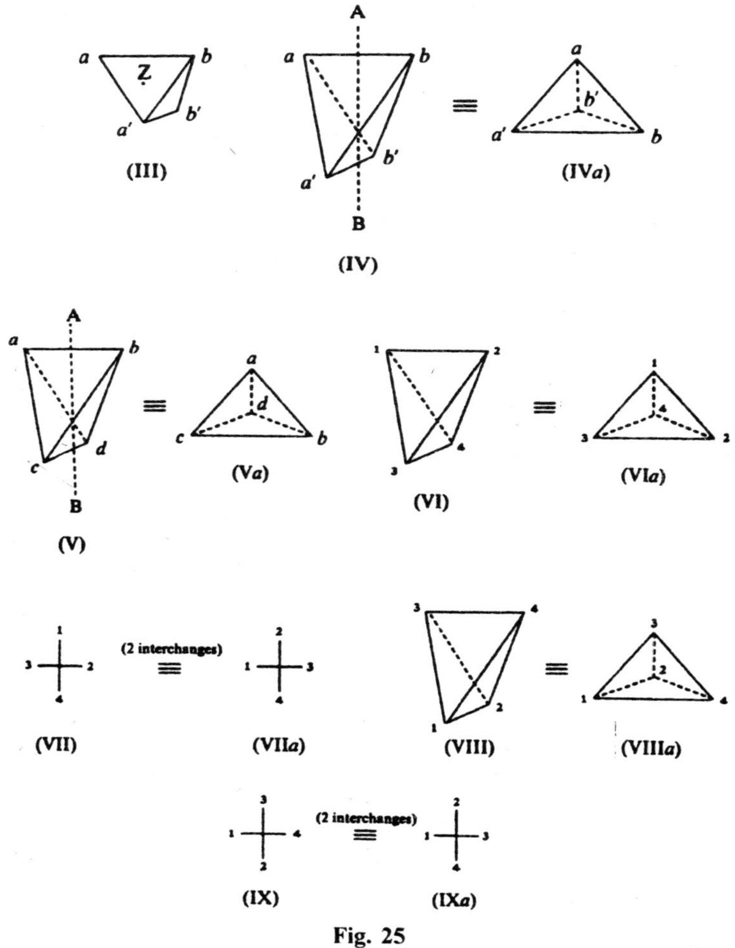

Fig. 25

If (IV) is viewed from point A, then the pair a–b, being nearer A, precedes the pair a′–b′; and in (V), pair a–b precedes pair c–d. If the order of priority is a > b and a′ > b′ (for IV), and a > b and c > d (for V), then both of these models give the final order of priority shown in (VI). If now, in accordance with the conversion rule, (VI) is viewed from the side remote from 4, then (VII) is obtained. This, by two interchanges gives (VIIa), and since the sequence 1 → 2 → 3 is clockwise, (IV) and (V) are (R)-configurations. Had (IV) and (V) been viewed from point B on the chiral axis, then pairs a-b, a′–b′ and c–d will be 1 and 2, and the pair a–b will be 3 and 4. This gives (VIII), which in turn gives (IX), and this, by two interchanges, gives (IXa), which is still the (R)-configuration.

Let us now consider some substituted biphenyls. First, the four ortho-substituent groups are inspected. If they are different, as pairs (2–6 and 2′–6′), then they are used. Thus, in (X), NO_2 = a,

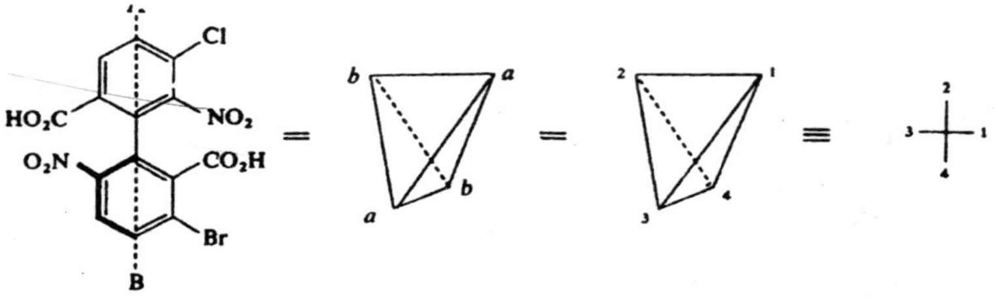

(X) (*S*)-form

Fig. 26

and CO_2H = b (priority a > h), and this molecule is therefore the (S)-form (the last diagram has 1 → 2 → 3 anticlockwise). In (XI), since the upper ring has Cl in both ortho-positions, groups H and

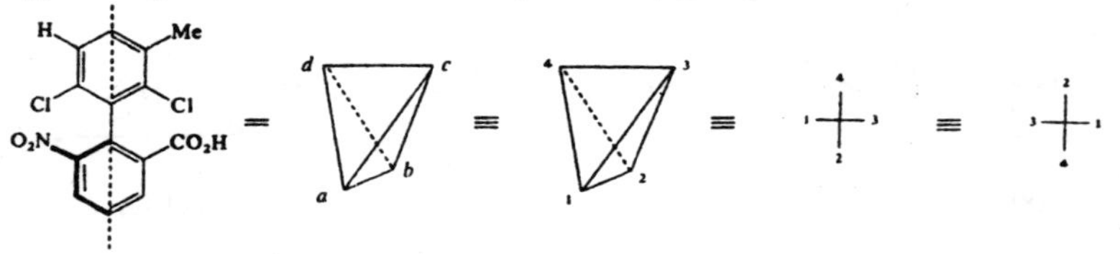

(XI) (*S*)-form

Fig. 27

Me are therefore selected; NO_2 = a, CO_2H = b, Me = c, and H = d. Hence, (XI) is the (S)-form (two interchanges in the last diagram give 1 → 2 → 3 in an anticlockwise direction).

Compound (I) is the (+)-ketone described above; NO_2 = a, $-CH_2CO$ = h, and so (I) is the (S)-form.

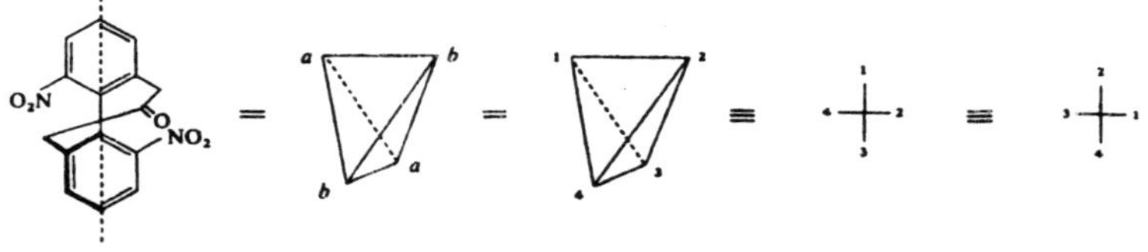

(I) (*S*)-form

Fig. 28

OTHER EXAMPLES OF ATROPISOMERISM

In addition to the biphenyl compounds, there are many other examples where optical activity in the molecule is produced by restricted rotation about a single bond which may or may not be one that joins two rings. The following examples are only a few out of a very large number of compounds that have been resolved.

(i) Adams *et al.*, (1931) have resolved the following N-phenylpyrrole and N,N′-bipyrryl.

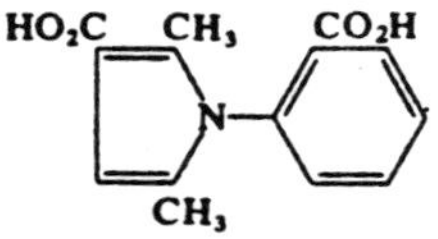

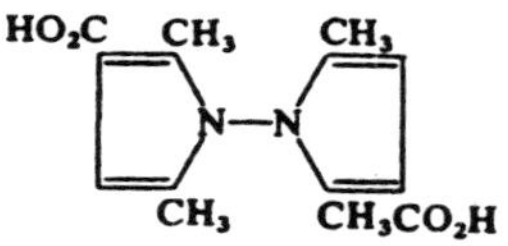

Adams *et al.*, (1932) have also resolved the 3,3′-bipyridyl

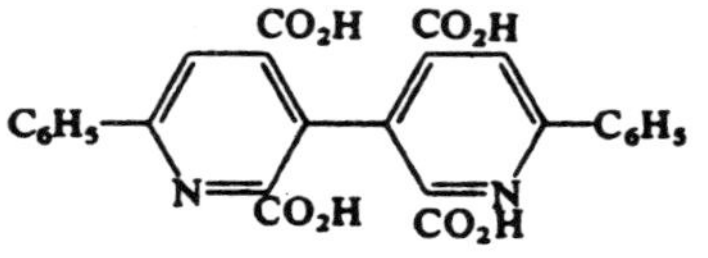

(ii) 1,1′-Binaphthyl-8,8′-dicarboxylic acid has been obtained in optically active forms by Stanley (1931).

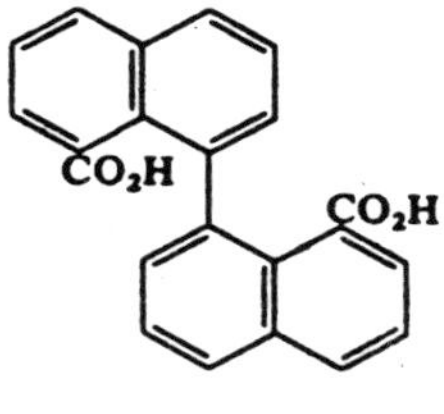

This compound gives rise to asymmetric transformation; resolution with brucine gave 100 per cent of either the (+)- or (–)-compound. Other compounds similar to the binaphthyl which have been obtained in optically active forms are 1,1′-binaphthyl-5,5′-dicarboxylic acid (I) (Bell *et al.,* 1951), the bianthryl derivatives, (II) and (III) (Bell *et al.,* 1949), and the 4,4′- and 5,5′-biquinolyls, (IV) and (V) (Crawford *et al.,* 1952).

(iii) Mills and Elliott (1928) obtained N-benzenesulphonyl-8-nitro-l-naphthylglycine (VI) in optically active forms; these were optically unstable, undergoing asymmetric transformation with

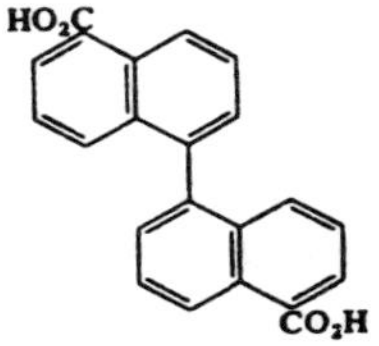

HO_2C CO_2H (I) CO_2H CO_2H (II)

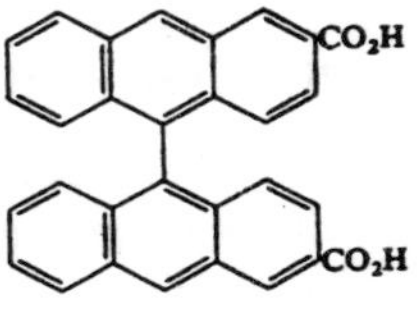

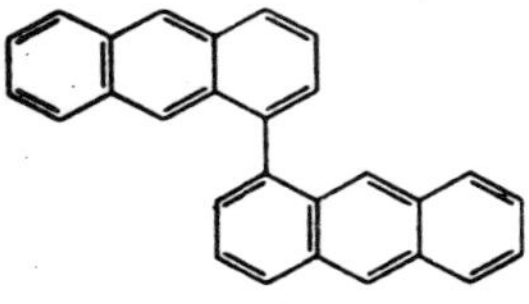

(III) 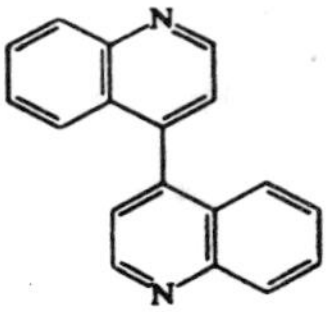 N N (IV) N N (V)

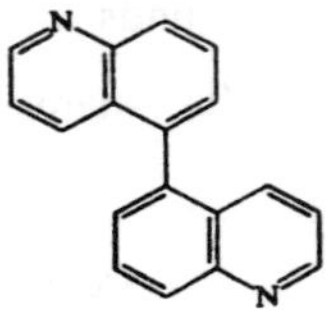

Fig. 29

brucine. Mills and Kelham (1937) also resolved N-acetyl-N-methyl p-toluidine-3-sulphonic acid (VII) with brucine, and found that it racemised slowly on standing. In both (VI) and (VII) the optical activity arises from the restricted rotation about the C–N bond (the C being the ring carbon to which the N is

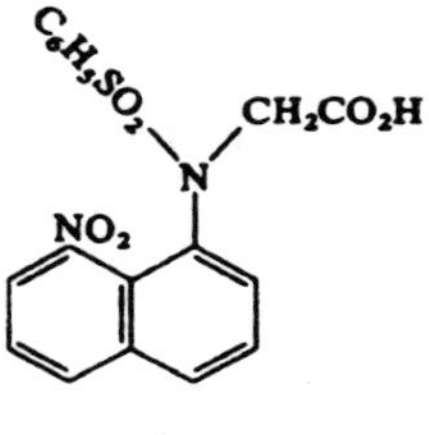

$C_6H_5SO_2$ CH_2CO_2H N NO_2 (VI) H_3C $COCH_3$ N SO_3H CH_3 (VII)

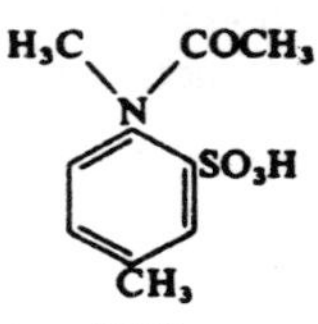

Fig. 30

attached). Asymmetry arising from the same cause is also shown by 2-acetomethylamido-4′,5-dimethyldiphenylsulphone (VIII); this was partially resolved by Buchanan *et al.*, It is also interesting to note in this connection that Adams *et al.*, (1950) have isolated pairs of geometrical isomers of compounds of the types (IX) and (X); here geometrical isomerism is possible because of the restricted rotation about the C–N bonds.

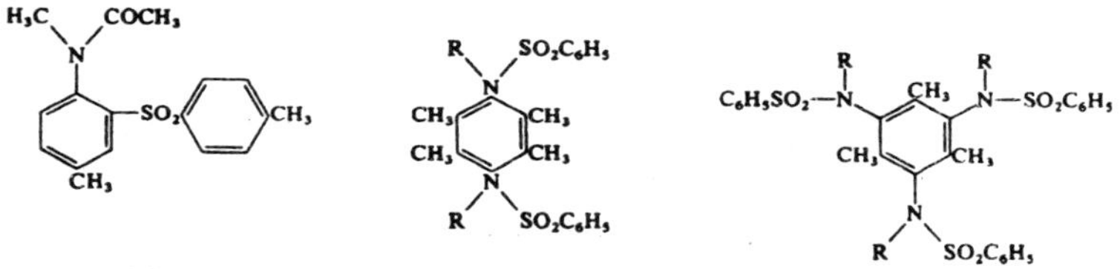

(VIII) (IX) (X)

Fig. 31

(iv) Luttringhaus *et al.*, (1940, 1947) isolated two optically active forms of 4-bromogentisic acid decamethylene ether (XI). This belongs to the group known as 'ansa' compounds, and the methylene ring is perpendicular to the plane of the benzene

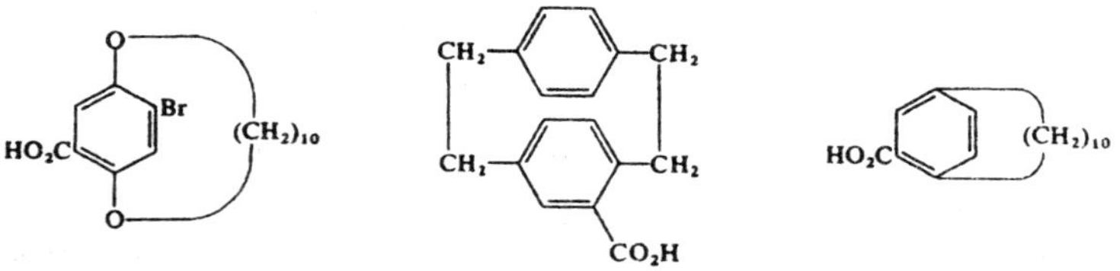

(XI) (XII) (XIII)

Fig. 32

ring; the two substituents, Br and CO_2H, prevent the rotation of the benzene nucleus inside the large ring. Cram *et al.*, (1955) have obtained paracyclophanes in optically active forms, *e.g.* (XII). In this molecule, the planes of the two benzene rings are approximately parallel (and the carboxyphenyl ring cannot rotate to give the enantiomer). When the bridges each contained four methylene groups, the compound could not be resolved (the carboxyphenyl ring can now rotate to give the enantiomers). On the other hand, Blomquist *et al.*, (1961) have resolved the simple paracyclophane (XIII).

(v) Terphenyl compounds can exhibit both geometrical and optical isomerism when suitable substituents are present to prevent free rotation about single bonds, *e.g,* Shildneck and Adams (1931) obtained (XIV) in both the cis- and trans-forms. Interference of the methyl and hydroxyl groups in the ortho-positions prevents free rotation and tends to hold the two outside rings perpendicular to the centre ring. Inspection of these formulae shows that if the centre ring does not possess a vertical plane of symmetry, then optical activity is possible. Thus, Browning and Adams (1930) prepared the dibromo cis- and trans-forms of (XV) and resolved the cis-isomer; the trans-isomer is not resolvable since it has a centre of symmetry.

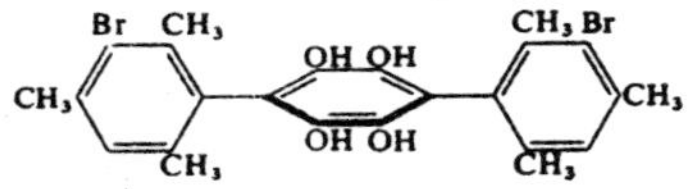

cis- (XIV) 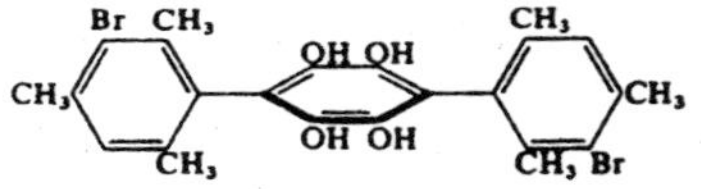 *trans*-(XIV)

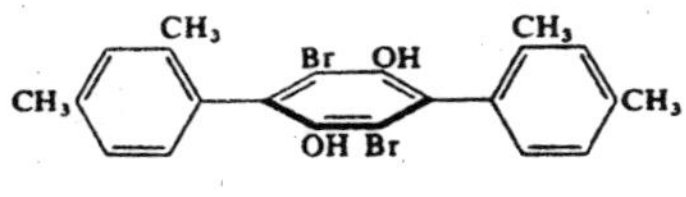

cis-(XV) 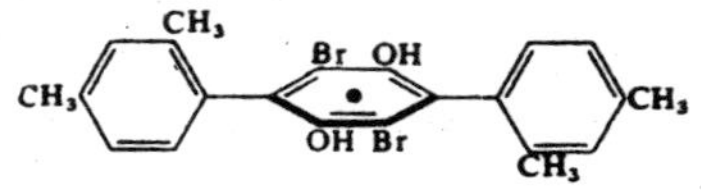 *trans*-(XV)

Fig. 33

It can be seen from the terphenyl compounds discussed that atropisomerism does not necessarily imply enantiomerism. The different forms may be related to each other as diastereoisomers, but whether optical isomerism is also exhibited depends on the substitution pattern.

(vi) A very interesting case of restricted rotation about a single bond is afforded by the compound 10-m-aminobenzylideneanthrone (XVI). This was prepared by Ingram (1950), but he failed to resolve it. He did show, however, that it was optically active by the mutarotation of its camphorsulphonate salt, and by the preparation of an active hydriodide. Thus the molecule is asymmetric, and this asymmetry can only be due to the restricted rotation of the phenyl group about the C-phenyl bond, the restriction being brought about by hydrogen atoms in the ortho-positions. The two hydrogen atoms labelled $\overset{\times}{H}$ overlap in space, and consequently the benzene ring cannot lie in the same plane as the 10-methyleneanthrone skeleton. Another example is the substituted cinnamic acids (XVII) (R = Cl, Me, OMe) [Adams *et al.*, 1940, 1941]. The benzene ring and the ethylenic double bond cannot become coplanar, and it was found that the order of stability to racemisation was Cl > Me > OMe.

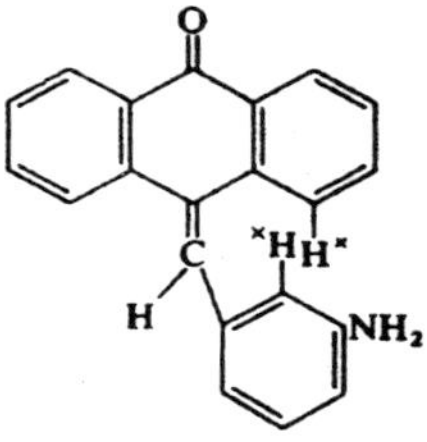

(XVI) 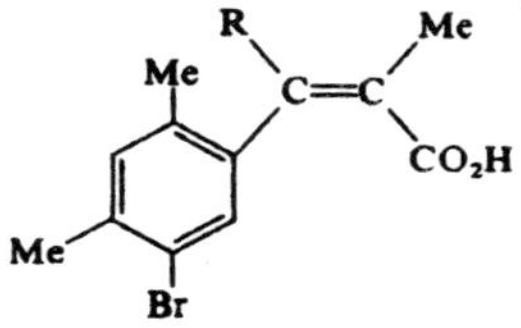 (XVII)

Fig. 34

Molecular Overcrowding

All the cases discussed so far owe their asymmetry to restricted rotation about a single bond. There is, however, another way in which steric factors may produce molecular asymmetry. It has been found that, in general, non-bonded carbon atoms cannot approach closer to each other than about 3.0Å. Thus, if the geometry of the molecule is such as to produce 'intramolecular overcrowding', the molecule becomes distorted. An example of this type is 4,5,8-trimethyl-l-phenanthrylacetic acid (I). The phenanthrene nucleus is planar and substituents lie in this plane. If, however, there are fairly large groups in positions 4 and 5, then there will not be enough room to accommodate both groups in the plane of the nucleus.

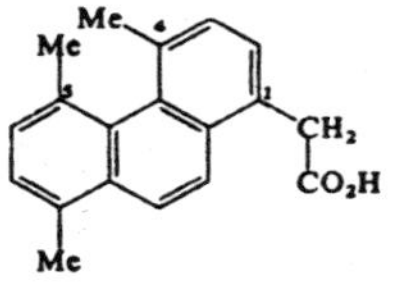

(I) 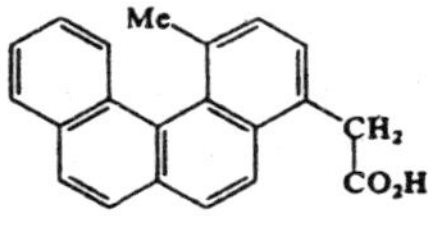 (II) 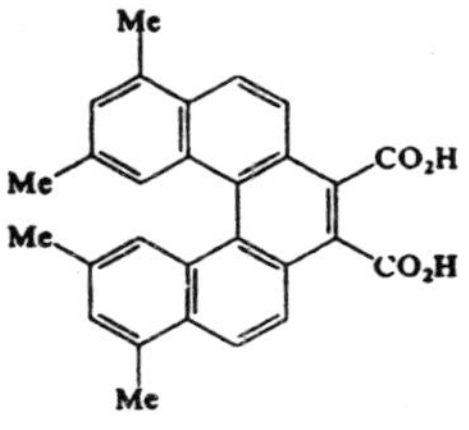 (III)

Fig. 35

This leads to strain being produced by intramolecular overcrowding, and the strain may be relieved by the bending of the substituents out of the plane of the nucleus, or by the bending (buckling) of the aromatic rings, or by both. Thus the molecule will not be planar and consequently will be asymmetric and therefore (theoretically) resolvable. Newman *et al.*, (1940, 1947) have actually resolved it, and have also resolved (II). Bell *et al.*, (1949) resolved (III), and it was these authors who introduced the term 'intramolecular overcrowding'. Theilacker *et al.*, (1953) resolved (IV), a heterocyclic analogue of phenanthrene. All of these compounds were found to have low optical stability, but Newman *et al.*, (1955, 1956) have prepared (V) and (VI; hexahelicene) which, so far, are the most optically stable compounds of the intramolecular overcrowding type.

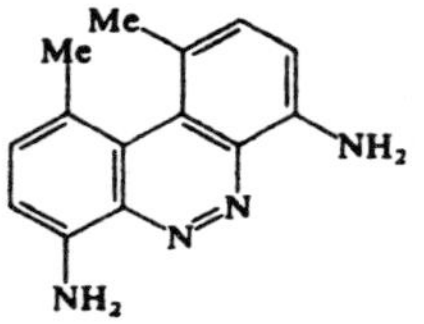

Me Me NH_2 N N NH_2

(IV)

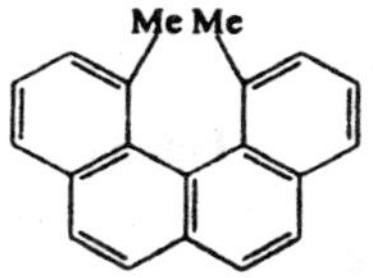

Me Me

(V)

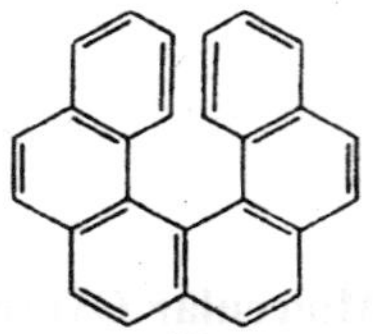

(VI)

Fig. 36

It will be noticed that in (VI) the only way in which out-of-plane distortion can occur is through buckling of the molecule. The simplest molecule exhibiting overcrowding and consequent out-ofplane buckling of the molecule is 3,4-benzophenanthrene (VII); this has been shown to be nonplanar by X-ray analysis (Schmidt *et al.*, 1954). Similarly, Robertson *et al.*, (1954) have shown that (VIII) exhibits out-of-plane buckling.

Another point to note in connection with out-of-plane buckling is that the buckling is distributed over all the rings in such a manner as to cause the minimum distortion in any one ring. This distortion, which enables non-bonded carbon atoms to avoid being closer together than 3.0Å (marked with dots in VII and VIII), forces some of the other carbon atoms to adopt an almost tetrahedral valency arrangement (the original hybridisation is trigonal), and this affects the physical and chemical properties of the molecule, *e.g*, Coulson *et al.*, (1955) have calculated that the deformation in (VIII) produces a loss of resonance energy of about 75 kJ mol^{-1}. We may now summarise the problem of molecular overcrowding as follows. A molecule is said to be overcrowded if, when the standard values are assigned to the bond lengths and bond angles, at least one pair of non-bonded atoms are closer to each other than the sum of their accepted van der Waals radii.

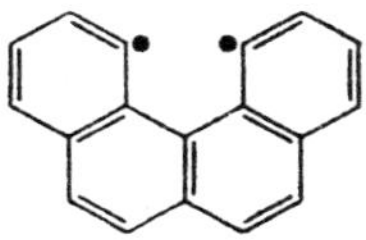

(VII) 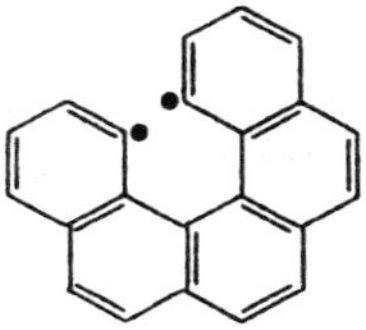 (VIII) 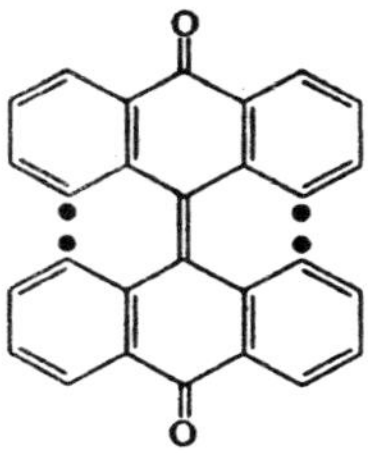 (IX)

Fig. 37

If these atoms were to remain very close to each other and the rest of the molecule remained unchanged, there would be a large steric strain. If, however, the molecule were deformed (buckled) so that the overcrowded atoms became separated to the sum of their van der Waals radii there would now be a considerable strain energy in the molecule. Since a large amount of energy is required to stretch bonds and much less energy is required to bend them in overcrowded molecules the geometry of the molecule adjusts itself so that the energy is a minimum by mainly bending various bonds, with the bond lengths not much changed from the unstrained analogue. Thus overcrowded molecules are non-coplanar and have the form of a segment of a helix. Mason *et al.,* (1965) have measured the circular dichroism spectra of (–)-(VIIa) and (+)-(VIIb) (these are derivatives of VII), and analysis of these spectra led the authors to conclude that (–)-(VIIa) has the M- (minus, left-handed) and (+)-(VIIb) the P-(plus, right-handed) helical configuration viewed in the direction perpendicular to the mean molecular plane.

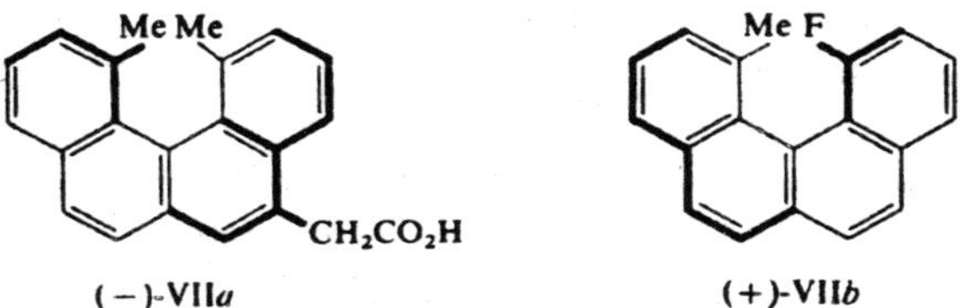

Fig. 38

Helical molecules are those in which the arrangement of the atoms or groups is an imaginary helix. Such molecules are optically active due to the presence of helical dissymmetry or helicity. Thus, helicity is a particular type of chirality. Just as benzene rings may suffer distortion, so can a molecule which owes its planarity to the presence of a double bond. Such an example is dianthronylidene (IX). The carbon atoms marked with dots are overcrowded (the distance between each pair is 2.9Å), and the strain is relieved by a rotation of about 40° around the olefinic double bond (Schmidt *et al.*, 1954). Even in such simple molecules as tiglic acid (X) the two methyl groups give rise to molecular overcrowding with the result that the β-methyl group appears to be displaced from the molecular plane, thereby relieving overcrowding which is also partly relieved by small distortions in bond angles. These results were obtained by Robertson *et al.*, (1959) from X-ray studies, and these authors also showed similar distortions in angelic acid (XI).

$(Me)(H)C{=}C(Me)(CO_2H)$

(X)

$(Me)(H)C{=}C(HO_2C)(Me)$

(XI)

In polynuclear aromatic hydrocarbons in which the strain tends to be overcome by out-of-plane displacements of substituents and out-of-plane ring buckling, these effects cause changes in the ultraviolet spectra, but it is not yet possible to formulate any correlating rules. NMR studies by Reid (1957) have shown a shift for the hydrogen atoms in positions 4 and 5 in phenanthrene itself. A similar phenomenon has been detected by Brownstein (1958) in 2-halogenobiphenyls, and the explanation offered is that the shift is due to the steric effect between the 2-halogen and the 2′-hydrogen atom.

Although, molecular overcrowding is normally confined in the polynuclear type to systems containing three or more rings, nevertheless various substituted benzenes may also exhibit out-of-plane displacements of the substituents. Electron-diffraction studies of polyhalogenobenzenes suggest that such molecules are non-planar (Hassel *et al.*, 1947), whereas X-ray studies indicate that in the solid state such molecules are very closely or even exactly planar (Tulinsky *et al.*, 1958; Gafner *et al.*, 1960). Ferguson *et al.*, (1959, 1961) have examined, by X-ray analysis, polysubstituted benzenes containing not more than one halogen atom, *e.g*, o-chloro- and bromobenzoic acid, and 2-chloro5-nitrobenzoic acid.